Nanostructured Catalysts

Nanostructure Science and Technology

Series Editor: David J. Lockwood, FRSC
National Research Council of Canada
Ottawa, Ontario, Canada

Current volumes in this series:

Nanostructured Catalysts
Edited by Susannah L. Scott, Cathleen M. Crudden, and Christopher W. Jones

Polyoxometalate Chemistry for Nano-Composite Design
Edited by Toshihiro Yamase and Michael T. Pope

Self-Assembled Nanostructures
Jin Z. Zhang, Zhong-lin Wang, Jun Liu, Shaowei Chen, and Gang-yu Liu

Nanostructured Catalysts

Edited by

Susannah L. Scott

University of Ottawa
Ottawa, Ontario, Canada

Cathleen M. Crudden

Queen's University
Kingston, Ontario, Canada

and

Christopher W. Jones

Georgia Institute of Technology
Atlanta, Georgia

Kluwer Academic / Plenum Publishers
New York   Boston   Dordrecht   London   Moscow

Library of Congress Cataloging-in-Publication Data

Nanostructured catalysts/edited by Susannah L. Scott, Cathleen M. Crudden and
Christopher W. Jones.
        p.   cm. (Nanostructure science and technology)
    Includes bibliographical references and index.
    ISBN 0-306-47484-0
      1. Catalysts. 2. Nanostructure materials. I. Scott, Susannah L. II. Crudden, Cathleen M.
III. Jones, Christopher W. IV. Series.

TP159.C3 N36 2003
660'.2995—dc21

2002040782

Top cover illustrations:

Quantum dot: Created by Jordon Johnson and David Lockwood.
(Used by permission of David Lockwood. National Research Council of Canada.)

Carbon nanotube: Created by Professor J.-C. Charlier.
(Used by permission of Prof. J.-C. Charlier. Université Catholique de Louvain, Belgium.)

ISBN 0-306-47484-0

©2003 Kluwer Academic / Plenum Publishers, New York
233 Spring Street, New York, New York 10013

http://www.wkap.nl/

10 9 8 7 6 5 4 3 2 1

A C.I.P. record for this book is available from the Library of Congress

CONTRIBUTORS

Takuzo Aida, University of Tokyo, Japan

Daryl Allen, University of New Brunswick, Canada

Reiner Anwander, Technishe Universität München, Germany

Gino Baron, Universiteit Brussel, Belgium

Knut Børve, University of Bergen, Norway

Daniel Brunel, Ecole National Supérieure de Chimie de Monpellier, France

Cathleen Crudden, Queen's University, Canada

Eric Deguns, University of Ottawa, Canada

Dirk De Vos, Katholieke Universiteit Leuven, Belgium

Robbert Duchateau, Dutch Polymer Institute, The Netherlands

Øystein Espelid, University of Bergen, Norway

Pierre Jacobs, Katholieke Universiteit Leuven, Belgium

Andreas Jentys, Technishe Universität München, Germany

Christopher Jones, Georgia Institute of Technology, USA

Jun Liu, Sandia National Laboratories, USA

Rasmita Raval, University of Liverpool, UK

Gopinathan Sankar, Royal Institution, UK

Susannah Scott, University of Ottawa, Canada

Rick Schroden, University of Minnesota, USA

Andreas Stein, University of Minnesota, USA

Kasuke Tajima, University of Tokyo, Japan

FOREWORD

With the recent advent of nanotechnology, research and development in the area of nanostructured materials has gained unprecedented prominence. Novel materials with potentially exciting new applications are being discovered at a much higher rate than ever before. Innovative tools to fabricate, manipulate, characterize and evaluate such materials are being developed and expanded. To keep pace with this extremely rapid growth, it is necessary to take a breath from time to time, to critically assess the current knowledge and provide thoughts for future developments. This book represents one of these moments, as a number of prominent scientists in nanostructured materials join forces to provide insightful reviews of their areas of expertise, thus offering an overall picture of the state-of-the art of the field.

Nanostructured materials designate an increasing number of materials with designed shapes, surfaces, structures, pore systems, *etc*. Nanostructured materials with modified surfaces include those whose surfaces have been altered via such techniques as grafting and tethering of organic or organometallic species, or through various deposition procedures including electro, electroless and vapor deposition, or simple adsorption. These materials find important applications in catalysis, separation and environmental remediation. Materials with patterned surfaces, which are essential for the optoelectronics industry, constitute another important class of surface-modified nanostructured materials. Other materials are considered nanostructured because of their composition and internal organization. For example, organic-inorganic nanocomposites, which may be prepared by various self-assembly processes or by inclusion of organic species such as surfactants and polymers within pores or between layers, are an important group of nanostructured materials. These materials are being used in a wide range of applications from catalysis to the automotive industry. Materials with designed regular pore systems such as zeolites, metallophosphates, periodic mesoporous materials, nanoporous organic and organometallic materials are also members of the large family of nanostructured materials. These materials are of paramount importance, particularly in catalysis.

One of the most distinctive characteristics of this book is the fact that it is all-inclusive, since most of the materials listed above have been dealt with in a concise and informative manner. The subject of periodic porous materials is discussed in great detail. In an authoritative report, Stein and Schroden tackle the issue of designing and synthesizing inorganic materials with controlled porosity in terms of size and architecture, *i.e.*, shape, connectivity, *etc*. Synthesis strategies for making micro-, meso-, macro- and bimodal porous materials are described. These are based on templating routes using molecules (amines, phosphines), cations (ammonium, phosphonium), supramolecular assemblies (surfactants, polymers) and other structures such as emulsions, colloidal crystals, bacteria and porous carbons.

Several chapters are devoted to surface functionalization of amorphous and periodic nanoporous silicas. In a concise contribution, Scott deals with the synthesis by design of complex, yet well-characterized multicomponent active sites on amorphous silica surfaces. Strategies involving reactions of siloxane and silica surface hydroxyl groups with one or more molecular complexes are presented. Surface modifications of nanoporous silica are the subject of contributions by Jones, Anwander, Brunel and Laspéras, Crudden *et al.* and Liu *et al.*, which describe a full range of techniques and strategies to impart these materials with controlled catalytic activity. Anwander and Crudden *et al.* focus their reviews on surface immobilization of rare earth and late transition metal complexes, respectively. Of particular significance is the use of chiral ligands for the development of novel, reusable enantioselective catalysts. Likewise, Brunel and Laspéras deal with chiral functionalization of periodic mesoporous silica with particular emphasis on their use as enantioselective epoxidation and alkylation catalysts. Jones devotes a sizeable part of his contribution to molecular imprinting of silica to generate pores with special shapes, which may contain organic functions in predetermined spatial arrangements. In a further development, Liu *et al.* combine grafting and imprinting techniques to create "surface micropores" with different shapes and densities depending on the structure and amount of the molecules used to generate such pores.

Instead of immobilizing chiral transition metal complexes on surfaces to induce asymmetry as described by Brunel and Laspéras and Crudden *et al.*, heterogeneous enantioselective catalysts may be obtained by adsorbing chiral modifiers on achiral solid surfaces. One of the most prominent examples of such materials is the enantioselective hydrogenation catalyst based on nickel modified by optically pure tartaric acid. Raval combines Reflection Absorption Infrared Spectroscopy (RAIRS) with a range of powerful surface analytical tools to gain in-depth understanding of enantioselectivity at the molecular scale. She studies model modified surfaces using a Cu(110) single crystal in the presence of adsorbed R,R- or S,S-tartaric acid.

With the same purpose of achieving detailed understanding of silica and transition metal modified-silica surfaces and catalytic processes taking place thereon, Duchateau provides a masterful contribution using silsesquioxanes and metallasilsesquioxanes, particularly suitable as model compounds for polymerization catalysts. In another contribution dealing with polymerization, Borve and Espelid use theoretical tools to study the Phillips-type Cr/silica catalyst for ethylene polymerization.

Two contributions are devoted to catalysis in the presence of nanoporous silica. Tajima and Aida demonstrate that polymerization in the confined space of MCM-41 silica affords polymers with unique properties and structures such as nanofibers and coaxial nanocables. Jentys and Vinek describe novel NO_x reduction catalysts based on MCM-41 silica doped with Pt and tungstophosphoric acids. Two additional contributions with particular emphasis on catalysis are fully devoted to crystalline microporous materials, namely zeolites and aluminophosphates. Jacobs *et al.*, address the important problem of surface hydrophobicity and its effect on catalytic activity for organic reactions in the presence of modified zeolites. Sankar and Raja review the literature related to cobalt-substituted aluminophosphates in great detail with particular emphasis on the relationship between the local environment of cobalt as determined by EXAFS and the catalytic performance of the corresponding materials.

With fourteen up-to-date reports on the design, synthesis and catalytic properties of nanostructured materials, this book sets the stage for things to come in this area. The development of novel catalysts, taking advantage of the many interesting attributes of periodic nanoporous materials, and the discovery of innovative materials such as polymer-silica nanocomposites and confined nanoparticles, are some of the leading ideas for future work in the increasingly important field of nanostructured materials.

Abdelhamid Sayari
Ottawa, 2002

PREFACE

The focus of this book is the convergence of smart materials design and powerful new surface characterization techniques, which together are transforming the field of heterogeneous catalysis from a "black box" tool of the chemical industry into a knowledge-intensive research field. Control of nanoscale molecular architecture in catalyst synthesis, which encompasses support materials such as zeolites and mesoporous materials, and the preparation of supported catalysts based on metal crystallites and organometallic fragments, are being realized by chemists and chemical engineers and applied to the design of active sites for applications in catalysis.

The importance of catalysis is immense. Catalysts are used to grow the food we eat (through the synthesis of fertilizers and pesticides) and to make it more appealing (through the manufacture of flavours and sweeteners). The clothes we wear contain synthetic fibres made using catalysts, which come in beautiful colors because of catalysts used to make longlasting dyes in an infinite variety of shades. The cars we drive depend on catalysis, from the sophisticated fuels they demand to the engine lubricants, brake fluids, antifreeze, durable interior appointments and high impact polymer components, as well as for the environmentally-important catalytic convertors which remove a large fraction of the troublesome combustion by-products which pollute the air we breathe. Overall, catalysis is involved in some 90% of modern chemical processes, cutting a broad swath across the fuels, plastics, materials and pharmaceutical industries.

One compelling reason for the dominance of catalytic technologies in modern chemical processes is clearly economic. Catalytic processes are usually less capital-intensive and require lower costs to operate than conventional non-catalytic approaches. Our standard of living and economic competitiveness are closely linked to the use of catalytic chemistry. We can also benefit from using our natural resources to manufacture in a less wasteful way. Better catalytic processes generate higher purity products with fewer by-products, thereby minimizing the environmental impact of chemical manufacturing. They effect chemical transformations with less energy input, which has lately become an area of urgent environmental concern. Finally, new catalysts, especially biocatalysts, are being sought to develop the chemistry of alternative raw materials, in order to eventually reduce our dependence on non-renewable natural resources.

The first objective of this book is to show some of the innovative ways in which researchers are developing new catalytic processes and better understanding of existing ones, in order to make the kinds of improvements that will raise our collective ability to maximize the returns for our investments of energy and resources. Nominally a catalyst is any substance which participates in a chemical reaction and causes its rate acceleration, but which can (in principle) be recovered in its original form after the reaction, to be reused. Catalysts are traditionally classified into two monolithic categories: homogeneous and heterogeneous, depending on whether or not they are soluble. Research into the former has been the largely the preserve of chemists, because soluble, molecular complexes tend to be inherently more tractable to spectroscopic and diffraction analyses. However, practical catalysts are solid metals or metal oxides, whose applications are most often studied by chemical engineers. The oft-lamented cultural divide between the two research communities of chemistry and chemical engineering has undoubtedly hindered progress in catalyst design and application. "How do you improve a reaction if you don't understand how it works?", asks the chemist, and the engineer replies: "What is the point of designing a new catalyst without any regard to its practicality?" Fortunately, this mutual *méfiance* is no longer universal, as more chemists and chemical engineers learn to communicate in each others' language, abandon conventional allegiances and recognize the importance of both kinds of knowledge. The second objective of this book is therefore to place the contributions of chemists and chemical engineers in close proximity, in the hope that readers too will appreciate the complementarity of their approaches.

Catalyst discovery at the dawn of the 21st century still involves a high degree of serendipity, and heterogeneous catalysis cannot yet be described as a truly predictive science. Hence, more fundamental knowledge about the intrinsic nature of active sites is critical to the rational development of better catalysts. There is first a need to define and then control the atomic structure of the active sites, which in the realm of heterogeneous catalysis involves the preparation of materials with well-defined architectures on length scales somewhat longer than the molecular. Catalysis is still largely a localized phenomenon; ensemble effects are generally transmitted over small distances of several to several dozen atoms. In addition, a materials science revolution is in progress: methods for the preparation and characterization of macroscopic materials and prediction of their properties have now been realized. The next challenge in catalysis also lies here, specifically, in the relationship between supramolecular structure, mechanism and activity. Many of the innovative accomplishments are the product of cross-disciplinary fertilization in the minds of young researchers who have sought experiences across traditional boundaries. A showcase of the latest progress in such emerging fields is the third and final objective of this book.

Cathy, Chris and I hope that this book will be of interest to our colleagues, from junior graduate students to independent researchers, both academic and industrial. We trust its breadth will appeal to both chemists and chemical engineers. As a series of topical discussion papers, it could serve as a textbook for a graduate course in catalysis in either discipline. Finally, it may be useful as an overview of current directions of thought in curiosity-driven research and as an introduction to the best young researchers in this field, whose ideas we have chosen to be represented here.

Susannah Scott
Ottawa, 2002

CONTENTS

Chapter 1

MULTIFUNCTIONAL ACTIVE SITES ON SILICA SURFACES BY GRAFTING OF METAL COMPLEXES

Susannah L. Scott,* Eric W. Deguns
Centre for Catalysis Research and Innovation, Department of Chemistry, University of Ottawa, Ottawa ON Canada K1N 6N5

Keywords: silica, supported metal catalysts, active sites, siloxane cleavage, silanols, multifunctional catalysts

Abstract: Supported metal catalysts containing more than one functionality can be created by routes involving simultaneous or sequential grafting of one or more molecular complexes. The resulting active sites are more complex than single component heterogeneous or homogeneous analogues, and may offer significant benefits in terms of catalytic selectivity and stability towards deactivation processes. Non-hydrolytic methods for the preparation of heterometallic active sites are potentially a powerful route to finely-tuned multicomponent heterogeneous catalysts. Strategies for using siloxane and surface hydroxyl sites to create such active sites are presented, with an emphasis on particularly well-characterized systems.

1.   INTRODUCTION

Multifunctionality in catalyst design is achieved by combining two or more components with different properties to create a catalyst system which is more effective or versatile than either component independently, and more efficient than running two catalytic processes sequentially. In tandem catalysis, the two catalyst components, present together, may have complementary functions, such as the clever homogeneous mixture of Ti and Ni "single-site" catalysts which generates oligomers from ethylene and then combines them into an ethylene copolymer.[1] Alternately, complementary functions may be performed by catalysts derived from the same precursor at different times during the catalytic process, such as the soluble Ru carbenes which can first induce olefin metathesis (*e.g.*, ring-opening or ring-closing), then hydrogenate the products;[2] switching is achieved by a modest change in the reaction conditions.

Combining multiple functions in a single homogeneous catalyst system requires considerable synthetic effort and luck in order to maintain the activity of the individual components while minimizing undesirable interactions. In contrast, tandem catalysis and multicomponent systems are the norm in heterogeneous catalysis, although the interactions of the constituent components are often poorly controlled. An example of a tandem heterogeneous catalyst is $Cr_2(CH_2SiMe_3)_8$ supported on silica, which performs simultaneous oligomerization and copolymerization of ethylene.[3] In multicomponent catalysis, the presence of Ti promotes the activity of Cr/SiO_2 polymerization catalysts[4] and enhances the stability of V/SiO_2 deNO$_x$ catalysts,[5] although the nature of the Cr-Ti and V-Ti interactions is far from clear.

Methods for generating well-defined heterogeneous active sites incorporating multiple functionality are emerging. In general, they build on the known chemistry of homogeneous catalysts and their surface organometallic chemistry.[6] Since purification of supported catalysts (analogous to the isolation of a homogeneous catalyst) is generally impossible, surface reactions must be quantitative, preferably generating no or only volatile, inert side-products. In this chapter, techniques for effecting clean, complex modifications of silica surfaces are described, utilizing either siloxane bonds or surface hydroxyls as grafting sites for one or more molecular species. Both one-step grafting reactions and sequential deposition of metal complexes are discussed.

2.   GRAFTING ON STRAINED SILOXANE BONDS

In principle, cleavage of a siloxane bond can result in the grafting of two identical or different chemisorbed groups on adjacent surface binding sites. However, because of the low reactivity of unstrained, nonpolar siloxane sites on silica,[7] this approach is usually limited to complexes of the highly oxophilic early transition metals, or grafting on the highly strained siloxane sites ("defects") created by extreme thermal treatments ($\geq 600°C$).[8,9] For example, highly dehydroxylated Aerosil silica pretreated at *ca.* 1100°C under vacuum, such that the OH density is reduced to *ca.* 0.4 per nm^2,[10] contains 0.15 highly strained four-membered siloxane rings per nm^2, formed by the condensation of vicinal hydroxyls, eq. 1.[11,12]

$$(1)$$

Such sites are reactive enough to dissociatively chemisorb H_2O, NH_3, *etc.*[8,13,14] It therefore comes as no surprise that they can also react with mixed alkylalkoxysilanes to generate silylated silicas containing adjacent alkoxide and alkylsilane sites, eq. 2.[15-17]

$$(2)$$

In contrast, silylating agents such as hexamethyldisilazane which react exclusively with surface hydroxyl sites generate silylated silicas with no spatial relationship between the silanol sites which are modified, eq. 3.

$$2 \equiv\!SiOH + (Me_3Si)_2NH \rightarrow 2 \equiv\!SiOSiMe_3 + NH_3 \qquad (3)$$

Early reports of siloxane cleavage reactions used to graft metal complexes to create supported catalysts include the deposition of organometallic reagents as diverse as $AlMe_3$[18-20] and $Ni(allyl)_2$[21] on silica. The evidence initally proposed for siloxane cleavage was the uptake of more metal and/or ligand than could be accounted for by the hydroxyl population of a highly dehydroxylated silica. However, the first direct observation of siloxane cleavage was reported during grafting of $Cp^*_2ThMe_2$ on highly dehydroxylated silica, eq. 4. The reaction was described as Th-Me addition to a Si-O bond.[22]

$$\equiv\!SiOSi\!\equiv\ +\ Cp^*_2ThMe_2 \rightarrow\ \equiv\!SiOThCp^*Me\ +\ \equiv\!SiMe \qquad (4)$$

In an early example of the use of high resolution solid state NMR to characterize surface organometallic complexes, the ^{13}C CP/MAS NMR spectrum of this material revealed the presence of three kinds of methyl groups. Signals at 59, 9 and −5 ppm were assigned to Cp-Me, Th-Me and $\equiv\!Si$-Me, respectively. Analogous ^{13}C and ^{29}Si solid state NMR spectroscopic evidence for methylation of porous silicas by $AlMe_3$ has now been published,[23,24] although in this system methyl transfer may be much less important than previously believed.[25]

$Zr(CH_2CMe_3)_4$ is also chemisorbed by a siloxane cleavage mechanism on $1000°C$-pretreated silica. Although reaction with residual surface hydroxyls does occur, resulting in the liberation of neopentane, the Zr loading exceeds the hydroxyl content of this silica by a factor of 2.[26] Furthermore, 25% of the neopentyl groups on the modified surface resist protonolysis by HCl, indicating that they are no longer bound to Zr. This suggests the formation of direct Si-C bonds during grafting, eq. 5.[27]

(5)

This alkylation of silicon is similar to the proposed mechanism of hydride transfer to silica which occurs during the transformation of ($\equiv$SiO)MR$_3$ (M is Ti, Zr, Hf) to ($\equiv$SiO)$_3$MH and $\equiv$SiH when heated in the presence of H$_2$.[28]

Since the maximum metal loading on ZrNp$_4$-modified 1000°C-pretreated silica exceeds the combined hydroxyl and "defect" site densities, siloxane cleavage is not restricted to highly strained four-membered rings. This observation raises the question of whether siloxane cleavage might be occurring during grafting of ZrR$_4$ (R = Np, Ns, allyl, *etc.*) and similar early transition metal organometallic reagents on silicas pretreated at temperatures below the threshold for formation of the defect sites (600°C). However, measurement of alkane yields suggests that grafting occurs on exclusively on isolated hydroxyl sites on 500°C-silica[31,38] and on pairs of hydroxyl sites on 200°C-silica.[29,31] Upon complete reaction of the hydroxyl sites, the surface is completed obscured by the neopentyl ligands, thereby rendering the unstrained siloxanes inaccessible to attack by the metal complex.[39] The remarkable catalytic activity of ZrR$_4$-modified low temperature silicas for olefin isomerization, hydrogenation[29,30] and polymerization,[31-34] as well as alkane hydrogenolysis and H-D exchange,[35-37] has been intensively investigated. In contrast, high temperature silica modified by MR$_4$ complexes, such as in eq. 5, have surfaces that are both active (*e.g.*, presence of ZrNp$_3$ fragments) and hydrophobic (*e.g.*, presence of SiNp fragments). The importance of catalyst hydrophobicity for substrate adsorption and subsequent conversion are emphasized by De Vos *et al.* in Chapter 14. Enhanced hydrophobicity may be particularly advantageous for early transition metal supported catalysts, which are highly moisture-sensitive.

Siloxane cleavage was recently reported to be a minor pathway in the modification of a partially dehydroxylated silica surface (T=450°C) with Cp*TaMe$_4$.[40] However, much more silica methylation was observed when highly dehydroxylated silica (T=950°C) was treated with Cp$_2$TaMe$_3$. In this case, though, the organotantalum complex does not become attached to the silica surface through an oxo bridge. Instead, it appears to be associated as a cationic complex with an anionic site on the silica, eq. 6.[40]

$$\equiv\text{SiOSi}\equiv \ + \ \text{Cp}_2\text{TaMe}_3 \ \rightarrow \ \equiv\text{SiO}^- + \ \text{Cp}_2\text{TaMe}_3^+ + \ \equiv\text{SiMe} \qquad (6)$$

The literature contains few examples of silica acting as a counteranion to a supported metal complex, in the reaction of grafted $\equiv\text{SiORh(PMe}_3)_3$ with H_2 in the presence of excess PMe_3, eq. 7,[41] and the reversible reaction of grafted $\equiv\text{SiORh(PMe}_3)_2(\text{CO})$ with excess PMe_3, eq. 8.[42]

$$\equiv\text{SiORh(PMe}_3)_3 + H_2 + PMe_3 \ \rightarrow \ \equiv\text{SiO}^- + \ cis\text{-Rh(PMe}_3)_4(\text{H})_2^+ \qquad (7)$$

$$\equiv\text{SiORh(PMe}_3)_2(\text{CO}) + PMe_3 \ \rightleftarrows \ \equiv\text{SiO}^- + \ \text{Rh(PMe}_3)_3(\text{CO})^+ \qquad (8)$$

Although cationic species are associated with increased activity in a number of homogeneous catalytic reactions, in the examples cited above the steric saturation of the metal coordination sphere required to maintain cation-anion separation at the surface appears to be incompatible with substrate binding.

Siloxane cleavage reactions have also been reported in surface coordination chemistry, where the driving force creating a new Si-C bond is absent. The gas-solid reaction of Re_2O_7 with the siloxane sites of highly dehydroxylated silica is shown in eq. 9.[43]

$$(9)$$

This surface reaction is analogous to the molecular siloxane cleavage which occurs during the elegantly atom-efficient synthesis of $Me_3SiOReO_3$. Cleavage of one Re-O bond of Re_2O_7 and one Si-O bond of $Me_3SiOSiMe_3$ is compensated by the concerted formation of a new Re-O bond and a new Si-O bond, eq. 10.[44]

$$O_3\text{ReOReO}_3 + Me_3\text{SiOSiMe}_3 \ \rightarrow \ 2 \ Me_3\text{SiOReO}_3 \qquad (10)$$

Siloxane cleavage is the only grafting reaction accessible to Re_2O_7, which does not react with hydroxyl sites on silica. Its reaction with siloxanes generates a modified silica containing adjacent grafted perrhenate esters. Reaction of Re_2O_7 at its sublimation temperature, 350°C, in the presence of 200 Torr O_2, with the four-membered defect rings generates vicinal silyl perrhenate esters, $[\equiv\text{SiOReO}_3]_2(\mu\text{-O})$. The reaction is characterized by the disappearance of IR vibrations characteristic of the strained rings (908, 888 cm^{-1}) and the appearance of bands assigned to $2v_{as}(ReO_3)$ at 1950 cm^{-1} and $v_{as}(ReO_3) + v_s(ReO_3)$ at 1989 cm^{-1}.[43] The analogous reaction of $Bu_3SnOSnBu_3$ with 1000°C-pretreated silica was reported to yield vicinal $\equiv SiOSnBu_3$ sites by siloxane cleavage.[27]

The maximum loading of Re achieved in the siloxane reaction of Re_2O_7 was ca. 1 per nm^2, which greatly exceeds the reported density of four-membered siloxane rings on 1100°C-treated silica, 0.15 per nm^2.[8] We inferred that grafting also occurs on less strained siloxane sites, as a result of the high reaction temperature (350°C) required for sublimation. The resulting perrhenates are not vicinal but adjacent, since the single oxo bridge linking the silicon atoms of the unstrained siloxane sites on which they are grafted is cleaved during the reaction. In contrast, the more volatile $Me_3SiOReO_3$, sublimed at room temperature onto 1100°C-pretreated silica, reacts selectively with four-membered siloxane rings, eq. 11.[6]

$$(11)$$

The maximum perrhenate loading from reaction of $Me_3SiOReO_3$ is only 0.14 per nm^2 (equal to the density of $=Si_2O_2$ sites), thus each perrhenate site is grafted vicinal to a bulky, hydrophobic trimethylsilylated silicon atom.

3.   GRAFTING ON ADJACENT HYDROXYL SITES

The phenomenon of a single metal complex being grafted onto adjacent hydroxyl sites, accompanied by displacement of two of its ligands, is well-established and has been suggested as evidence for the predominantly paired nature of hydroxyl groups on silicas treated at low dehydroxylation temperatures (<400°C), eq. 12.[31,39]

$$(12)$$

An important goal in the development of multifunctional catalysts is the ability to graft, concurrently or sequentially, two similar or different moieties on adjacent hydroxyl sites. Such a situation arose spontaneously when silica pretreated at 200°C was exposed to mononuclear Rh(C$_3$H$_5$)(CO)$_2$, resulting in association of Rh sites and formation of dinuclear dicarbonyl Rh complexes, eq. 13.[45]

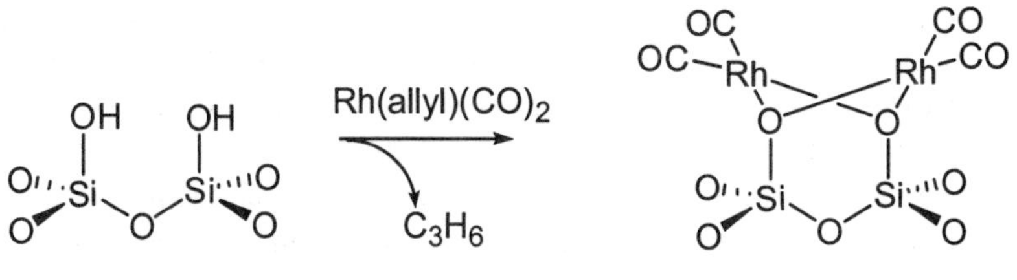

$$(13)$$

The dimeric sites were identified by their characteristic three-band υ(CO) pattern.[45] Since the carbonyl ligands are all terminal, bridging by vicinal surface siloxide ligands was inferred to be responsible for dimer formation. The same product was obtained upon treatment of silica modified with Rh(allyl)$_3$ (containing 18-electron bis(allyl)rhodium sites associated with surface hydroxyl groups)[46] with CO.[47] However, in this case, mononuclear ≡SiORh(CO)$_2$ was initially detected, and evolved over several minutes to the dinuclear product, Scheme 1. This observation of a long-lived intermediate suggests that monomers which are not initially adjacent to one another can migrate across the silica surface. Mobility of grafted organometallic fragments is probably mediated by residual surface hydroxyls.[48]

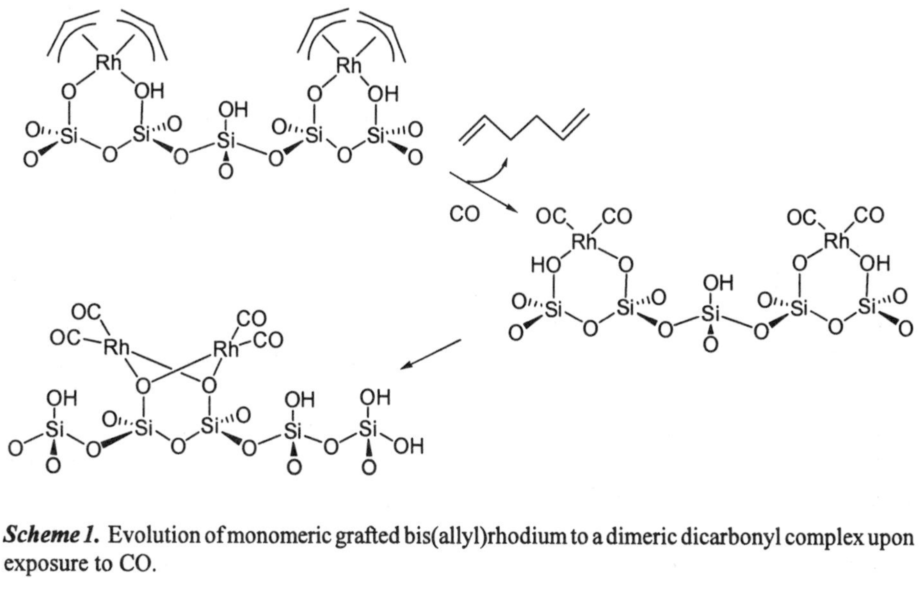

Scheme 1. Evolution of monomeric grafted bis(allyl)rhodium to a dimeric dicarbonyl complex upon exposure to CO.

Residual hydroxyl groups which persist after grafting of metal complexes on the silica surface have been blamed for deactivation of "site-isolated" supported catalysts, since aggregation can be mediated by the facile migration of surface protons.[49-51] In cases where such mobility is not desired, it may be greatly impeded by cofunctionalization of the catalyst, for example, modification of the Rh(CO)$_2$–modified silica surface with trimethylsilyl groups in order to cap unreacted hydroxyls.[52] The resulting site isolation is reflected in both higher activity and stability of the catalyst. Capping can also stabilize supported catalysts towards hydrolysis, since hydroxyl-terminated silica surfaces are rendered hydrophobic by the chemical modification. Ideally, each active site should be surrounded by hydrophobic, non-displaceable trialkylsilyl sites which prohibit its migration. The capping reaction usually involves a silylating agent such as SiCl$_x$R$_{4-x}$, Si(OR)$_x$R$_{4-x}$ or (Me$_3$Si)$_2$NH (see eq. 3 above). Displacement of HCl, ROH or NH$_3$, respectively, leads to chemisorbed alkylsilane fragments, although the reactions are complicated by self-condensations and chlorination of silica.[7,53,54] The effectiveness of this procedure depends on the efficiency of the capping or silylating agent and its selectivity towards reaction with the silica hydroxyl groups rather than with the supported metal complexes.

Recently, the modification of MCM-41 with rare-earth bis(dimethylsilylamides) was shown to result in a one-pot grafting/capping reaction involving transfer of dimethylsilyl groups to the surface via condensation on surface hydroxyl groups of the HN(SiHMe$_2$)$_2$ liberated during grafting, eq. 14.[55]

$$3 \equiv\!SiOH + Y\{N(SiHMe_2)_2\}_3(THF)_2 \rightarrow$$
$$\equiv\!SiOY\{N(SiHMe_2)_2\}_2 + 2 \equiv\!SiOSiHMe_2 + NH_3 + 2 THF \quad (14)$$

The capping process resembles the well-known reaction of silica hydroxyls with hexamethyldisilazane,[7] and is described in more detail in the chapter by Anwander.

We discovered that the adhesion mechanism for metallasiloxanes on silica diverges from the simple ligand metatheses observed for metal halides and alkoxides. Reaction of (Me$_3$SiO)$_2$Mo(=N^{t}Bu)$_2$ with the hydroxyl groups of a silica surface was expected to result in grafting of bis(*tert*-butylimido) molybdenum(VI) fragments with liberation of Me$_3$SiOH. Surprisingly, although the metal complex was indeed irreversibly adsorbed on the surface, and the hydroxyl groups were shown to be consumed by *in situ* IR spectroscopy, no volatiles were generated. In fact, elemental analysis of the resulting material showed that all of the carbon originally present in the molecular complex was still present on the surface. Furthermore, the ^{29}Si CP/MAS NMR spectrum showed evidence of trimethylsilylation of the silica surface. These results were interpreted in terms of a two-step grafting mechanism, Scheme 2. Initial reaction

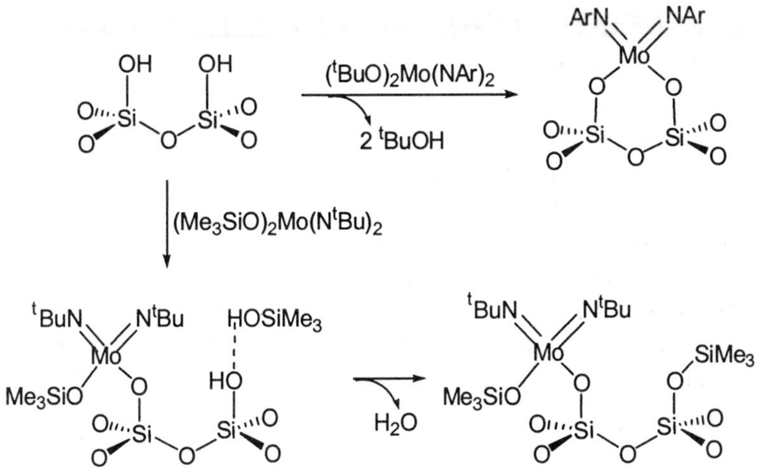

Scheme 2. Comparison of grafting modes of molybdenum alkoxide and siloxide complexes on silica.

on the silanol sites presumably results in the liberation of the expected Me₃SiOH, which is not released from the surface but undergoes condensation with an adjacent silanol. This condensation is apparently faster than self-condensation of Me₃SiOH, intermolecular reaction of the adjacent silanol with another equivalent of (Me₃SiO)₂Mo(=N'Bu)₂, or intramolecular reaction of the adjacent silanol with the grafted Mo fragment. Its efficiency may be due to retention of the Me₃SiOH on the nearest hydroxyl site by H-bonding, combined with activation of that hydroxyl site by the neighboring grafted organometallic fragment. In contrast, the reaction of the analogous alkoxide complex Mo(NAr)₂(O'Bu)₂ (Ar is 2,6-diisopropylphenyl) with 200°C-activated silica occurs by straightforward disubstitution at the metal with liberation of 'BuOH, Scheme 2.[56] Unlike trimethylsilanol, alcohols condense with silica hydroxyl groups to form ≡SiOR and H₂O only at elevated temperatures (> 400 K).

The ability of metallasiloxanes to generate trialkylsilanols *in situ* during grafting suggests that this class of single-source precursors can be used to effect a designed dual modification of the surface. The reaction is apparently general for metal trimethylsiloxides, whose grafting leads to local environments for the supported catalysts which are modified by the presence of adjacent Me₃Si groups adsorbed on the surface *via* nonhydrolyzable siloxane or "glass bonds". Thus grafting of Ti(OSiMe₃)₄ also generates trimethylsilylated silica,[57] unlike the analogous reaction of Ti(O'Pr)₄ where the 'PrOH product is not retained on the surface.[58]

4.   GRAFTING ON METAL-MODIFIED SILICA SURFACES

A third method for creating complex active site architectures in supported catalysts involves the chemisorption of a metal complex on an already modified silica surface.   An early, serendipitious example involved the reaction of $Bu_2Sn(OMe)_2$ with silica, which was reported to generate a material containing 1.4 times more Sn than the hydroxyl groups originally present.[59] IR evidence for μ-methoxy ligands suggested a trinuclear structure. Furthermore, one-third of the Sn was displaced by extraction with MeOH, eq. 15, suggesting that the central Sn complex is labile while the outer Sn complexes are strongly chemisorbed.

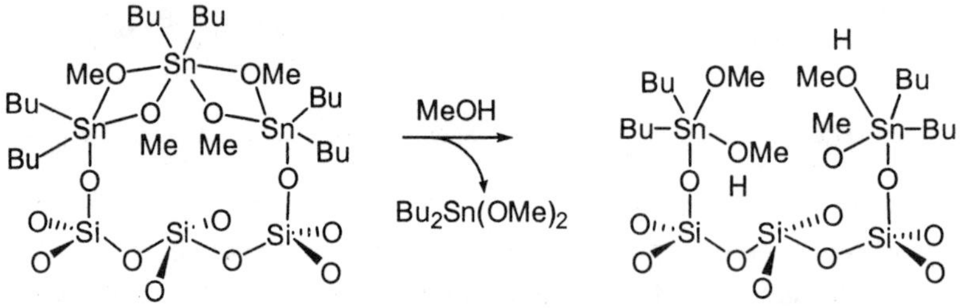

$$(15)$$

If the displacement reaction is reversible, it suggests the possibility of introducing a different metal complex between the outer chemisorbed Sn complexes, in order to create a heterobimetallic active site.

The spontaneous association of metal alkoxides was also observed during grafting of $Ti(O^iPr)_4$ onto partially dehydroxylated fumed silica surfaces.[58] Alkoxide disproportionation ensues, Scheme 3.

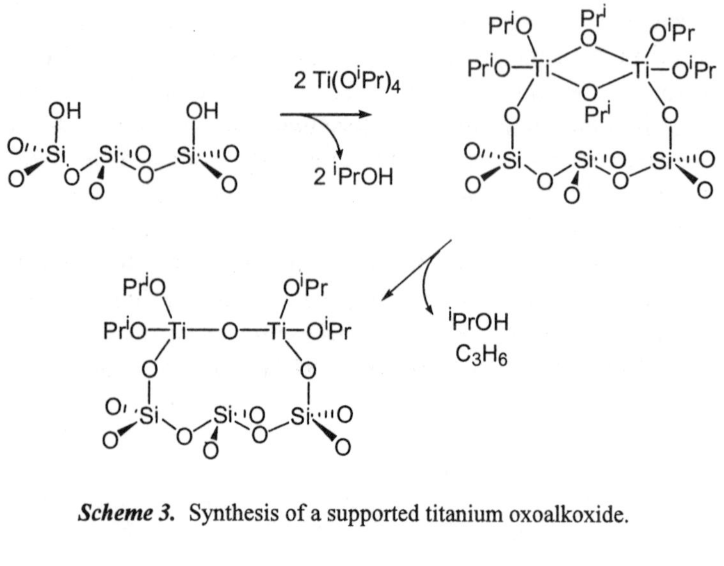

Scheme 3. Synthesis of a supported titanium oxoalkoxide.

The presence of Ti-O-Ti appears to be essential for catalytic activity in olefin epoxidation in these systems.[60]   The formation of dinuclear oxoalkoxide complexes is apparently promoted by silica, since the reactions of Ti(OR)$_4$ (R is Me, Et and ^{i}Pr) with partially condensed silsesquioxanes do not lead to alkoxide disproportionation,[61] although alkoxide-bridged dititanium complexes were isolated in some cases.[62-64] These observations suggest that there is an important electronic difference between silica surfaces and the silsesquioxane model compounds, discussed in detail in the chapter by Duchateau.   Curiously, complexes featuring the Ti-O-Ti structure formed spontaneously when silsesquioxanes reacted with titanocene dihalides, despite the absence of an explicit oxygen source.[64]

Condensation reactions analogous to those shown in Scheme 3 can be induced to occur between different metal complexes, creating heterobimetallic active sites. For example, reaction of a VO(O^{i}Pr)$_3$-modified silica surface with Ti(O^{i}Pr)$_4$ generates a material containing equal amounts of each metal.[65] The IR spectrum indicates that the SiO-V bond is displaced by SiO-Ti, and the observation of alkoxide disproportionation products implies the formation of Ti-O-V, eq. 16.

$$\equiv\text{SiOVO(O}^i\text{Pr)}_2 + \text{Ti(O}^i\text{Pr)}_4 \rightarrow \equiv\text{SiOTiOVO(O}^i\text{Pr)}_4 + {}^i\text{PrOH} + \text{C}_3\text{H}_6 \quad (16)$$

The resulting Ti-anchored oxidation catalyst is more stable under reaction conditions than the V-anchored catalyst, presumably due to the greater strength of the Ti-O bond to the silica surface. It is possible to envisage such reactions taking place with other pairs of metal alkoxide partners to create a variety of bimetallic sites with interesting catalytic properties.

5.   OUTLOOK

With methods for grafting single component homogeneous catalysts onto oxide (most often siliceous) supports now well-established, attention will naturally focus on more complex active site architectures. The placement of hydrophobic alkylsilyl groups proximal to the active site may slow catalyst deactivation due to leaching and/or surface migration of metal fragments. Positioning of chiral organosilyl auxiliaries at adjacent/vicinal sites by one-step concurrent metal and ligand grafting will accomplish the same goals and, in addition, may influence the enantioselectivity of the catalytic reaction at neighboring active sites.

Deposition methods involving hydrolysis of mixtures of metal complexes, with little or no control over the interactions between the components, will be complemented by cleaner, non-hydrolytic techniques (essential for sensitive organometallic catalyst components). Controlled sequential deposition of two simple metal complexes will prove to be a more versatile approach than the time-consuming synthesis of heterometallic molecular precursors. Furthermore, post-grafting ligand incorporation through simple metathetical exchange will prove highly useful, avoiding synthesis of discrete molecular catalyst precursors and lending itself to combinatorial catalyst screening.

ACKNOWLEDGEMENTS

Portions of this work were funded by NSERC (Canada), Union Carbide's Innovation Recognition Program and the Cottrell Scholar program of Research Corporation. The author gratefully acknowledges the financial support of the Province of Ontario (PREA) and the Canada Research Chairs program.

REFERENCES

1.   Komon, Z. J. A.; Bu, X.; Bazan, G. C. *J. Am. Chem. Soc.* **2000**, *122*, 1830.
2.   Louie, J.; Bielawski, C. W.; Grubbs, R. H. *J. Am. Chem. Soc.* **2001**, *123*, 11312; Drouin, S. D.; Zamanian, F.; Fogg, D. E. *Organometallics*, **2001**, *20*, 5495.
3.   Benham, E. A.; Smith, P. D.; McDaniel, M. P. *Polym. Eng. Sci.* **1988**, *28*, 1469.
4.   McDaniel, M. P.; Welch, M. B.; Dreiling, M. J. *J. Catal.* **1983**, *82*, 118.
5.   Rajadhyaksha, R. A.; Hausinger, G.; Zeilinger, H.; Ramstetter, A.; Schmelz, H.; Knözinger, H. *Appl. Catal.* **1989**, *51*, 67.
6.   Scott, S. L.; Basset, J. M. *J. Mol. Catal.* **1994**, *86*, 5.
7.   Morrow, B. A. *Stud. Surf. Sci. Catal.* **1990**, *57A*, 161.
8.   Morrow, B. A.; Cody, J. A. *J. Phys. Chem.* **1976**, *80*, 1998.
9.   Bunker, B. C.; Haaland, D. M.; Michalske, T. A.; Smith, W. L. *Surf. Sci.* **1989**, *222*, 95.
10.   Curthoys, G.; Davydov, V. Y.; Kiselev, A. V.; Kiselev, S. A.; Kuznetsov, B. V. *J. Coll. Interface Sci.* **1974**, *48*, 58.
11.   Bunker, B. C.; Haaland, D. M.; Ward, J. K.; Michalske, T. A.; Smith, W. L.; Binkley, J. S.; Melius, C. F.; Balfe, C. A. *Surf. Sci.* **1989**, *210*, 406.
12.   Bunker, B. C.; Haaland, D. M.; Michalske, T. A.; Smith, W. L. *Surf. Sci.* **1989**, *222*, 95.
13.   Morrow, B. A.; Cody, I. A. *J. Phys. Chem.* **1976**, *80*, 1995.
14.   Morrow, B. A.; Cody, I. A.; Lee, L. S. M. *J. Phys. Chem.* **1976**, *80*, 2761.
15.   Dubois, L. H.; Zegarski, B. R. *J. Phys. Chem.* **1993**, *97*, 1665.
16.   Dubois, L. H.; Zegarski, B. R. *J. Am. Chem. Soc.* **1993**, *115*, 1190.
17.   Blümel, J. *J. Am. Chem. Soc.* **1995**, *117*, 2112.
18   Yates, D. J. C.; Debinski, G. W.; Kroll, W. R.; Elliott, J. J. *J. Phys. Chem.* **1969**, *73*, 911.
19.   Kunawicz, J.; Jones, P.; Hockey, J. A. *Trans. Farad. Soc.* **1971**, *67*, 848.
20.   Peglar, R. J.; Murray, J.; Hambleton, F. H.; Sharp, M. J.; Parker, A. J.; Hockey, J. A. *J. Chem. Soc. (A)* **1970**, 2170.
21.   Ermakov, Y. I.; Kuznetsov, B. N.; Karakchiev, L. G.; Derbeneva, S. S. *Kinet. Catal.* **1973**, *14*, 611.
22.   Toscano, P. J.; Marks, T. J. *Langmuir* **1986**, *2*, 820.
23.   Kratochvila, J.; Kadlc, Z.; Kazda, A.; Salajka, Z. *J. Non-cryst. Solids* **1992**, *143*, 14.

24. Anwander, R.; Palm, C.; Groeger, O.; Engelhardt, G. *Organometallics* **1998**, *17*, 2027.
25. Tao, T.; Maciel, G. E. *J. Am. Chem. Soc.* **2000**, *122*, 3118.
26. Scott, S. L.; Basset, J. M., unpublished results.
27. Adachi, M.; Lefebvre, F.; Basset, J.-M. *Chem. Lett.* **1996**, 221.
28. Corker, J.; Lefebvre, F.; Lécuyer, C.; Dufaud, V.; Quignard, F.; Choplin, A.; Evans, J.; Basset, J.-M. *Science* **1996**, *271*, 966.
29. Schwartz, J.; Ward, M. D. *J. Mol. Catal.* **1980**, *8*, 465.
30. Zakharov, V. A.; Ryndin, Y. A. *J. Mol. Catal.* **1989**, *56*, 183.
31. Ballard, D. G. H. *Adv. Catal.* **1973**, *23*, 263.
32. Candlin, J. P.; Thomas, H. *Adv. Chem. Ser.* **1974**, *132*, 212.
33. Zakharov, V. A.; Yermakov, Y. I. *Catal. Rev.-Sci. Eng.* **1979**, *19*, 67.
34. Vasnetsov, S. A.; Nesterov, G. A.; Zakharov, V. A.; Thiele, K.-H.; Scholtz, I. *React. Kinet. Catal. Lett.* **1988**, *36*, 383.
35. Quignard, F.; Lécuyer, C.; Choplin, A.; Olivier, D.; Basset, J.-M. *J. Mol. Catal.* **1992**, *74*, 353.
36. Niccolai, G. P.; Basset, J.-M. *Appl. Catal. A: Gen.* **1996**, *146*, 145.
37. Casty, G. L.; Matturro, M. G.; Myers, G. R.; Reynolds, R. P.; Hall, R. B. *Organometallics* **2001**, *20*, 2246.
38. Quignard, F.; Lécuyer, C.; Choplin, A.; Olivier, D.; Basset, J.-M. *J. Mol. Catal.* **1992**, *74*, 353.
39. Amor Nait Ajjou, J.; Scott, S. L. *Organometallics* **1997**, *16*, 86.
40. Ahn, H.; Marks, T. J. *J. Am. Chem. Soc.* **2002**, *124*, 7103.
41. Scott, S. L.; Dufour, P.; Santini, C. C.; Basset, J.-M. *J. Chem. Soc., Chem. Commun.* **1994**, 2011.
42. Scott, S. L.; Szpakowicz, M.; Mills, A.; Santini, C. C. *J. Am. Chem. Soc.* **1998**, *120*, 1883.
43. Scott, S. L.; Basset, J.-M. *J. Am. Chem. Soc.* **1994**, *116*, 12069.
44. Schmid, M.; Schmidbauer, H. *Inorg. Synth.* **1967**, *9*.
45. McNulty, G. S.; Cannon, K.; Schwartz, J. *Inorg. Chem.* **1986**, *25*, 2919.
46. Dufour, P.; Houtman, C.; Santini, C. C.; Nédez, C.; Basset, J. M.; Hsu, L. Y.; Shore, S. G. *J. Am. Chem. Soc.* **1992**, *114*, 4248.
47. Scott, S. L.; Dufour, P.; Santini, C. C.; Basset, J.-M. *Inorg. Chem.* **1996**, *35*, 869.
48. Santini, C. C.; Scott, S. L.; Basset, J.-M. *J. Mol. Catal.* **1996**, *107*, 263.
49. Basu, P.; Panayotov, D.; J. T. Yates, J. *J. Am. Chem. Soc.* **1988**, *110*, 2074.
50. Dufour, P.; Scott, S. L.; Santini, C. C.; Lefebvre, F.; Basset, J.-M. *Inorg. Chem.* **1994**, *33*, 2509.
51. Santini, C. C.; Scott, S. L.; Basset, J.-M. *J. Mol. Catal. A: Chem.* **1996**, *107*, 263.
52. Drago, R. S.; Pribich, D. C. *Inorg. Chem.* **1985**, *24*, 1983.
53. Feher, F. J.; Newman, D. A. *J. Am. Chem. Soc.* **1990**, *112*, 1931.
54. Tripp, C. P.; Hair, M. L. *Langmuir* **1991**, *7*, 923.
55. Anwander, R.; Runte, O.; Eppinger, J.; Gerstberger, G.; Herdtweck, E.; Spiegler, M. *J. Chem. Soc., Dalton Trans.* **1998**, 847.
56. Li, F.; Scott, S. L., unpublished results.
57. Roveda, C.; Church, T. L.; Alper, H.; Scott, S. L. *Chem. Mater.* **2000**, *12*, 857.
58. Bouh, A. O.; Rice, G. L.; Scott, S. L. *J. Am. Chem. Soc.* **1999**, *121*, 7201.
59. Ballivet-Tkatchenko, D.; dos Santos, J. H. Z.; Malisova, M. *Langmuir* **1993**, *9*, 3513.
60. Bouh, A. O.; Hassan, A.; Scott, S. L. In *Catalysis of Organic Reactions*, Morrell, D. Ed., Dekker: New York, 2002.
61. Maschmeyer, T.; Klunduk, M. C.; Martin, C. M.; Shephard, D. S.; Thomas, J. M.; Johnson, B. F. G. *Chem. Commun.* **1997**, 1847.
62. Crocker, M.; Herold, R. H. M.; Orpen, A. G. *Chem. Commun.* **1997**, 2411.

63.   Crocker, M.; Herold, R. H. M.; Orpen, A. G.; Overgaag, M. T. A. *J. Chem. Soc., Dalton Trans.* **1999**, 3791.
64.   Edelmann, F. T.; Gießmann, S.; Fischer, A. *J. Organomet. Chem.* **2001**, *620*, 80.
65.   Rice, G. L.; Scott, S. L. *Chem. Mater.* **1998**, *10*, 620.

Chapter 2

NANOSTRUCTURED RARE EARTH CATALYSTS VIA ADVANCED SURFACE GRAFTING

Reiner Anwander

Anorganisch-chemisches Institut, Technische Universität München, D-85747 Garching, Lichtenbergstraße 4, Germany

Keywords: rare earth elements (group 3 elements, lanthanides), nanostructuring, nano-sized particles, nanoporous materials, periodic mesoporous silica (PMS), lanthanide silylamides, surface organometallic chemistry (SOMC), organic/inorganic composite materials, catalysts, surface confinement, pore confinement, nanoenvironment, molecular oxo-surfaces

Abstract: Mesoporous silicas of the M41S family are discussed as attractive host materials in catalysis and materials science. Their intrinsic zeolite-like pore architecture with tunable pore sizes and narrow pore size distributions provides a unique platform (model support) to study the grafting of highly reactive organometallic compounds. In contrast to their alkyl and alkoxide derivatives, silylamide complexes of oxophilic and electrophilic metal centers display favorable surface reactions featuring (i) mild reaction conditions, (ii) the formation of thermodynamically stable metal siloxide bonds, (iii) concomitant surface silylation, (iv) favorable atom economy, and (v) the absence of any insoluble by-products. Various conceptual approaches to this novel heterogeneously performed silylamine elimination are presented for the lanthanide elements, which are promising components of nanostructured catalysts. Particular emphasis is put on the application of conclusive methods of characterization of the hybrid materials, elaborating the importance of spectroscopic probe ligands and nitrogen physisorption measurements. It is shown that the Lewis acidity (*i.e.*, cation size), pore diameter, surface silylation and confinement are critical parameters influencing the catalytic performance of lanthanide(III) surface species, for example, in the hetero Diels-Alder and Meerwein-Ponndorf-Verley reactions. Moreover, pore entrapment and surface confinement seem to be important factors which govern the stabilization and reactivity patterns of the surface species involved in the samarium(II)-mediated reduction of ketones.

1.   INTRODUCTION

The nanosize regime, approximately 1 to 100 nm, is of fundamental importance in materials science technology.[1-4] The modification of chemical and physical properties, using size effects, is the ultimate inspiration for the development of nanostructured catalysts[5] and quantum-confined materials.[6] Principally, nanostructuring of catalytic species can occur either through nano-sized bulk particles[7] or via a nanoporous matrix, hosting the catalytic species as a so-called nanoreactor.[8,9] Surface reactivity is the pivotal criterion of such cluster/nanophase materials. Nanostructured bulk catalysts feature intrinsically different adsorption and impart peculiar surface reactivity which is ascribed to unusual surface morphology, unusual surface defects, and unusual electronic sites.[10] Unlike larger crystallites, their increased number of edges and corners produces a high concentration of coordinatively unsaturated sites of enhanced reactivity. Stabilization of such nanoparticles can be achieved by, for example, a "protective" ligand shell L.[11] On the other hand, nanocavities of porous host materials can serve as templates for the growth of nano-particles[12] or as a nanoreactor for immobilized catalysts.[13] The "control-led" generation of "artificial" surface defects via post-treatment methods such as surface grafting is a prominent approach toward enhanced surface reactivity of inorganic materials.[14] Following this classification, Figure 1 outlines various manifestations of nanostructured rare earth catalysts.

The present chapter will focus on the advanced synthesis and catalytic relevance of rare earth-modified mesoporous materials with special emphasis on surface organometallic chemistry at periodic mesoporous silica (SOMC@PMS).[15] Moreover, the following section pays tribute to the expanding area of nanosized rare earth bulk catalysts by highlighting some recent developments.

2.   NANO-SIZED RARE EARTH PARTICLES

A variety of known chemical and physical preparative methods has been applied to produce rare earth materials with nanometer structures. The SMAD (*s*olvated *m*etal *a*tom *d*ispersion) procedure or cryo-chemical synthesis,[16] involving low-temperature solvent trapping (77 K) of vaporized metal, was used

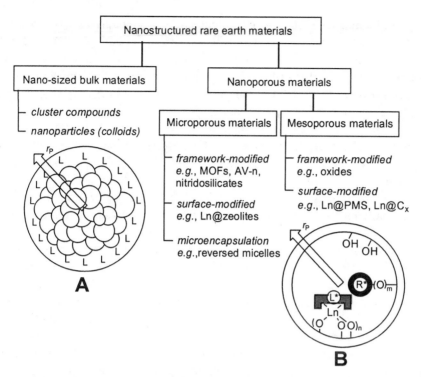

Figure 1. Schematic presentation of various forms of nanostructuring: **A**, bulk nanoparticle embedded in (protective/reactive) ligand shell (r_P = particle diameter); **B**, functionalized periodic nanoporous materials (r_P = pore diameter and R*/L* are additional parameters determining the local, micro- and mesoenvironments of the immobilized species); MOFs = metal–organic open frameworks; AV-n = Aveiro microporous solid no. n; PMS = periodic mesoporous silica; C_x = mesoporous tubular carbon.

for the generation of samarium and ytterbium nanoparticles.[17-19] THF, benzene, and methylenecyclohexane were employed as frozen matrices. The solvated lanthanide atoms Ln(S) are thermally labile and, upon warming, agglomerate to small spheroidal particles with sizes of 15-50 nm and BET (Brunauer–Emmett–Teller) surface areas of a_s = *ca.* 70 m^{2}g^{-1} (Figure 1, **A**).[18] Such cryo-suspensions or cryo-colloids exhibit unusual activity and selectivity in the hydrogenation of unsaturated hydrocarbons.[18]

Hydrothermal oxidation of cerium metal chips in 2-methoxyethanol at 200-250°C yielded a transparent colloidal solution of ultrafine, 2 nm-sized ceria particles.[20] Note that a spherical ceria particle with a diameter of 2 nm contains *ca.* 100 Ce atoms. Nanocrystalline nonstoichiometric CeO$_2$-based catalysts (particle size d_P = *ca.* 8 nm, a_s = *ca.* 50-65 m^{2}g^{-1}), obtained by magnetron sputtering and inert gas condensation,[21] exhibit increased catalytic acitivity in SO$_2$ reduction by CO, CO oxidation and methane oxidation compared with those of stoichiometric chemically precipitated catalysts.

Thermal treatment of organometallic or metalorganic compounds is another versatile method for the synthesis of catalytically relevant refractory lanthanide particles.[22] For example, lanthanide nitride nanoparticles were obtained via ammonolysis of molten silylamide complexes according to Scheme 1.[23] The presence of LiNH$_2$, formed via ammonolysis of Li[N(SiMe$_3$)$_2$], catalyzes the crystallization of LnN$_{1-x}$ at lower temperature.

Ammonolysis of Ln[N(SiMe$_3$)$_2$]$_3$ in hydrocarbon solvents also yielded precipitates of low ligand content which transform to the corresponding metal nitrides at *ca.* 400 °C.[24] The complex ScCl$_2$[N(SiMe$_3$)$_2$](THF)$_2$ was converted to ScN at 400°C *in vacuo*, with loss of THF and ClSiMe$_3$.[25] Borohydride complexes of europium(II) and ytterbium(II), (CH$_3$CN)$_x$Ln(BH$_4$)$_2$, produced crystalline single-phase metal borides EuB$_6$ and YbB$_4$ when heated at 700-1000°C/10^{-5} Torr.[26]

Gas phase pyrolysis of a structurally identified heterobimetallic neodymium aluminum alkoxide in a low-pressure chemical vapor deposition (LPCVD) apparatus at 500°C produced a nanoscaled NdAlO$_3$/Al$_2$O$_3$ ceramic–ceramic composite (Scheme 2).[27] Group 3 metal oxides and lanthanide oxides are classified as true "selective" catalytic reduction (SCR) systems, with the ability to reduce NO$_x$ in the presence of large excess of oxygen.[28] For example, nanocrystalline Y$_2$O$_3$, featuring particle sizes of $d_p = 9$-17 nm and BET surface areas of $a_s = 52$-101 m^{2}g^{-1}, exhibits a catalytic activity for the reduction of nitric oxide with methane comparable to that of Co-ZSM-5.[29] The highly

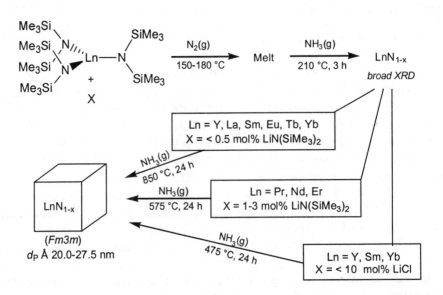

Scheme 1. Synthesis of crystalline LnN$_{1-x}$ nanoparticles. Adapted with permission from *Inorg. Chem.* **1992**, *31*, 1311. Copyright 1992 American Chemical Society.

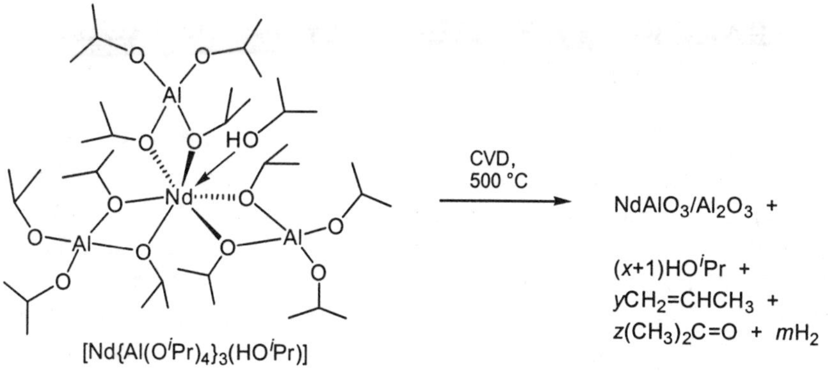

$$[Nd\{Al(O^iPr)_4\}_3(HO^iPr)] \xrightarrow{\text{CVD,}\ 500\ ^\circ C} NdAlO_3/Al_2O_3 +$$

$$(x+1)HO^iPr + yCH_2{=}CHCH_3 + z(CH_3)_2C{=}O + mH_2$$

Scheme 2. Synthesis of a NdAlO$_3$/Al$_2$O$_3$ ceramic–ceramic composite ($x/y \neq 1$, $x \gg y$, z, m and $x + y + z = 12$).[27]

dispersed yttrium oxide was prepared using a chemical precipitation technique, starting from an aqueous solution of yttrium nitrate and tetraethylammonium hydroxide.

A sol-gel preparation method was used, employing Ce(acac)$_3$ • nH$_2$O and Al(O^{s}Bu)$_3$ as ceria and alumina precursors, for the synthesis of efficient nanoscale CeO$_2$-based materials for deNO$_x$ applications.[30] CeO$_2$ contents of <25 wt% were necessary in order to obtain highly dispersed alumina at 1000°C ($d_P = ca.$ 30-40 nm; a$_S = ca.$ 100m^{2}g^{-1}). Moreover, functionalized emulsion and suspension polymer particles such as cross-linked acrylic emulsion polymers behave as nanoreactors for the synthesis of pure CeO$_2$ nanoparticles at temperatures as low as 150°C ($d_P = ca.$ 7 nm).[31] The versatility of this synthetic technique was demonstrated for the processing of supported catalysts such as Pt/CeO$_2$ and the coating of substrates with nanoparticulate inorganic materials. Water-in-oil microemulsions were also found to be ideal media for the preparation of ultrafine yttrium iron garnet (Y$_3$Fe$_5$O$_{12}$, YIB) nanoparticles ($d_P = ca.$ 3 nm),[32] ultrafine perovskites such as LaNiO$_3$ and La$_2$CuO$_4$,[33] and ultrafine ceria ($d_P = 2.6$-4.1 nm).[34] The reversed micelle technique was further exploited for the synthesis of polymer-based porous nanocomposites such as Eu(III) LLC-networks containing only water in the periodic nanochannels.[35] Cross-linking of the initially formed lyotropic (*i.e.*, amphiphilic) liquid crystals (LLCs), which adopt the inverted hexagonal phase, via photopolymerization is the crucial step.

Finally, microencapsulation was postulated for polystyrene-supported Sc(OTf)$_3$ (OTf = trifluoromethanesulfonato, "triflate", CF$_3$SO$_3$).[36] The Lewis acid is probably enveloped by the polymer thin film and stabilized by scandium-arene interactions. Such microencapsulated Lewis acids display higher activity than the monomeric Lewis acid in C–C bond-forming reactions including Michael and aldol reactions and Friedel-Crafts acylation, and are recoverable and reusable.

3.   RARE EARTH ELEMENTS IN NANOPOROUS ENVIRONMENTS

The incorporation of rare earth elements into nanoporous host materials is well-known to provide a beneficial effect for various catalyst systems. The main markets for rare earth catalysts are fluid cracking (FCC) and automotive post-combustion.[37] In FCC catalysts, exchanged rare earth cations promote the stability, activity and selectivity of the parent microporous host, in particular zeolite Y.[38] Consequently, the location of Ln(III) cations within the zeolitic framework and their mobility as a function of hydrolysis and heat treatment has attracted increased attention.[39] Novel developments in the field of rare earth-based microporous inorganic materials include the synthesis and structural characterization of rare earth silicates, *e.g.*, $Na_4K_2Tb_2Si_{16}O_{38} \cdot 10H_2O$,[40] and nitridosilicates, *e.g.*, $Ba_2Nd_7Si_{11}N_{23}$.[41] A completely different yet equally promising approach toward microporous networks is the controlled design of metal–organic open frameworks (MOFs).[42] Ln(III) cations were shown to preferentially assemble to such modular porous solids in the presence of strongly coordinating, difunctional organics such as oxalato,[43] dicarboxylato[44] and organodiphosphonato ligands.[45]

The synthesis of thermally stable periodic mesoporous main group metal and transition metal oxides is a major challenge for materials scientists.[46,47] Various identified rare earth-based and -modified oxidic mesophases are summarized in Tables 1 and 2 and are treated in more detail in the following sections.

3.1   Periodic Mesoporous Rare Earth (Mixed) Metal Oxides

Mesoporous rare earth oxides were obtained by homogeneous precipitation of rare earth nitrates $Ln(NO_3)_3 \cdot nH_2O$ using urea at 80°C in the presence of sodium dodecylsulfate [SDS, $CH_3(CH_2)_{11}OSO_3Na$] as a templating reagent.[48,49] Hydrothermal treatment of the reaction mixture at 80°C initially formed a layered structure with an interlayer spacing of *ca.* 3.6 nm which could be converted into a hexagonal MCM-41-analogous phase depending on the size of the rare earth cation. Stable, hexagonal phases formed only for the smaller cations Er, Ho, Tm, Yb, and Lu, while "middle"-sized Eu, Gd, Tb and Dy showed a phase transfer lamellar→hexagonal, and converted to amorphous powder upon prolonged hydrothermal treatment (30 hrs). The larger rare earth centers La, Ce, Pr, Nd, and Sm yielded layered structures exclusively, which converted partially to $La_2O(CO_3)_3(H_2O)_x$ and polycrystalline $Ce_2O(CO_3)_3(H_2O)_x$, $Pr(CO_3)(OH)$, $Nd(CO_3)(OH)$ and $Sm_2O(CO_3)_2(H_2O)_x$, respectively, upon prolonged hydrothermal treatment (30 hrs). For the stable carbonate-containing hexagonal phases, the anionic surfactant molecules were removed by exchange with acetate, yielding mesoporous materials (Table 1).

The mesophases show surface areas of a_s = 250-550 m^2g^{-1} which correspond to porous silica materials with surface areas as high as 900 m^2g^{-1}, taking into account the Ln_2O_3 to SiO_2 mass ratios. The effective pore diameters lie in the range of 2.5 to 3.0 nm, exhibiting broad pore size distributions (FWHM = full width at half maximum = *ca.* 1.0 nm). Interestingly, the formation of the hexagonal phases parallels the classification of the rare earth metal sesquioxides which form three different type of structures, namely La_2O_3 (A), Sm_2O_3 (B), and Sc_2O_3 (C), respectively, depending on the size of the rare earth cation. It is noteworthy, in contrast to the corresponding bulk oxides, that the magnetic susceptibility of these porous materials features a significant minimum near 23-25 K. This can be explained as a specific effect of the mesostructural arrangement of the rare earth cations, causing decrystallization of the paramagnetic ordering of the spins at low temperatures.

Table 1. Surface area, pore volume and pore diameter of rare earth (mixed) metal oxide mesophases.

Oxide[a]	Phase	a_s^e m^2g^{-1}	d_p^f nm	V_p^h cm^3g^{-1}	Ref.
Y_2O_3 (CO_3^{2-})	lamellar	–	–	–	48
	hexagonal	545	3.0		
Ln_2O_3 (CO_3^{2-})	lamellar	–	–	–	49
(Ln = Ho,Tm,Yb,Lu)	hexagonal	250-340	2.5-3.0		
Ln_2O_3 (CO_3^{2-})	lamellar	–	–	–	49
(Ln = Eu,Gd,Tb,Dy)	hexagonal	250-340	2.5-3.0	–	
Ln_2O_3 (CO_3^{2-})	lamellar		3.5^g	–	49
(Ln = La,Ce,Pr,Nd,Sm)	hexagonal			–	
$M^1M^2O_3$ (CO_3^{2-})	lamellar	798	1.8	–	50
($M^{1/2}$ = Y,Al)	hexagonal	714	2.0	–	51
($M^{1/2}$ = Y,Ga)					
Y_2O_3–ZrO_2 (YSZ)	lamellar				52
	hexagonal	245	–	0.3	
YSZ (56% Y)[b]	–	116	1.9	0.05	54
Pt–YSZ (20% Y)[b]	–	184	1.8	0.08	54
Ln-Al_2O_3	MSU-X[d]	350-530	5.3-10.8	0.65-1.31	55
(Ln = La,Ce)	("wormlike")				
CeO_2^c	fluorite-type	230	4.0-8.0	–	56

[a] If not otherwise noted, template removal was accomplished via extraction with sodium acetate. [b] YSZ = yttria-stabilized zirconia; calcination: 12 hr ramp to 600°C, held for 3 hr. [c] Calcination at 450°C. [d] MSU = Michigan State University; calcination at 500°C. [e] Specific BET surface area. [f] Pore diameter. [g] Interlayer spacing. [h] Cumulative pore volume.

Under similar reaction conditions, mixtures of metal salt precursors, such as $Y(NO_3)_3 \cdot 6H_2O/Al(NO_3)_3 \cdot 9H_2O$ ($Al/(Al+Y)=0.375$) or $Y(NO_3)_3 \cdot 6H_2O/GaCl_3$ ($Ga/(Ga+Y)=0.5$) produced high-quality mesoporous mixed metal oxide phases. However, the hexagonal structure of the acetate-exchanged materials completely disappeared upon calcination at 300°C in air for 5 hrs.[50,51] Typical molar ratios of $Y(NO_3)_3 \cdot 6H_2O/ZrO(NO_3)_2/SDS/urea$ were 1:1:4:60, for the synthesis of mesostructured yttrium–zirconium oxides.[52] Sonication of the synthesis gels was shown to favorably affect the formation of lamellar (after 1.5 hrs) and hexagonal phases (after 6 hrs). Although the hexagonal mesophase collapsed after extraction with sodium acetate or upon calcination at 400°C for 4 hours, a relatively high surface area of 245 m^2g^{-1} remained (Table 1).

Yttria-stabilized zirconia (YSZ) is the material of choice for use in solid oxide fuel cells (SOFC).[53] Binary and ternary high surface area YSZ and metal-YSZ, respectively, were also be obtained from synthesis gels of composition $Y(OCOCH_3)_3/Zr(OEt)_4/CTMABr/$ ethylene glycol/water/NaOH, using cetyltrimethylammonium bromide (CTMABr) as the structure-directing mesophase and preformed glycolate solutions of the metal precursors.[54] Nitrogen physisorption of the calcined materials revealed type-I isotherms indicating the presence of microporosity (<2.0 nm) rather than mesoporosity (2.0-50 nm). The template-free YSZ materials retained their structural integrity to around 800°C.

The incorporation of 1.0-5.0 mol% Ce(III) or La(III) ions in MSU-X alumina molecular sieves dramatically improved their thermal stability without altering the mesopore size or the wormhole channel motif.[55] The MSU-X materials were prepared through an $N^\circ I^\circ$ assembly pathway and exhibit BJH (Barret–Joyner–Halenda) pore diameters as high as 10 nm, depending on the surfactant size (N^0 = electrically neutral polyethylene oxide surfactant such as Tergitol® 15-S-12 ($C_{11-15}H_{23-31}O(CH_2CH_2O)_{12}H$), Pluronic®P65 (($PEO)_{19}(PPO)_{30}(PEO)_{19}$), P123 (= $(PEO)_{20}(PPO)_{70}(PEO)_{20}$); I° = aluminum alkoxide). The synthesis of Ln(III)-doped alumina was accomplished by dissolving the corresponding rare earth nitrate in a solution of the surfactant in warm butanol and subsequent addition of $Al(^sBu)_3$ and *sec*-butanol. The reaction mixture was treated in a reciprocating shaker bath at 45°C for a period of 40 hrs and was ultimately calcined at 500°C.

Drying and calcination of gelatinous hydrous ceria/surfactant mixtures, obtained from solutions of $CeCl_3 \cdot 7H_2O$ in aqueous ammonia in the presence of CTMABr, gave pure mesoporous high surface area fluorite-structured CeO_2.[56] This ceria, which exhibits a broad pore size distribution (2.0-8.0 nm; FWHM = *ca.* 5.0 nm, Table 1) and the absence of any long-range mesoscopic organization, shows enhanced textural and thermal resistance compared to ceria prepared by conventional routes. It was suggested that the cationic surfactant molecule does not act as a true templating reagent but rather as a surface area enhancer through interaction with Ce(IV) species of type $Ce(H_2O)_x(OH)_y^{(4-y)-}$ formed under such basic conditions.

3.2   Rare Earth-Modified Periodic Mesoporous Silica

Ion exchange of mesoporous silica Al-MCM-41 (Si/Al ratio = 39) in aqueous media with $Y(NO_3)_3 \cdot nH_2O$ afforded ion exchange levels of Y/Al = 0.39 (Scheme 3, Table 2).[57]   The presence of Y(III) increased the hydrothermal stability of the Al-MCM-41 material by 100°C to T_m = 1170 K, where with T_m is defined as the maximum temperature to which the sample can be heated for 2 hrs in flowing O_2 with 2.3 kPa of water vapor without decreasing the BET surface area more than 10%.

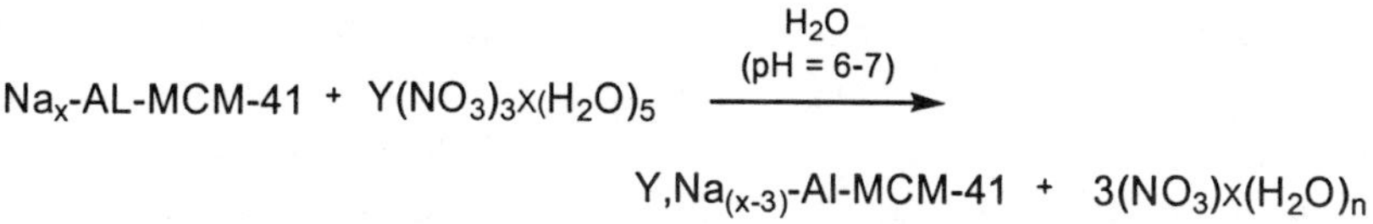

$$Na_x\text{-AL-MCM-41} + Y(NO_3)_3 x(H_2O)_5 \xrightarrow[\text{(pH = 6-7)}]{H_2O} Y,Na_{(x-3)}\text{-Al-MCM-41} + 3(NO_3)x(H_2O)_n$$

Scheme 3. Synthesis of yttrium-exchanged Al-PMS.

Lanthanum and cerium-modified MCM-41 were synthesized via a hydrothermal method using CTMABr as template and triethylammonium hydroxide as a mineralizer.[58] Cerium-incorporated MCM-41 was also obtained from synthesis gels of composition $0.08\ CeO_2/4SiO_2/1Na_2O/1CTMABr/200H_2O$ (Si/Ce = 50) using $CeCl_3 \cdot 7H_2O$ and fumed silica as cerium and silicon sources (Table 2). The Ce-MCM-41 material revealed higher structural ordering than the corresponding purely siliceous MCM-41 sample and showed medium and strong acid sites by thermo-gravimetric analysis of the *n*-butylamine thermodesorption.

Table 2. Surface area, pore volume, and pore diameter of rare earth (mixed) metal-modified silica mesophases.

Oxide[a]	Phase	a_s^b m^2g^{-1}	d_p^c nm	V_p^d cm^3g^{-1}	Ref.
Y-Al-MCM-41	hexagonal	995	ca. 2.6	–	57
La-MCM-41	hexagonal	950	3.3	–	58a,b
Ce-MCM-41	hexagonal	850	3.8	0.78	58c,d
LaCsO$_x$-Al-MCM-41	hexagonal	560-730	2.6-3.6	0.29-0.50	59
M-smectite (M = Ce,Al)	"pillared"	370	2.2[e]	ca. 0.6	62

[a] Calcination at 500-550°C. [b] Specific BET surface area. [c] Pore diameter. [d] Cumulative pore volume.
[e] Layer distance; calcination at 700°C.

The acidity was much higher for cerium than for lanthanum.[58c,d] Furthermore, incorporation of La(III) into MCM-41 (Si/La = 42, Table 2), like Fe(III) congeners, gave better thermal and hydrothermal stability than incorporation of Al(III) species.[58a,b] This can be ascribed to salt effects during the crystallization process and increase of the channel wall thickness. Binary cesium–lanthanum oxide supported MCM-41 and HMS (= hexagonal mesoporous silica) materials were alternatively synthesized from Al-MCM-41 (Si/Al ratio = ∞, 30, 15), CsOAc and La(NO$_3$)$_3$•nH$_2$O either by solid-state impregnation or by the incipient wetness approach and subsequent thermal treatment (Table 2).[59] The presence of both La(III) and a mesoporous framework affect the activity and product selectivity in the liquid-phase Michael addition of ethyl cyanoacetate to ethyl acrylate (Scheme 4a). The mildly basic and thermally recoverable CsLaO$_x$-MCM-41 catalyst was also employed for the Knoevenagel addition of enolates to benzaldehyde in aqueous media (Scheme 4b). Similar oxide-supported mesoporous aluminosilicate catalysts were shown to mediate a novel isomerization of ω-phenylalkanal to phenyl alkyl ketones, featuring an aldol condensation as a side-reaction (Scheme 4c).[60] A heterobimetallic substrate activation involving Cs$^+$---arene and La^{3+}---O interactions was discussed as a mechanistic detail. Note that such a coordination pattern was X-ray structurally proven in heterobimetallic aryloxide complexes of type CsLn(OC$_6$H$_3^i$Pr-2,6)$_4$.[61]

Pillared clays (PILCs) with large pores and good thermal stability were synthesized by intercalation of montmorillonites with polyhydroxy cations of Al/Ce; CeCl$_3$•7H$_2$O was used as the cerium source.[62] Depending on the composition of the pillaring solution, *i.e.*, hydroxide oligomerization, materials with specific surface areas > 300 m^{2}g^{-1} (Table 2) and varying micro- and mesoporous volumes were obtained. A ratio of Al/Ce of *ca.* 2.9 produced materials with a large portion of micropores, while high cerium concentrations (Al/Ce = 0.41) gave materials with a relatively high mesopore volume. Such Ce-Al-PILC materials show higher Brønsted to Lewis acidity ratios than corresponding Al-PILC samples. The presence of Ce in the pillars also markedly improves the conversion and selectivity for cracking of *n*-heptane by Pt-impregnated samples.[63]

3.3 Rare Earth Elements in Carbon-Based Mesoenvironments

Nanometer-scale carbon tubules can be used efficiently as high-surface area catalyst supports.[64] The mesopores of carbon nanotubes with internal diameters of *ca.* 3-6 nm were completely filled with crystalline rare earth metal oxides by treatment with lanthanide nitrates (Ln = Y, La, Ce, Pr, Nd, Sm, Eu) and subsequent calcination at 450°C.[65] Wet chemical methods involving the treatment of closed nanotubes with a nitric acid solution of the soluble metal nitrate resulted in pore-filling with discrete crystallites. Ln$_2$O$_3$@C$_x$ featuring

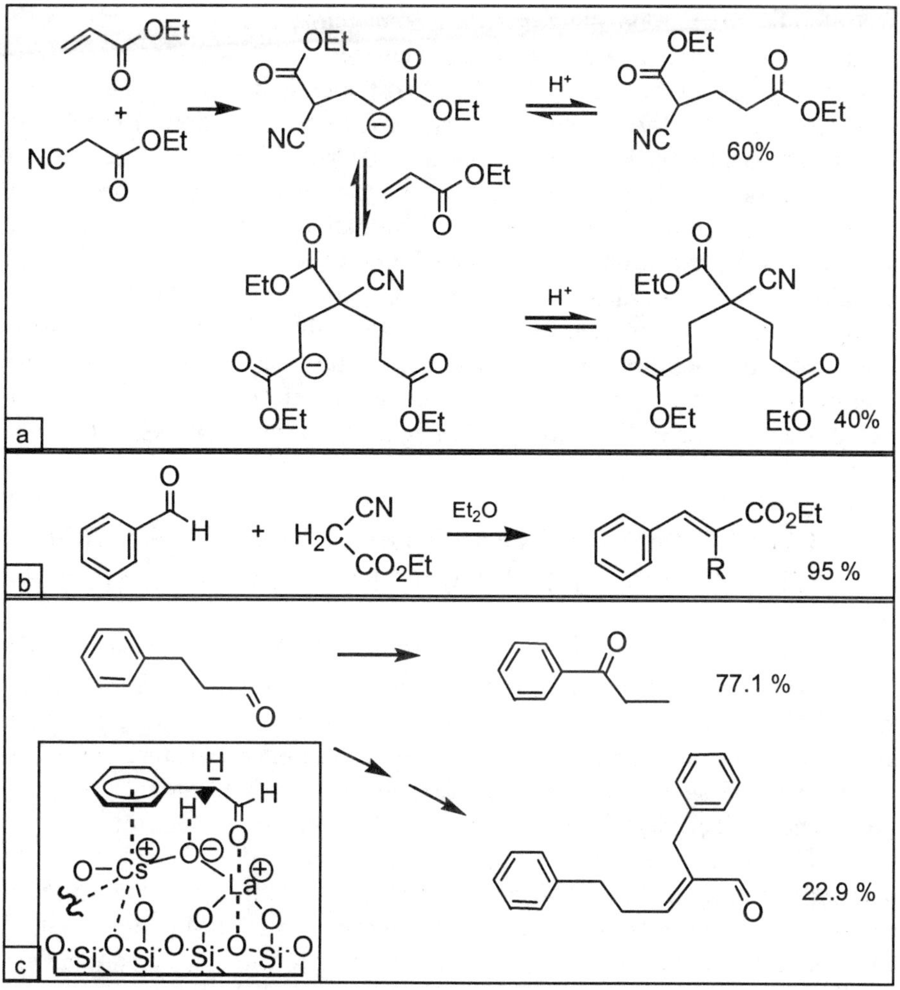

Scheme 4. LaCsO$_x$-Al-MCM-41-catalyzed Michael reaction (a; Si/Al = ∞), Knoevenagel addition (b; Si/Al = 30), and aldehyde–ketone–isomerization (c; Si/Al = ∞; proposed interaction of an arylalkenal with a supported heterobimetallic oxide catalyst).

long, continuous single crystals was obtained according to a molten media method, starting from a ground mixture of open carbon nanotubes and lanthanide halide. The latter method was also used to insert 1D lanthanide halide crystals into single-walled carbon nanotubes (SWNTs) at melt temperatures of 650-910°C.[66] High resolution transmission electron microscopy (HRTEM) revealed that the nanostructure of the encapsulated crystals varied with tubule diameter.

Extremely large, mesoporous activated carbon fibers have been obtained by steam invigoration of pitch fibers containing rare earth metal complexes such as Y(acac)$_3$ (acac = acetylacetonate) or LnCp$_3$ (Ln = Y, Sm, Yb, Lu).[67]

Accordingly, the carbon precursor was homogenized with 0.3 wt.% of the metalorganic compound in a large excess of quinoline, followed by airblowing of the resulting carbon material at 300°C/159 Torr, and final steam invigoration of spun fibers at 850°C for 45 min. The activated carbon fibers showed a mesopore ratio of 81%, a mesopore diameter of 4.4 nm and also a BET surface area of 1470 m^2g^{-1}. The rare earth metal complexes are assumed to assist the ordered construction of mesoporous fibers via the formation of thermally destabilized complexes with the aromatics contained in pitches. During this process, crystalline Ln_2O_3 particles (d_p = 4.5-9.5 nm) are dispersed homogeneously throughout the fiber. The adsorption capacity/capability of such activated carbon fibers was revealed for the effective purification of drinking water from carcinogenic humic acid (molecular weight at pH 7.2 = *ca.* 21000), γ-cyclodextrin (diameter 1.314 nm), and vitamin B_{12} (size 1.42×1.835×1.14 nm).

4.   SOLnC@PMS – SURFACE ORGANOLANTHANIDE CHEMISTRY AT PERIODIC MESOPOROUS SILICA

*S*urface *o*rgano*m*etallic *c*hemistry (SOMC), *i.e.*, derivatization of thermally robust condensed solid materials with molecularly well-defined organometallic compounds, has given enormous impetus to the field of organic-inorganic composite materials of relevance for catalysis.[14] The microstructure of surface-deposited metal species and the transformation of isolated atomic/molecular spot→mono(multi)layer/film→particle is of fundamental importance for a catalytic process.[68] SOMC represents a highly efficient method to postsynthetically incorporate surface metal centers, and is often superior to framework substitution methods, *i.e.*, isomorphous substitution via hydrothermal Si/M coprecipitation, due to better accessibility of the active centers. SOMC produces predominantly surface centers, the distribution of which is easily controllable over a wide range. Porous oxidic materials featuring rigidity and thermal stability as well as regular, adjustable, nano-sized cage and pore structures are commonly classified as promising catalysts supports.[13] The lack of "smaller-sized" organolanthanide compounds excludes extensive SOLnC in the intracrystalline space of traditional microporous materials such as zeolites,[8,13] zeotypes and pillared clays.[69] Only reactive solutions of ytterbium and europium metals in liquid ammonia have been successfully used so far for the impregnation of zeolite Y featuring 12R-windows of 0.74×0.74 nm diameter (Scheme 5).[70] Thermal degradation of the initially formed Eu(II) amide species produced Eu(II) imide and nitride species as evidenced by FTIR and XANES (*X*-ray *a*bsorption *n*ear-edge *s*tructure) spectroscopy. Depending on the type of alkaline metal cation M^+, such hybrid materials can display high catalytic activity for the isomerization of

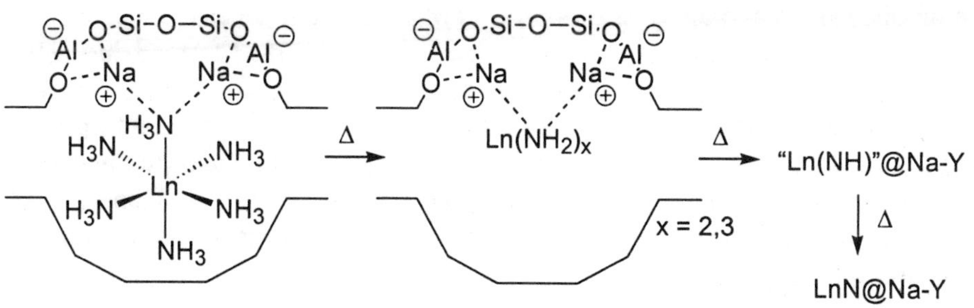

Scheme 5. Schematic presentation of the immobilization and thermal degradation of Ln(NH$_3$)$_6$ within a zeolitic 12R-window, *e.g.*, of zeolite Y.

2,3-dimethylbut-1-ene (proposed catalytically active species: "Eu(II)NH"), the Michael reaction of cyclo-2-enone with dimethylmalonate ("Eu(II)NH"), and the hydrogenation of ethene ("Eu(II)N").

4.1 Periodic Mesoporous Silica as a Catalyst Support

The intrinsic structural and morphological properties of semicrystalline PMS materials have been summarized in several review articles.[71] PMS materials are unique in so far as they combine favorable properties of crystalline microporous materials, such as structural variety,[72] with those of amorphous silica,[73] for example, high number of surface silanol groups. Due to their intrinsic zeolite-like pore architecture and thermal stability, MCM-41,[74] MCM-48[75] and SBA-15 silica[76] provide a unique platform for an efficient intrapore chemistry comprising complex grafting, surface-mediated ligand exchange, and catalytic applications. Surface areas as high as 1500 m^{2}g^{-1} and uniformly arranged mesopores with pore volumes as high as 3 cm^{3}g^{-1} ensure both a higher guest loading and a more detailed characterization by means of nitrogen adsorption/desorption, PXRD (powder X-ray diffraction), and HRTEM (high resolution transmission electron spectroscopy) compared to conventional silica support materials.

As a rule, SOMC is performed under anaerobic conditions on dehydrated support materials to avoid deactivation and hydrolysis and hence agglomeration of the molecular precursor, *e.g.*, by formation of metal hydroxide. Dehydration is commonly performed under vacuum at temperatures between 200 and 500°C yielding silanol populations in the range of 1.5-2.5 SiOH/nm^2. Various types of silanol groups, including isolated, vicinal, geminal, and hydrogen-bridged moieties, were assigned by FTIR spectroscopy.[77] The surface coverage of the predominantly isolated silanol groups formed at elevated temperature not only depends on the pretreatment temperature but also on the synthesis procedure and type of PMS, and pore diameter, *i.e.*, surface curvature. Like the commonly used amorphous silica and alumina supports,[14] PMS materials are capable of surface reactions via these silanol groups and strained siloxane bridges formed during the calcination and dehydration process at elevated temperature.[78] In particular, nitrogen physisorption measurements create a unique picture of the intraporous population, *i.e.*, of the steric bulk of the immobilized metal–ligand fragments, by representing of the surface area and the pore size distribution.[79] Hence, PMS materials also seem especially well-suited as *model* supports of the commonly used amorphous silica.[80]

The intracrystalline space of PMS materials, Figure 2, imposes a stabilizing environment which can be fine-tuned by variation of the pore diameter. Pore confinement is supposed to impart a stabilizing effect by better protecting the reactive site from deactivation and leaching effects (encapsulation phenomenon).[4] Moreover, implications of pore confinement for structure-reactivity relationships can be anticipated.

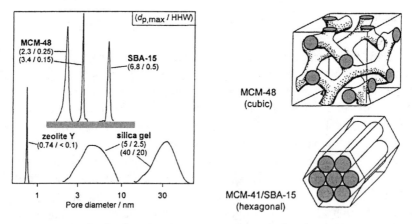

Figure 2. Pore size distribution of various porous materials and structural models of various PMS materials. Note that in contrast to MCM-41, the hexagonally arranged mesopores of SBA-15 are interconnected by micropores. Reproduced with permission from *Chem. Mater.* **2001**, *13*, 4421. Copyright 2001 American Chemical Society.

4.2 Advanced Surface Grafting – The "Heterogenized Silylamide Route"

SOLnC@PMS, starting from *pseudo*-organolanthanides, is a promising alternative to metal alkyl-based surface reactions. Such *pseudo*-organolanthanides, containing no direct metal-carbon linkage, feature otherwise readily hydrolyzable Ln–X bonds represented by amide and alkoxide ligands of enhanced basicity.[81] The basicity of the metal bonded ligand, or the pK_a value of

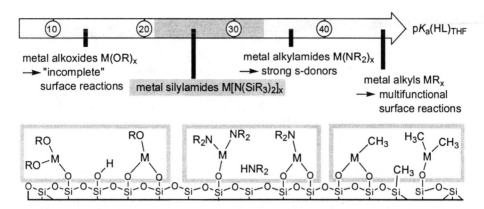

Scheme 6. The pK$_a$ criterion as a measure for surface reactivity. Reproduced with permission from *Chem. Mater.* **2001**, *13*, 4419. Copyright 2001 American Chemical Society.

the protonated ligand, gives a reasonable measure of the reactivity of a metal–X(ligand) bond toward a dehydrated PMS surface assuming a pK$_a$ value of ca. 5-7 similar to silica gel (Scheme 6).[82]

    Table 3 classifies the diversity of SOLnC@PMS according to the molecular precursor, the mesoporous support and aspects of characterization and application. Rare earth alkoxide complexes are the most inexpensive precursors for SOLnC and, depending on the substituents R, the pK$_a$ value of alcohols HOR can range from 5 to 20. Grafting of homoleptic derivatives Ln(OR)$_x$ revealed several disadvantages including incomplete surface silanol consumption (Scheme 6),[83,84] release of strongly surface-coordinating alcohols, agglomeration of

Table 3. Application of rare earth-grafted periodic mesoporous silica (PMS).

Organometallic Precursor	PMS	Applications	Ref.
Nd(OC^{t}Bu$_3$)$_3$	MCM-41	MPV reduction of TBCH	83
AlMe$_3$/*rac*-[Me$_2$Si(2-Me-C$_9$H$_5$)$_2$]Y[N(SiHMe$_2$)$_2$]	MCM-41	Olefin polymerization	89
Nd(N^{i}Pr$_2$)$_3$(THF)	Al-MCM-41	Surface morphology	86
Nd[N(SiMe$_3$)$_2$]$_3$	MCM-41	MPV reduction of TBCH, Nd$_2$O$_3$ overlayer	83
Nd[N(SiHMe$_2$)$_2$]$_3$(THF)$_2$	MCM-41	MPV reduction of TBCH	86
Sc[N(SiHMe$_2$)$_2$]$_3$(THF)	MCM-41	Diels-Alder cyclization (Scheme 15)	80
Ln[N(SiHMe$_2$)$_2$]$_3$(THF)$_2$ (Ln = Y,La)	MCM-41	Diels-Alder cyclization (Scheme 15)	80, 116
(L)Y[N(SiHMe$_2$)$_2$](THF)	MCM-41	Diels-Alder cyclization (Scheme 15)	96
Sm[N(SiHMe$_2$)$_2$]$_2$(THF)$_x$	MCM-41/48	Ketyl formation (Scheme 10)	97

MPV = Meerwein Ponndorf Verley. TBCH = *tert*-butylcyclohexanone. H$_2$L = *N,N'*-Bis-(3,5-di-*tert*-butyl-salicylidene)ethylenediamine,*trans*-1,2-Bis(2,4,6-triisopropyl-benzenesulfonamidato)cyclohexane.

alkoxide surface species,[85] and limited secondary ligand exchange. It has been shown that highly fluorinated alcohols such as $HOC(CF_3)_3$ (pK_a 5.7) readily "degraft" rare earth metal centers from a MCM-41 surface.[86] Calcination of the metal alkoxide-grafted materials is a popular approach for the synthesis of metal oxo overlayers.[83,87]

Metal alkyl precursor compounds display enhanced reactivity toward various silica sites, particularly toward silanol groups and strained siloxane bridges, favoring multifunctional surface reactions as shown the for material $AlMe_3$@MCM-41.[88] *Ansa*-lanthanidocene complexes synthesized in our laboratory, *e.g.*, *rac*-[$Me_2Si(2$-Me-$C_9H_5)_2$]Y[$N(SiHMe_2)_2$],[89] underwent partial protonation and detachment of the indenyl ancillary ligand when reacted with dehydrated PMS. Additionally, significant leaching of alkylated species was found when *rac*-[$Me_2Si(2$-Me-$C_9H_5)_2$]Y[$N(SiHMe_2)_2$] was reacted with $AlMe_3$@MCM-41 (Scheme 7).[89]

Until recently, there has been a dearth of data on SOMC involving metal alkyl amide derivatives $M(NR_2)_x$.[85,90] Similar to metal alkoxides, their surface reaction generates strongly donating amines (Scheme 6) and siloxane cleavage reactions can be an issue.[91,92] We found that rare earth silylamide complexes[93] give rise to an advanced surface grafting[86,94] featuring (i) mild reaction conditions, (ii) formation of thermodynamically stable metal siloxide bonds, (iii) the production of a unique hydrophobic platform due to concomitant surface silylation, (iv) favorable atom economy, (v) suppression of extensive complex agglomeration due to the steric bulk of the silylamide ligands, (vi) release of weakly coordinating and hence easily separable silylamines, and (vii) the absence of any insoluble by-products. Additionally, silylamide ligands provide a stabilizing environment (low coordination numbers, low metal oxidation states) for most of the main group and transition metals except for the noble metals.[93,95]

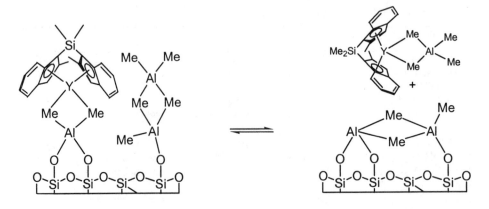

Scheme 7. Proposed surface species obtained via the reaction of *rac*-[$Me_2Si(2$-Me-$C_9H_5)_2$]Y[$N(SiHMe_2)_2$] with $AlMe_3$@MCM-41. Reproduced with permission from *Chem. Mater.* **2001**, *13*, 4419. Copyright 2001 American Chemical Society.

Various conceptual approaches in heterogeneous silylamine elimination were initially reported for the lanthanide elements as outlined in Scheme 8. Approach **A** is based on a two-step reaction sequence, comprising an initial grafting of silylamide complexes[94] and a subsequent surface-mediated secondary ligand exchange under fairly mild conditions.[80,83] Depending on the type of silylamide ligand (steric bulk) and the oxidation state of the metal center (number of ligands), approximately 1 to 2.5 mmol of homoleptic silyl-amide complex can be grafted onto 1 gram of dehydrated PMS material (surface area *ca.* 1000 m^2g^{-1}). This corresponds to a relatively high surface coverage of 0.8 to 1.4 "Ln"/nm^2. For comparison, the multi-functional grafting of AlMe$_3$ gave a metal surface population of *ca.* 2.7 Al/nm^2.[88]

Approach **B** utilizes tailor-made heteroleptic molecular precursors exhibiting reactive docking positions (a silylamide moiety) and strongly chelating ancillary ligands such as SALEN and *bis*-sulfonamides which disfavor protonolysis and counteract oligomerization reactions.[96] In addition to their favorable surface reaction, silylamide ligands can be easily decorated with spectroscopic probe moieties as shown for the Si–H group of the dimethylsilylamide ligand N(SiHMe$_2$)$_2$.

The IR spectra of a silylamide-modified hybrid material (**C**), shown in Figure 3 for the reaction of Y[N(SiHMe$_2$)$_2$]$_3$(THF)$_2$ (**B**) with dehydrated MCM-41 (**A**), feature the complete consumption of the strong band at 3695 cm^{-1} due to the isolated OH groups (**C**) as well as the characteristic SiH stretching vibrational modes for metal bonded-silylamide ligands (< 2100 cm^{-1}) and for ≡SiOSiHMe$_2$ surface sites (2145 cm^{-1}). The asymmetric and symmetric stretching vibrations of

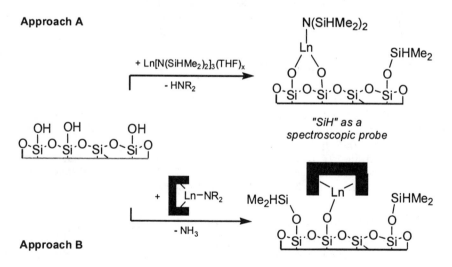

Scheme 8. Various options in the "heterogenized silylamide route". Adapted with permission from *J. Chem. Soc., Dalton Trans.* **1999**, 3611. Copyright 1999 Royal Society of Chemistry.

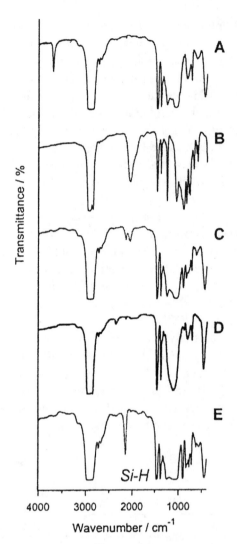

Figure 3. "Si–H" as a spectroscopic probe: The IR spectra (Nujol) of MCM-41$_{dehydrat.}$ (**A**, pretreatment temperature 250˚C, 3hrs, 10^{-3} Torr), Y[N(SiHMe$_2$)$_2$]$_3$(THF)$_2$ (**B**), [MCM-41]Y-[N(SiHMe$_2$)$_2$]$_x$(THF)$_y$ (**C**, pretreatment temperature 100˚C, 3hrs, 10^{-3} Torr), [AS200]Y-[N(SiHMe$_2$)$_2$]$_x$(THF)$_y$ (**D**, pretreatment temperature 100˚C, 3hrs, 10^{-3} Torr), [MCM-41]-SiHMe$_2$ (**E**, pretreatment temperature 250 ˚C, 3 hrs, 10^{-3} Torr).[94] Reproduced with permission from *Chem. Mater.* **2001**, *13*, 4419. Copyright 2001 American Chemical Society.

the"YNSi$_2$ " fragment are also visible at 900 and 796 cm^{-1}.[80] The IR spectrum of a material obtained under analogous conditions from a dehydrated non-porous Aerosil AS200 (**D**) reinforces the importance of a high surface-containing material and hence high immobilized complex/support ratios for easy and quick characterization. An independently performed silylation reaction, starting from excess tetramethyldisilazane, $HN(SiHMe_2)_2$ and dehydrated MCM-41, revealed the formation of a completely silylated material [MCM-41]$SiHMe_2$ (**E**) under fairly mild conditions. The presence of probably metal-bonded ammonia as a silylation co-product is evidenced by the appearance of the symmetric and asymmetric stretching vibrations of N–H (3380, 3330 cm^{-1}).

The profound importance of nitrogen physisorption for characterizing such intraporous SOMC is shown for the grafting of $Sm[N(SiHMe_2)_2]_2(THF)_x$ onto dehydrated MCM-48 (Figure 4, **B**),[97] the subsequent reaction with fluorenone (Figure 4, **C**), and final acidic work-up (Figure 4, **D**). Filling of the mesopores according to the approximate steric bulk of the immobilized species is impressively demonstrated and analysis of the Barret-Joyner-Halenda (BJH) pore size distribution suggests a regular distribution of the surface species accounting for reduced pore volumes and mean pore diameters. Note that use of pore-expanded PMS materials ensures the preservation of a mesoporous hybrid material after silylamide-grafting as indicated by the host-characteristic type-IV isotherm.[79]

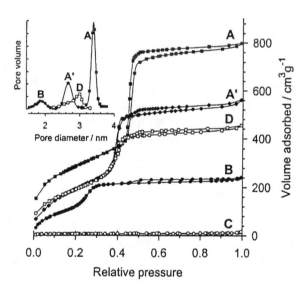

Figure 4. Nitrogen adsorption/desorption isotherms at 77.4 K and the corresponding BJH pore size distributions of the parent and modified MCM-48 materials **A** (■; 10^{-3} Torr, 4 hrs, 280°C; V_p = 1.21 cm^{3}g^{-1} / d_p = 3.3 nm), **A'** (♦; A + excess of $HN(SiH Me_2)_2$; 10^{-3} Torr, 3 hrs, 250°C; 0.77 / 2.7), **B** (●; A + {$Sm[N(SiHMe_2)_2]_2(THF)_x$}; 10^{-3} Torr, > 5 h, r.t.; 0.30 / 1.9), **C** (–○–; **B** + fluorenone; 10^{-3} torr, > 5 hrs, r.t.), and **D** (–□–; recovered material; 10^{-3} Torr, 3 hrs, 250°C; 0.65/ 3.0). V_p = BJH desorption cumulative pore volume of pores between 1.5 and 6.5 nm diameter. d_p = pore diameter according to the maximum of the BJH pore size distribution (d_p < 2.0 nm should be viewed critically). Adapted with permission from *J. Am. Chem. Soc.* **2000**, *122*, 1544. Copyright 2000 American Chemical Society.

4.3   Fine-Tuning of Organolanthanide Surface Species

Agglomeration, dispersion and coordination phenomena are well-known to govern the catalytic activity of surface-deposited metal species.[12-14] In contrast to salt impregnation methods utilizing metal salts such as nitrates ("incipient wetness approach"), the grafting of oxophilic and Lewis acidic main group and early transition metal centers via SOMC produces highly reactive surface species without thermal post-treatment. SOMC on ordinary silica and alumina materials revealed that such surface species can exhibit unforeseen reactivity patterns.[98] Such enhanced reactivity can originate from sterically unsaturated metal centers featuring a highly distorted coordination environment and from a strongly electron-withdrawing effect of the support. Moreover, surface confinement implies a "small-ligand" chemistry of the metal centers. Factors such as the podality Π^n of the grafted metal center, *i.e.*, the number of covalent M–O(support) bonds, the connectivity of the metal center, *i.e.*, the type of M–O–X(support) linkage, and the remaining (chiral) ligand environment $L^{(*)}$ are important factors which govern the reactivity of the hybrid material (Figure 1, **B**).[15] Π^n depends on the silanol population, which can be controlled by thermal pretreatment, *i.e.*, dehydration of the support material and its surface curvature. The intrinsic monodisperse pore structure of PMS materials facilitates the variation of the inner surface curvature, *i.e.*, the concavity of the pore walls, via the pore diameter.[99] The micro- and meso-environment imposed by surface-attached (chiral) groups R* and the pore diameter d_P is found to additionally tailor the reaction behavior. Group R*-decorated pore walls not only modify the reactivity of the active metal site via various steric effects including metal spacing and stereochemical void space formation but also via hydrophobicity effects.

4.3.1   Surface Confinement

Surface confinement also seems to have an effect on the reductive behavior of samarium(II) centers. The rate and selectivity of these one-electron reduction reactions, usually induced by the standard reagent SmI_2, are markedly affected by the type of solvent, additives such as strong Lewis bases, *e.g.*, HMPA, or metal salts,[100] and ancillary anionic ligands bonded to the Sm(II) center, such as amide, alkoxide, or cyclopentadienyl ligands.[101]

Sm(II) surface species obtained by the heterogeneously performed silylamide route via reaction of bis(dimethylsilyl)amide $Sm[N(SiHMe_2)_2]_2(THF)_x$ with pore-expanded MCM-41 and MCM-48 (Scheme 9) form stable "storable" ketyl surface radicals and yield selectively the alcoholic product in the fluorenone/fluorenol transformation.[97] This contradicts findings in homogeneous solutions where organosamarium(II) complexes treated with equimolar amounts of ketone hydrolyze to give the pinacol product exclusively.[102] The presence of

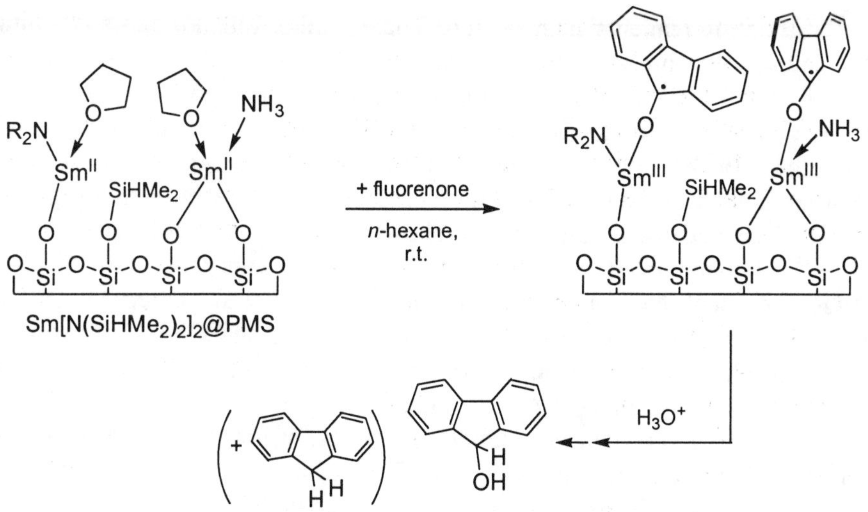

Scheme 9. Formation of surface-confined ketyl radicals via Sm(II)@PMS. Adapted with permission from *J. Am. Chem. Soc.* **2000**, *122*, 1544. Copyright 2000 American Chemical Society.

sterically separated, surface-bonded Sm(II) species seems to drastically modify the selectivity of this transformation. Further evidence for the formation of surface-confined and sterically unsaturated Sm(II) centers was obtained by FTIR spectroscopy: metal-bonded silylamide ligands featuring agostic Sm---SiH interactions were unequivocally indicated by a broad band for the SiH stretching vibration in the region from 2030 to 1920 cm^{-1}.[97]

4.3.2 Effect of Local and Nanoenvironments

Variation of the inner coordination sphere of the active metal site, *i.e.*, change of the local environment, is crucial for the design of homogeneous single-site catalysts,[103] and, consequently, is examined for the development of more efficient and specialized catalysts of the category discussed in this chapter. Reports have accumulated revealing that the micro- and mesoenvironment implied by PMS materials can have important implications for catalytic processes ("nanovessels", "inclusion polymerization").[104] In particular, surface morphology, surface polarity and pore confinement are important factors which define the micro- and mesoenvironment in PMS materials. Silylation and alkylation reactions are of fundamental importance for examination of the surface morphology and for the manipulation of siliceous support materials, including zeolites.[105,106] Fine-tuning the catalytically active interface via adjustment of both the dispersion and accessibility of the catalytically active sites by steric and hydrophobicity effects is known to be crucial.[107]

Silylating reagents such as chlorosilanes, alkoxysilanes and silylamines have found widespread application in post-silylation experiments and their reactions with surface silanol functionalities has been reported in detail.[73,108] Silazane-based silylation reactions and SOMC seem to optimally complement each other. Both *consecutive* and *competitive* silylation reactions revealed important details of the kinetics of this special silylation reaction and support a controlled surface functionalization, as shown for vinyl moieties (Scheme 10).[109]

Surface-mediated secondary ligand exchange seems particularly viable to change the local environment of surface-bonded metal centers.[110] Surface confined metal amide species give access to "small-ligand-SOLnC" of oxophilic and electrophilic metal centers, a chemistry unknown in solution due to agglomeration phenomena.[83] As a standard example, the class of lanthanide alkoxide complexes produces polymeric solid state structures for OMe and OEt ligands, cluster compounds such as $Ln_5O(O^iPr)_{13}$ for medium-sized ligands, and mononuclear complexes for the bulky ligand OC^tBu_3 ("tritox").[111] Surface-mediated chemistry was shown to generated "mononuclear" Ln-alkoxide sites independent of the size of the ligand (Scheme 11). The size of the ligand is reflected in the nitrogen physisorption data as the pore volume and pore diameter decrease in the order $[MCM-41]_{dehydrated}$ ($V_p = 0.89$ cm^3g^{-1}; $d_p = 2.8$ nm) > $[MCM-41]Nd(OMe)_x$ (0.48 cm^3g^{-1}; 2.2 nm) > $[MCM-41]Nd(OC^tBu_3)_x$ (0.35 cm^3g^{-1}; 1.8 nm). Such rare earth alkoxide surface complexes display subtly differentiated catalytic behavior in the MPV reduction of *tert*-butylcyclohexanone (TBCH).[83]

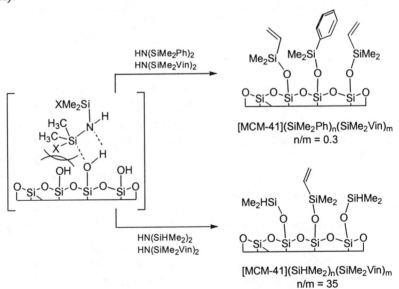

Scheme 10. Competitive surface silylation utilizing disilazane reagents. Reproduced with permission from *J. Phys. Chem. B* **2000**, *104*, 3532. Copyright 2000 American Chemical Society.

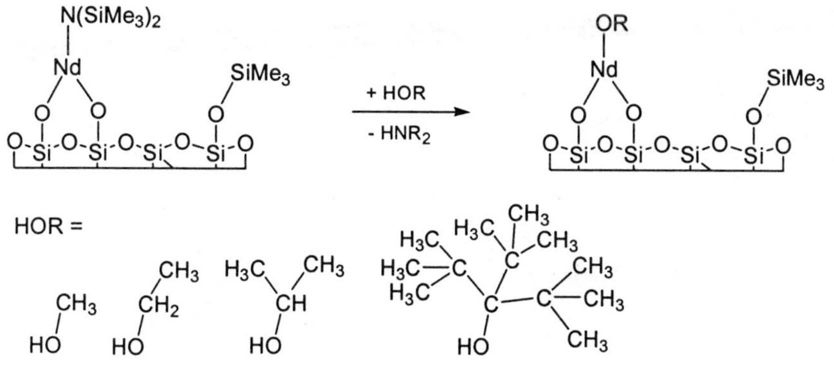

Scheme 11. "Small-ligand-SOLnC" of alkoxide derivatives.

Under the prevailing reaction conditions, commercially available $Al(O^iPr)_3$ exhibits no significant catalytic activity after 5 hrs (Table 4). The neodymium alkoxide supported materials designated $[MCM-41]Nd(OR)_x$ show catalytic conversions higher than 90% after 5 hrs reaction time at ambient temperature, comparable to that of monomeric $Nd(OC^tBu_3)_3$. Apparently, the metal centres are readily accessible to the substrate molecules, maintaining the catalytic cycle given in Scheme 12.

Table 4. Catalytic activity of grafted rare earth alkoxide species in the MPV reduction of *tert*-butylcyclohexanone.

Run	Catalyst Precursor	mol% M	Conversion % 5 hrs	(*trans:cis*) 24 hrs
1[b]	$[Al(O^iPr)_3]$	5	<1	7(4.0)
2	$[Al(O^iPr)_3]$	5	12 (2.1)	59 (2.1)
3[c]	$Nd(OC^tBu_3)_3$	1.4	96 (2.7)	>99 (2.7)
4	$[Nd(OEt)_3]$	1.5	96 (2.7)	>99 (2.7)
5[d]	$[MCM-41]_{dehydrated}$	-	-	<1
6[d]	$[MCM-41]SiMe_3$	-	-	<1
7	$[MCM-41]Nd(O^iPr)_x$	1.9	90 (3.3)	>99 (3.3)
8	$[MCM-41]Nd(O^iPr)_x$	2	34 (3.9)	82 (3.9)
9	$[MCM-41]Nd(OC^tBu_3)_x$	1.7	34 (3.9)	82 (3.9)
10	$[MCM-41]Nd(OEt)_x$	2	29 (4.2)	68 (4.2)
11	$[MCM-41]Nd(OMe)_x$	2.1	33 (4.1)	70 (4.2)
12	$[MCM-41]Nd(OC^tBu_3)_y$	1.3	63 (3.3)	> 99 (3.3)

[a] Conditions: 25 g HO^iPr, 0.1 g *n*-nonane, 0.78 g ketone, *ca.* 0.10 g of precatalyst, 25 °C (mol% M = 100 $n_{metal}/n_{substrate}$). [b] HO^iPr not predried. [c] 95% conversion after 50 min. [d] 0.20 g mesoporous material. [e] Recovered material (no activity of the centrifugate during 48 hrs). Reprinted with permission from *Stud. Surf. Sci. Catal.* **1998**, *117*, 413. Copyright 1998 Elsevier Science.

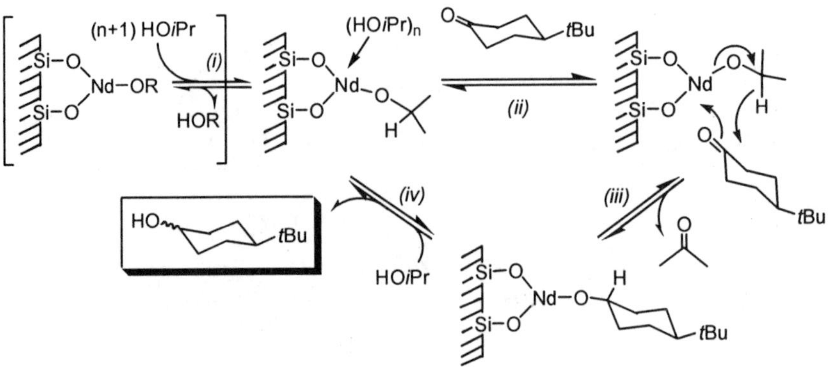

Scheme 12. Mechanistic scenario for rare earth-mediated MPV reduction of TBCH. Reprinted with permission from *Stud. Surf. Sci. Catal.* **1998**, *117*, 413. Copyright 1998 Elsevier Science.

The similar catalytic activities of $[MCM-41]Nd(O^iPr)_x$ and $[MCM-41]Nd(OC^tBu_3)_x$ suggest that the OC^tBu_3/HO^iPr ligand exchange [step (i)] and adduct formation [step (ii)] are fast compared to acetone elimination [step (iii)], involving a six-membered transition state and subsequent C_α-hydrogen transfer. Hence, both systems should produce the same catalytically active species. This can be explained by the structural and electronic peculiarities of the tritox ligand, featuring enhanced Brønsted basicity, increased steric bulk and lack of C_α-hydrogen transferability. The solid catalysts can easily be separated from the reaction mixture *via* centrifugation and can be reused. The decreased catalytic activity of materials $[MCM-41]Nd(OMe)_x$ and $[MCM-41]Nd(OEt)_x$ is plausible from the decreased susceptibility to oxidation of methoxide and ethoxide ligands. Although the more Brønsted acidic primary alcohols should additionally depress the alcohol exchange reactions (i) and (iv), the observed catalytic activity is quite remarkable, probably due to the pronounced ligand exchange ability of rare earth alkoxide complexes. Surprisingly, *insoluble* $[Nd(OEt)_3]$ also proved to be quite active in this MPV reduction. The reduced catalytic activity of the material $[MCM-41]Nd(OC^tBu_3)_y$, obtained from the reaction of dehydrated MCM-41 and $Nd(OC^tBu_3)_3$, (*i.e.*, the "detour" over the heterogenized silylamide route was left out), might be due to the lower concentration of active metal centers, changed hydrophobicity of the catalyst and/or metal-coordinated THF.

The peculiar reaction behavior of molecular lanthanide alkoxide and (silyl)amide complexes with trimethylaluminum to yield Lewis acid base adduct complexes $Ln(OR)_3(AlMe_3)_3$ and permethylated derivatives $Ln(AlR_4)_3$[81] has been exploited for the synthesis of "heteroleptic" amide/siloxide surface species. For example, scandium methyl surface species were generated by consecutively treating dehydrated, pore-enlarged MCM-41 silica with $Sc[N(SiHMe_2)_2]_3(THF)$, trimethylaluminum and tetrahydrofuran under mild conditions (Scheme 13).[112]

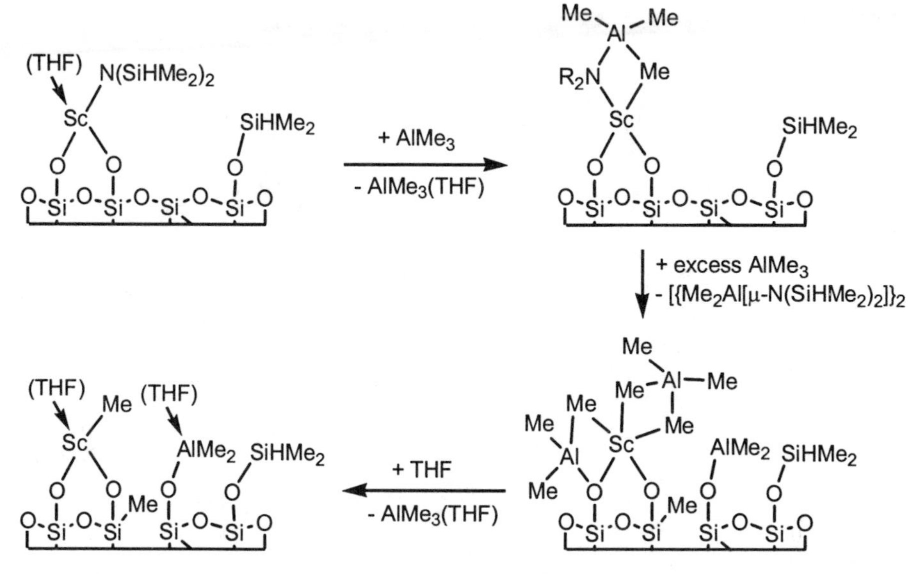

Scheme 13. Synthesis of scandium alkyl surface species.

Such lanthanide(III) alkyl species stay immobilized in the presence of excess trimethylaluminum, and the variously formed inorganic/organometallic hybrid materials denoted as [MCM-41]Sc[N(SiHMe$_2$)$_2$]$_x$L)$_y$, [MCM-41]Sc(AlMe$_4$)$_x$(AlMe$_3$)$_y$, and [MCM-41]ScMe(THF)$_z$(AlMe$_3$)$_y$ were successfully characterized by solid state NMR spectroscopy (Figure 5). The ^{1}H MAS NMR spectra clearly show SiH groups at 4.6-4.9 ppm and a broad resonance centered at 0.2 ppm assignable to Si-CH$_3$, Al-CH$_3$, "Sc(μ-CH$_3$)$_2$Al", and Sc-CH$_3$ groups.[109,113]

^{13}C CP/MAS NMR spectroscopy additionally revealed the presence/absence of coordinated THF molecules (25.6 and 72.5 ppm) and a more conclusive picture of the surface alkyl groups. A broad signal centered at 0.5 ppm attributable to ≡SiCH$_3$ and ≡SiOSiH(CH$_3$) sites exhibits two well-resolved higher-field shoulders at -6.5 and -13.7 ppm assignable to various metal methyl groups (Figure 5b/**B**).[88,113] A pronounced broadening at the lower-field side of this resonance is also visible. The disappearance of the lower-field shoulder and the significantly decreased relative intensity of the higher-field shoulders in the ^{13}C CP/MAS spectrum of material [MCM-41]ScMe(THF)$_z$(AlMe$_3$)$_y$ (Figure 5b/**C**) is in accordance with the loss of AlMe$_3$(THF). The higher-field shoulder at -11.0 ppm can be assigned to terminally bonded M-CH$_3$ moieties.

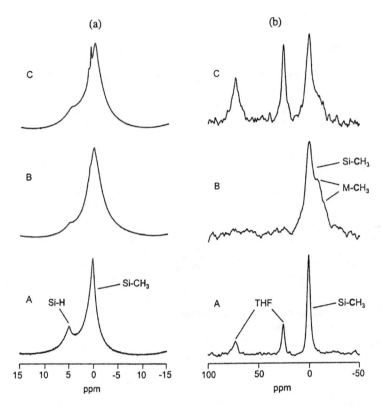

Figure 5. (a) ¹H MAS NMR (a) and ¹³C CP MAS NMR spectra (b) of samples [MCM-41]Sc[N(SiHMe₂)₂]ₓ(L)ᵧ (**A**), [MCM-41]Sc(AlMe₄)ₓ(AlMe₃)ᵧ (**B**), and [MCM-41]ScMe(THF)ᵤAlMe₃)ᵧ (**C**). [MCM-41]Sc[N(SiHMe₂)₂]ₓ(L)ᵧ and [MCM-41]Sc(AlMe₄)ₓ(AlMe₃)ᵧ were washed several times with *n*-hexane and evacuated for 5 hrs at r.t./10⁻² Torr; [MCM-41]ScMe(THF)ₓ(AlMe₃)ᵧ was washed with THF and evacuated for 5 hrs at r.t./10⁻² Torr. Adapted with permission from *Micropor. Mesopor. Mater.* **2001**, *44-45*, 311. Copyright 2001 Elsevier Science.

The reaction of homoleptic alkyl complexes (LnR₃(THF)ₓ, R = CH₂SiMe₃, CH(SiMe₃)₂) with surface silanol groups offers an alternative route to rare earth alkyl surface species,[89] however, this immobilization sequence is hampered by synthesis conditions for the preparation of pure alkyl precursor complexes[114] and is restricted to bulky alkyl ligands exclusively. In contrast, the silylamide route implicates a "small–ligand–SOMC" for the entire lanthanide series based on commercially available precursor compounds. The generation of metal alkyl grafted silica materials by SOMC gives access to highly reactive catalyst precursor species for olefin transformation including oxidation and polymerization processes.[98] In particular, rare earth alkyl moieties belong to the most reactive organometallic fragments as evidenced by, for example, the CH activation of alkanes.[115]

The ligand-modified PMS materials [MCM-41]Y(fod)$_x$ obtained according to the heterogeneously performed silylamide route provide a unique hydrophobic platform for the handling and isolation of sensitive substrate molecules and reaction intermediates. Such species show highly selective reaction behavior in the Danishefsky transformation, a special hetero-Diels-Alder reaction starting from Danishefsky's diene (*trans*-1-methoxy-3-trimethylsilyloxy-1,3-butadiene) and benzaldehyde, to form product **A** exclusively, as shown in Scheme 14 (TOF = 70: Ln = Y, 50 °C).[80,96,116] This was ascribed to an *in situ* silylation by the released silylamine ensuring the complete "end-capping" of all of the Brønsted acidic surface silanol groups. Supportive of this was the catalytic behavior of a hybrid material which was obtained by contacting a dehydrated MCM-41 sample directly with Y(fod)$_3$ (fod = 1,1,1,2,2,3,3-heptafluoro-7,7-dimethyl-4,6-octanedione). Although the latter material displayed enhanced initial catalytic activity comparable to molecular complexes, conversion of product **A** into **B** was observed from the beginning.

Silylamide grafting seems to be applicable for the high-throughput synthesis and screening of structurally-defined rare earth metal catalysts on PMS materials.[116] Thus, by means of parallel synthesis, any desired number of intraporous ligand exchange reactions can be conducted simultaneously under identical conditions, resulting in a large and structurally diverse array of promising substances.

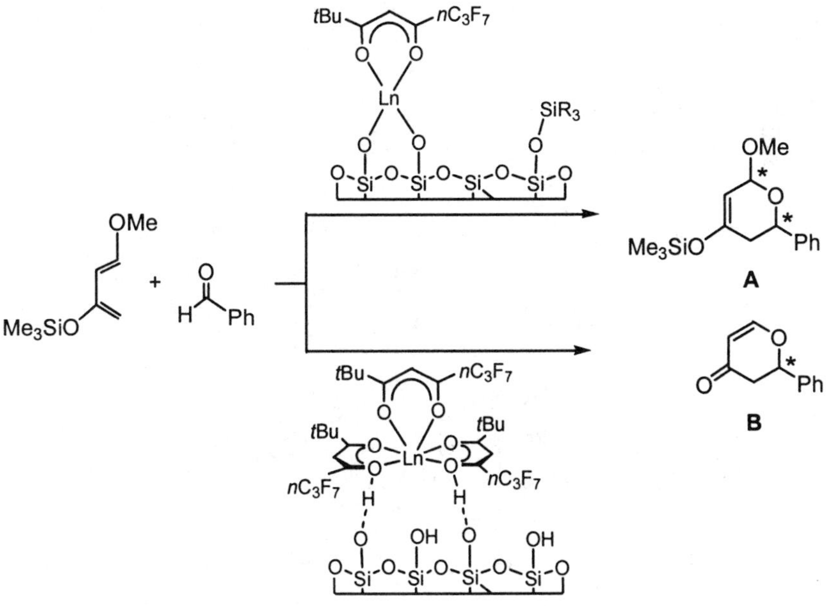

Scheme 14. The "Danishefsky" transformation mediated by [MCM-41]Ln(fod)$_x$. Reproduced with permission from *Chem. Mater.* **2001**, *13*, 4419. Copyright 2001 American Chemical Society.

Mesoporous inorganic/metalorganic hybrid materials combining grafted rare earth metal (Y) centers with different chiral ligands including: **(a)** L-(-)-3-(perfluorobutyryl)-camphor **(b)** 2,2'-methylenebis-[(4*S*)-4-*tert*-butyl-2-oxazoline] **(c)** (*S*)-(-)-2-(diphenyl-hydroxymethyl)pyrrolidine **(d)** (*R*)-(+)-1,1'-*bis*-2,2'-naphthol **(e)** (1*R*,2*S*)-(-)-*N*-methylephedrine **(f)** and (1*S*,2*R*,5*S*)-(+)-menthol **(g)** were generated *via* surface-mediated ligand exchange reactions from silylamide modified MCM-41 materials (Scheme 15). The immobilized catalysts display markedly increased catalytic activity (65-85% yield after 10 hrs) in the above-mentioned Danishefsky transformation compared to their molecular congeners. The L-(-)-3-(perfluorobutyryl)-camphor derivative [MCM-41]Y((-)-hfc)$_x$(THF)$_y$ produced the highest diastereomeric and enantiomeric excesses [67% *de*, 37% *ee* (-35 °C); compared to Er[(-)-hfc]$_3$: 68% *de*, 55% *ee*).

Pore-confinement of the catalytic center can have a favorable effect in some respects including (i) protection of the reactive site from deactivation processes, (ii) stabilization of labile reaction intermediates, and (iii) control of product selectivity (regioselectivity, enantioselectivity) and morphology. For example, the long-time stability of a confined catalyst was demonstrated for the

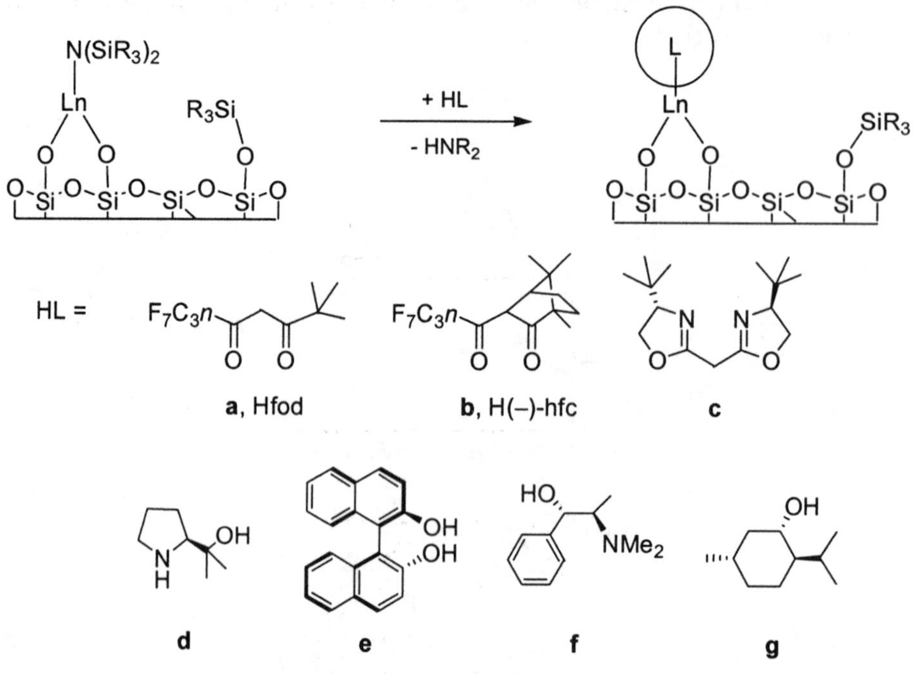

Scheme 15. "High-throughput synthesis" of chiral rare earth surface species. Adapted with permission from *Micropor. Mesopor. Mater.* **2001**, *44-45*, 303. Copyright 2001 Elsevier Science.

above-mentioned $Y(fod)_x$-catalyzed Danishefsky transformation (Scheme 14).[80] The homogeneous catalyst $Y(fod)_3$ was more active at the beginning of the reaction, however, the conversion comes to an halt after approx. 1 hr (80%) conversion, and no further activity was observed upon addition of another equivalent of substrate. In contrast, although the initial activity of the PMS-confined catalyst was slightly decreased compared to its molecular congener due to diffusion limitation, almost 95% conversion was obtained after 50 hrs, and new substrates were converted as quickly as the first time, revealing no marked decrease in activity towards the end of the reaction.

Ketyl radicals are key intermediates in the reduction of carbonyl compounds and their stabilization markedly influences product selectivities. The unequivocal formation of surface ketyl radicals for the Sm(II)-mediated fluorenone/fluorenol transformation (Scheme 5) was revealed by their X-band EPR spectra (Figure 6, A).[97] The rigid bonding of the radical to the surface was indicated by the absence of any qualitative changes in the spectra recorded at temperatures between 130 and 323 K. In contrast, ketyl radicals produced on an ordinary dehydrated silica material (Aerosil-200, Degussa-Hüls, surface area 200 m^2g^{-1}) displayed a relatively weak EPR signal (Figure 6, **B**). Although non-porous

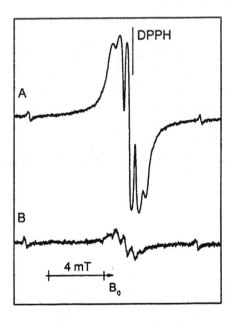

Figure 6. X-band EPR spectra (T = 293 K) observed for $Sm[N(SiHMe_2)_2]_2(THF)_x$@MCM-41 (A) and $Sm[N(SiHMe_2)_2]_2(THF)_x$@AS200 (B) under argon atmosphere (EPR settings: n = 9.05 GHz, microwave power 2 mW, modulation frequency 100 kHz, modulation amplitude 0.05 mT; the two small satellites are due to the Mn^{2+} standard).[97] Adapted with permission from *J. Am. Chem. Soc.* **2000**, *122*, 1544. Copyright 2000 American Chemical Society.

Aerosil-200 gave similar Sm(II) surface species, showing qualitatively analogous reaction behavior, the ketyl radicals seem to be increasingly prone to hydrogen radical abstraction, as indicated by a significant amount of fluorenol in the supernatant solution. This ketyl destabilization was ascribed to the lack of a protective effect of the intrapore arrangement and the changed morphology of the Aerosil-200 silica material (presence of an increased number of bulk silanol groups). The analysis of the BJH pore size distribution was in agreement with such a stabilization via pore confinement, indicating apparent loss of total pore volume and hence pore blocking (Figure 4).

Sm(II)@PMS was also employed for extrusion polymerization studies. For example, polymerization of methylmethacrylate (MMA) produced inorganic–organic composite materials featuring a completely filled mesopore system.[117]

5.   MODELING OF RARE EARTH SURFACE SITES

Modern methods of surface characterization contribute powerfully to a better understanding of the formation, appearance and catalytic action of active surface sites.[118] Moreover, due to their structural order and mesoporosity PMS materials give access to more diverse physico-chemical characterization which makes them especially well-suited as a model support for more ordinary forms of silica.[80] However, important information about the interaction of molecular precursors with reactive surface sites often remains elusive. In order to shed more light on supported catalysts, so-called molecular oxo–surfaces are increasingly being investigated (Figure 7).[119] Accordingly, ordinary silanols such as triphenylsilanol (not shown in Figure 7) are employed for the modeling of a monopodal (II^1) surface attachment of metal ligand fragments to isolated surface hydroxy groups.[120] Polyhedral oligosilsesquioxane (POSS)[121] and calixarene

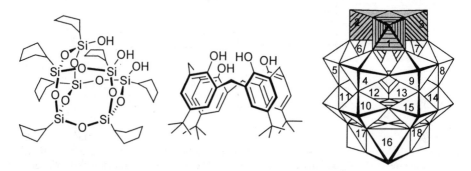

Figure 7. Examples of molecular oxo–surfaces. Reproduced with permission from *Chem. Mater.* **2001**, *13*, 4419. Copyright 2001 American Chemical Society.

derivatives[122] represent oxo–surfaces with various sets of OH groups. Polyoxyanion-supported catalysts such as $[Cp^*RhP_2W_{15}Nb_3O_{62}]^{7-}$ and $[(Cp_2U)_2(\mu\text{-}\kappa^2O\text{-}TiW_5O_{19})_2]^{4-}$ may also provide models of atomically dispersed metal complexes.[123]

Combined examination of the soluble model complexes via high-resolution solution spectroscopy and diffraction methods, such as X-ray structure analysis, assists in developing a detailed qualitative understanding of immobilization processes and the appearance of surface species. Such investigations are usually hampered by agglomeration processes in solution, *e.g.*, monomer–dimer equilibria, which produce higher-coordinated metal centers and hence imply changed (decreased) reactivity. Incompletely condensed silsesquioxanes of type $(c\text{-}C_5H_9)_7Si_7O_9(OH)_3$ (cyclopentyl-$T_7(OH)_3$: according to the conventional silicone nomenclature, T denotes a trifunctional unit of siloxane structure)[124] present a "realistic" electronic and steric situation as revealed by the strongly electron-withdrawing character of the trisilanol "SiO-framework". Of particular significance are the short-range structural similiarities between $T_7(OH)_3$ and geometrically comparable trisilanol sites available on idealized surfaces of SiO_2 polymorphs such as β-cristobalite and β-tridymite.[121]

Yttrium surface sites [≡Si–O]Y(SALEN) and [≡Si–O]Y-(disulfonamide), obtained according to route **B** of our "heterogenized silylamide route" (Scheme 8, Table 3) were modeled by reacting the corresponding heteroleptic silylamide complexes with triphenylsilanol, HOSiPh$_3$ (Scheme 16).[96] In order to probe the stability of the Y–SALEN and Y–disulfonamide moieties towards silanolysis, the silylamide precursor complexes were also reacted with excess HOSiPh$_3$. ^{1}H-NMR

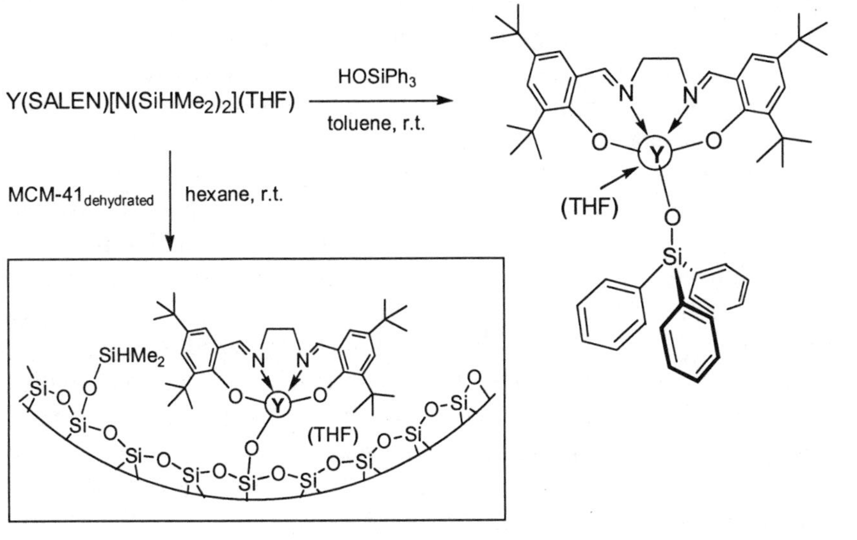

Scheme 16. Triphenylsilanol, HOSiPh$_3$, as a model for isolated surface silanol groups.

spectroscopy revealed both silylamide/siloxide ligand exchange and the stability of the Y–SALEN and –disulfonamide fragments under the prevailing conditions. Only prolonged treatment (>10 days) of complex Y–disulfonamide with HOSiPh$_3$ results in ligand redistribution and formation of [Y(OSiPh$_3$)$_3$(THF)$_3$].[124] The siloxide complexes [Ph$_3$Si–O]Y(SALEN) and [Ph$_3$Si–O]Y(disulfonamide) were isolated and fully characterized and can be regarded as model compounds of the predominant metal surface species of the hybrid materials. For example, the vibration v(C=N) of both the SALEN-derived hybrid material and the corresponding model complex is located at 1618 cm^{-1} (free SALEN ligand: 1626 cm^{-1}). The molecular structure of complex [Ph$_3$Si–O]Y(SALEN)(CH$_3$CN)(THF) features a seven-coordinate yttrium center in a slightly distorted pentagonal bipyramidal geometry. The SALEN and acetonitrile ligands are located in the equatorial plane, while the siloxide and the THF ligand occupy the apical positions. A remarkable coordinative flexibility of the SALEN ligand is documented by its significant bending (Y(SALEN)[N(SiHMe$_2$)$_2$](THF): ∠aryl/aryl=74°) and almost flat coordination mode ([Ph$_3$Si–O]Y(SALEN)(CH$_3$CN)(THF):∠aryl/aryl = 19°). This coordinative flexibility seems to be highly desirable during the grafting reaction onto the curved walls of the MCM-41 material. Considering the lateral extension of the SALEN (*ca.* 1.0X1.6 nm) and disulfonamide ligands (*ca.* 1.0X2.0 nm) and the pore dimensions, orientation of the bulky ligands parallel to the internal walls is suggested.

Tetravalent calix[4]arene ligands seem to qualify as superior model systems to mimic the reactivity of metal silylamide complexes on real oxo–surfaces, *i.e.*, a scenario involving both metal oxygen bond formation and *in situ* silylation (Scheme 17).[126] Reaction of Y[N(SiHMe$_2$)$_2$]$_3$(THF)$_2$ with a slurry of calix[4]arene in toluene gave complete ligand exchange via a homogeneous silylamine elimination. [Y{*p-tert*-butylcalix[4]arene(SiHMe$_2$)}(THF)]$_2$ was reproducibly isolated in high yield (>85%). As established by X-ray structure analysis, its solid state structure features a centrosymmetric aryloxide dimer with the yttrium centers placed in a distorted trigonal pyramidal coordination geometry. The appearance of a multifunctional surface reaction was also clarified by the spectroscopic details of the additionally formed "OSiHMe$_2$" silylether moieties showing a strong SiH stretching vibration at 2194 cm^{-1} and a singlet at 4.73 ppm for the SiH protons. The analogous Y[N(SiMe$_3$)$_2$]$_3$ reaction gave the corresponding product in low yield (17%). The presence of adjacent yttrium aryloxide and arylsilylether moieties impressively parallels our findings on mesoporous silica materials.[94] We propose an initial fast amide/aryloxide exchange reaction involving three phenolic positions of the tetravalent model oxo surface. The forth phenolic position seems to be not readily accessible to another bulky silylamide complex. This was further examined by the reaction of calix[4]arene with an excess of the complex Y[N(SiHMe$_2$)$_2$]$_3$(THF)$_2$, from which

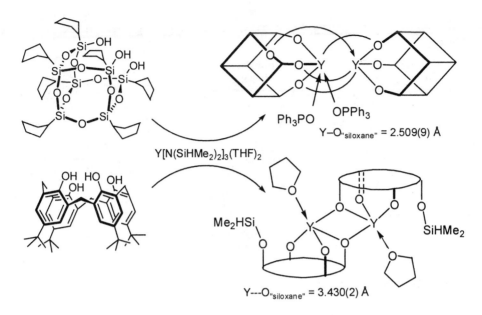

Scheme 17. Reactivity of yttrium silylamide toward various multipodal molecular oxo-surfaces.

the same product was isolated. Under these conditions, the released silylamine favorably competes for this remaining site as otherwise pointed out by the ready formation of [*p*-tert-butylcalix[4]arene(SiHMe$_2$)$_4$].[126]

Reaction of Y[N(SiMe$_3$)$_2$]$_3$ with a slurry of cyclopentyl-T$_7$(OH)$_3$ in toluene also gave complete ligand exchange to afford Y(cyclopentyl-T$_7$O$_3$)(THF) in quantitative yield.[127] Interestingly, this molecular oxo–surface produced metal siloxide bonds exclusively, *i.e.*, an *in situ* silylation did not occur (Scheme 17). Instead, this complex is capable of mimicking another essential feature of supported metal catalysts, namely the interaction of metal center with adjacent oxygen donors. X-ray structure analysis of a triphenylphosphine oxide adduct revealed a dinuclear composition and Y–O("SiO-framework") distances of 2.509(9) Å. However, both of the model systems shown in Scheme 17 emphasize the kinetic and thermodynamic limitations of multivalent molecular model oxo–surfaces. In particular, in the presence of oxophilic, electron-deficient metal centers, agglomeration or rearrangement to dimeric and hence higher-coordinated metal centers, is a predominant reaction pathway in solution.

The modeling of mononuclear, non-agglomerated surface metal sites featuring reactive metal ligand moieties seems to be feasible after initial manipulation of the model oxo–surface. Disilazane silylation was shown to be an effective tool to manipulate the calix[4]arene surface for subsequent derivatization reactions (Scheme 18).[109,126]

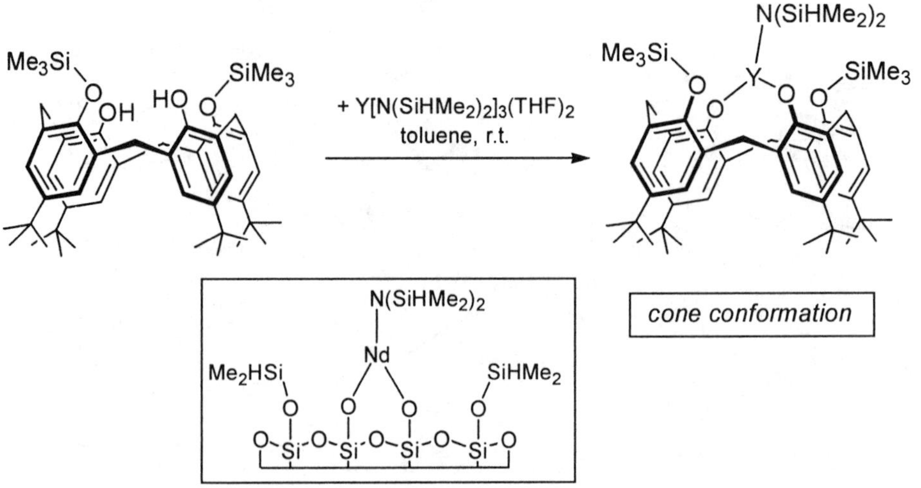

Scheme 18. Mononuclear model complexes via metallation of a "manipulated" oxo-surface.

For example, reaction of [*p-tert*-butylcalix[4]arene(SiMe$_3$)$_2$]H$_2$ with an equimolar amount of Y[N(SiHMe$_2$)$_2$]$_3$(THF)$_2$ in toluene proceeds under fairly mild conditions to give the complex {Y[*p-tert*-butylcalix[4]arene-(SiMe$_3$)$_2$][N(SiHMe$_2$)$_2$]}, the cone conformation of which was evidenced by X-ray structure analysis. The yttrium-bonded bis(dimethylsilyl)amide ligand was also clearly indicated by a strong SiH stretch vibration at 2069 cm^{-1} and a SiH proton at 5.45 ppm, considerably down-shifted compared to its synthetic precursor Y[N(SiHMe$_2$)$_2$]$_3$(THF)$_2$ (δ_{SiH} = 5.02 ppm). Heteroleptic complexes of the type {Ln[*p-tert*-butylcalix[4]arene(SiMe$_3$)$_2$][N(SiHMe$_2$)$_2$]} can be considered as promising compounds to study secondary surface-mediated ligand (silylamide) exchange reactions.

Partly-silylated silsesquioxanes,[128] *e.g.*, (c-C$_5$H$_9$)$_7$Si$_7$O$_9$(OH)$_2$-(OSiMe$_2$ ^{t}Bu)[129] and (c-C$_5$H$_9$)$_7$Si$_7$O$_9$(OH)$_2$(OSiMe$_3$),[130] were also reported as precursors for functional models for doped silicate materials. Examples of X-ray structurally characterized POLnSS derivatives are presented in Figure 8. The silyl substituents impart high solubility and partially block the missing vertex of the POSS trisilanol. Thus, "corner-capping" reactions yielding *polyhedral oligo-metallasilsesquioxanes* (POMSS)[120] are markedly affected. For example, access of the "SiO-framework" to the primary coordination sphere of the rare earth metal center seems to be sterically prevented.

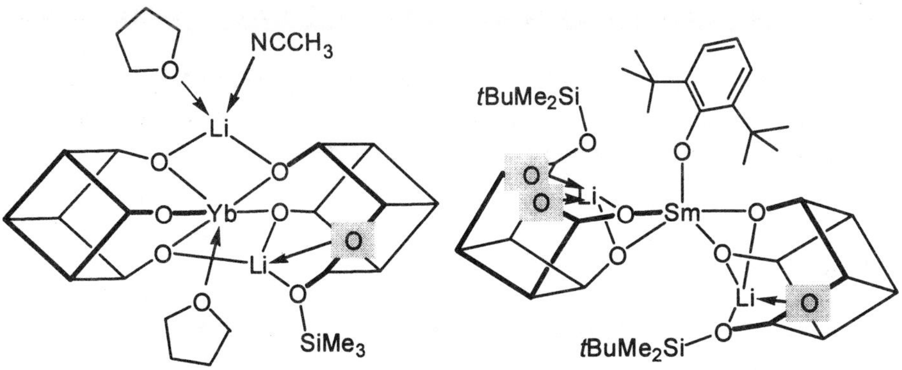

Figure 8. POLnSS complexes featuring silylated corner oxygen atoms.

6. CONCLUSIONS

Their abundant availability, non-toxicity and intrinsic stereoelectronic properties position the rare earth elements as increasingly valuable elements for applications in homogeneous and heterogeneous catalysis. The uniqueness of nanostructuring has been revealed by the changed (increased) surface reactivity of a series of heterogeneous rare earth catalysts. Advanced chemical and physical preparative methods will further improve their microstructural control and bring forth new catalytic applications. Surface organometallic chemistry (SOMC) in confined spaces adds a new dimension to the nanostructuring of rare earth catalysts. The uniform composition and intrinsic zeolite-like pore architecture of semicrystalline periodic mesoporous silicas (PMS) produce well-defined nanoparticle replicas for the design of new chemicals and inorganic–organic composite materials. Nanosized cavities make PMS materials of the M41S and SBA families ideal for use as catalyst supports and as precise reaction chambers accessible to an extended intraporous chemistry ("molecular nanofactories"). Encapsulation and surface grafting produce confined non-agglomerated metal ligand species or nanoparticles, primarily by steric effects.

Novel reaction pathways, arising from sterically highly unsaturated metal centers, surpass the phenomenology of contemporary molecular organometallic chemistry. Advanced methods of surface grafting, as evidenced by the versatility of a novel "heterogenized silylamide route," have important implications for the design of the nanoenvironment of immobilized metal species. Conclusive methods of characterization, elaborating the importance of spectroscopic probe ligands and nitrogen physisorption, reinforce the excellent capacity of PMS materials as a model support, for example, of amorphous silica. The importance of connecting novel molecular chemistry and surface chemistry is emphasized by the application of tailor-made grafting reagents and model oxo–surfaces. The application of SOMC@PMS to emulate the incredible cooperativity between metal site, protein tertiary structure, and substrate molecule in natural enzymes by "mesozymes" is certainly challenging. Note that the parallels of microporous zeolitic materials to natural metalloenzymes have been investigated for almost 20 years.[131]

ACKNOWLEDGMENTS

The author would like to thank Gisela Gerstberger, Iris Nagl, Clemens Palm and Markus Widenmeyer, who worked on this research topic, for their shared enthusiasm and their patience and technical assistance during the writing of this article. Our research is generously supported by the Deutsche Forschungsgemeinschaft.

REFERENCES

1.   (a) Hu, J.; Odom, T. W.; Lieber, C. M. *Acc. Chem. Res.* **1999**, *32*, 435. (b) El-Sayed, M. A. *Acc. Chem. Res.* **2001**, *34*, 257.

2.   (a) Ozin, G. A.; Kuperman, A.; Stein, A. *Angew. Chem. Adv. Mater.* **1989**, 373; *Angew. Chem. Int. Ed. Engl., Adv. Mat.* **1989**, 359. (b) Ozin, G. A.; Kuperman, A.; Stein, A. *Angew. Chem.* **1989**, *101*, 373; *Angew. Chem. Int. Ed. Engl.* **1989**, *28*, 359. (c) Ozin, G. A.; Özkar, S. *Adv. Mater.* **1992**, *4*, 11. (d) Ozin, G. E. *Adv. Mater.* **1992**, *4*, 612. (e) Ozin, G. *Adv. Mater.* **1994**, *6*, 71. (f) Mulvaney, P.; Liz-Marzán, L. M.; Giersig, M.; Ung, T. *J. Mater. Chem.* **2000**, *10*, 1259.

3.   (a) Tenne, R.; Homyonfer, M.; Feldman, Y. *Chem. Mater.* **1998**, *10*, 3225. (b) Edelmann, F. T. *Angew. Chem.* **1999**, *111*, 1473; *Angew. Chem. Int. Ed.* **1999**, *38*, 1381.

4.   (a) Meldrum, F. C.; Wade, V. J.; Nimmo, D. L.; Heywood, B. R.; Mann, S. *Nature* **1991**, *349*, 684. (b) König, B. *Angew. Chem.* **1997**, *109*, 1919; *Angew. Chem. Int. Ed.* **1997**, *36*, 1833. (c) Stewart, S.; Liu, G. *Angew. Chem.* **2000**, *112*, 348; *Angew. Chem. Int. Ed.* **2000**, *39*, 340. (d) Crooks, R. M.; Zhao, M.; Sun, L.; Chechik, V.; Yeung, L. K. *Acc. Chem. Res.* **2001**, *34*, 181.

5.   Moser, W. R., Ed. *Advanced Catalysts and Nanostructured Materials*, Academic Press: San Diego, 1996.

6.   Interrante, L. V.; Hampden-Smith, M. J., Eds. *Chemistry of Advanced Materials*, Wiley-VCH: New York, 1998.

7.   (a) Rieke, R. D. *Acc. Chem. Res.* **1977**, *10*, 301. (b) Davis, S. C.; Klabunde, K. J. *Chem. Rev.* **1982**, *82*, 153.

8.   (a) Stucky, G. D.; MacDougall, J. E. *Science* **1990**, *247*, 669. (b) Stucky, G. D. *Prog. Inorg. Chem.* **1992**, *40*, 99. (c) Behrens, P.; Stucky, G. D. in *Comprehensive Supramolecular Chemistry*, Alberti, G.; Bein, T., Eds. Elsevier: Oxford, **1996**, *7*, 721. (d) Schüth, F. *Chem. Zeit* **1995**, *29*, 42.

9. For recent review articles on the modification of mesoporous silica materials, see: (a) Clark, J. H.; Macquarrie, D. J. *Chem. Commun.* **1998**, 853. (b) Moller, K.; Bein, T. *Chem. Mater.* **1998**, *10*, 2950. (c) Ying, J. Y.; Mehnert, C. P.; Wong, M. S. *Angew. Chem.* **1999**, *111*, 58; *Angew. Chem. Int Ed.* **1999**, *111*, 56.

10. (a) Bönnemann, H.; Brijoux, W.; Joussen, T. *Angew. Chem.* **1990**, *102*, 324; *Angew. Chem. Int. Ed. Engl.* **1990**, *29*, 273. (b) Itoh, H.; Utamapanya, S.; Stark, J. V.; Klabunde, K. J.; Schlup, J. R. *Chem. Mater.* **1993**, *5*, 71.

11. Steigerwald, M. L.; Brus, L. E. *Acc. Chem. Res.* **1990**, *23*, 183.

12. (a) Iwasawa, Y.; Gates, B. C. *CHEMTECH* **1989**, 173. (b) Ozin, G. E. *Adv. Mater.* **1992**, *4*, 612. (c) Sachtler, W. M. H.; Zhang, Z. *Adv. Catal.* **1993**, *39*, 129. (d) De Vos, D. E.; Thibault-Starzyk, F.; Knops-Gerrits, P. P.; Parton, R. F.; Jacobs, P. A. *Macromol. Symp.* **1994**, *80*, 157.

13. (a) Ozin, G. A.; Özkar, S.; Prokopowicz, R. A. *Acc. Chem. Res.* **1992**, *25*, 553. (b) Bein, T. in *Comprehensive Supramolecular Chemistry*, Alberti, G.; Bein, T., Eds. Elsevier: Oxford, **1996**, *7*, 579. (c) Thomas, J. M. *Angew. Chem.* **1999**, *111*, 3800; *Angew. Chem. Int. Ed.* **1999**, *38*, 3588. (d) Barton, T. J.; Bull, L. M.; Klemperer, W. G.; Loy, D. A.; McEnaney, B.; Misono, M.; Monson, P. A.; Pez, G.; Scherer, G. W.; Vartuli, J. C.; Yaghi, O. M. *Chem. Mater.* **1999**, *11*, 2633.

14. (a) Robinson, A. L. *Science* **1976**, *194*, 1261. (b) Hartley, F. R.; Vezey, P. N. *Adv. Organomet. Chem.* **1977**, *15*, 189. (c) Bailey, D. C.; Langer, S. H. *Chem. Rev.* **1981**, *81*, 109. (d) Schwartz, J. *Acc. Chem. Res.* **1985**, *18*, 302. (e) Iwasawa, Y. *Adv. Catal.* **1987**, *35*, 187. (f) Basset, J.-M.; Gates, B. C.; Candy, J. P.; Choplin, A.; Leconte, H.; Quignard, F.; Santini, C., Eds., *Surface Organometallic Chemistry: Molecular Approaches to Surface Catalysis*, Kluwer: Dordrecht, **1988**. (g) Lamb, H. H.; Gates, B. C.; Knözinger, H. *Angew. Chem.* **1988**, *100*, 1162; *Angew. Chem. Int. Ed. Engl.* **1988**, *27*, 1127. (h) Gates, B. C. *J. Mol. Catal.* **1989**, *52*, 1. (i) Marks, T. J. *Acc. Chem. Res.* **1992**, *25*, 57. (j) Zecchina, A.; Areán, C. O. *Catal. Rev.-Sci. Eng.* **1993**, *35*, 262.

15. Anwander, R. *Chem. Mater.* **2001**, *13*, 4419.

16. For a review, see: Klabunde, K. J.; Li, Y.-X.; Tan, B.-J. *Chem. Mater.* **1991**, *3*, 30.

17. Evans, W. J.; Bloom, I.; Engerer, S. C. *J. Catal.* **1983**, *84*, 468.

18. (a) Imamura, H.; Ohmura, A.; Haku, E.; Tsuchiya, S. *J. Catal.* **1985**, *96*, 139. (b) Imamura, H.; Kitajima, K.; Tsuchiya, S. *Chem. Lett.* **1988**, 249. (c) Imamura, H.; Kitajima, K.; Tsuchiya, S. *J. Chem. Soc., Faraday Trans. 1* **1989**, *85*, 1647.

19. Zagorskii, V. V.; Kondakov, S. E.; Kosolapov, A. M.; Sergeev, G. B.; Solovev, V. N. *Organomet. Chem. USSR* **1992**, *5*, 255.

20. Inoue, M.; Kimura, M.; Inui, T. *Chem. Commun.* **1999**, 957.

21. Tschöpe, A.; Liu, W.; Flytzani-Stephanopoulos, M.; Ying, J. Y. *J. Catal.* **1995**, *157*, 42.

22. For examples, see: (a) Hussein, G. A. M. *J. Phys. Chem.* **1994**, *98*, 9657. (b) Okabe, K.; Sayama, K.; Kusama, H.; Arakawa, H. *Bull. Chem. Soc. Jpn* **1994**, *67*, 2894. (c) Jiang, J. C.; Pan, X. Q.; Graham, G. W.; McCabe, R. W.; Schwank, J. *Catal. Lett.* **1998**, *53*, 37.

23. LaDuca, R. L.; Wolczanski, P. T. *Inorg. Chem.* **1992**, *31*, 1311.

24. Baxter, D. V.; Chisholm, M. H.; Gama, G. J.; DiStasi, V. F.; Hector, A. L.; Parkin, I. P. *Chem. Mater.* **1996**, *8*, 1222.

25. Karl, M.; Seybert, G. Massa, W.; Dehnicke, K. *Z. Anorg. Allg. Chem.* **1999**, *625*, 375.

26. White III, J. P.; Deng, H.; Shore, S. G. *Inorg. Chem.* **1991**, *30*, 2337.

27. Veith, M.; Mathur, S.; Lecerf, N.; Bartz, K.; Heintz, M.; Huch, V. *Chem. Mater.* **2000**, *12*, 271.

28. (a) Zhang, X.; Walters, A. B.; Vannice, M. A. *J. Catal.* **1995**, *155*, 290. (b) Otsuka, K.; Zhang, O.; Yamanaka, I.; Tono, H.; Hatano, M; Kinoshita, H. *Bull. Chem. Soc. Jpn* **1996**, *69*, 3367.

29.   (a) Fokema, M. D.; Ying, J. Y. *J. Catal.* **2000**, *192*, 54. (b) Fokema, M. D.; Chiu, E.; Ying, J. Y. *Langmuir* **2000**, *16*, 3154.

30.   Guizard, C. G.; Julbe, A. C.; Ayral, A. *J. Mater. Chem.* **1999**, *9*, 55.

31.   Manziek, L.; Langenmayr, E.; Lamola, A.; Gallagher, M.; Brese, N.; Annan, N. *Chem. Mater.* **1998**, *10*, 3101.

32.   Vaqueiro, P.; López-Quintela, M. A.; Rivas, J. *J. Mater. Chem.* **1997**, *7*, 501.

33.   Gan, L. M.; Zhang, L. H.; Chan, H. S. O.; Chew, C. H.; Loo, B. H. *J. Mater. Sci.* **1996**, *31*, 1071.

34.   Masui, T.; Fujiwara, K.; Machida, K.; Adachi, G.; Sakata, T.; Mori, H. *Chem. Mater.* **1997**, *9*, 2197.

35.   Deng, H.; Gin, D. L.; Smith, R. C. *J. Am. Chem. Soc.* **1998**, *120*, 3522.

36.   Kobayashi, S.; Nagayama, S. *J. Am. Chem. Soc.* **1998**, *120*, 2985.

37.   (a) Rosynek, M. P. *Catal. Rev.* **1977**, *16*, 111. (b) Sauvion, G. N.; Ducros, P. *J. Less-Common. Met.* **1985**, *111*, 23. (c) Trovarelli, A. *Catal. Rev.* **1996**, *38*, 439.

38.   (a) Breck, D. W. *Zeolite Molecular Sieves*, Wiley: New York, 1974. (b) Rabo, J. A. *Zeolite Chemistry and Catalysis, ACS Monograph 171*, A.C.S.: Washington, D.C. 1976.

39.   For examples, see: (a) Cheetham, A. K.; Eddy, M. M.; Thomas, J. M. *J. Chem. Soc., Chem. Comm.* **1984**, 1337. (b) Gaare, K.; Akporiaye, D. *J. Phys. Chem. B* **1997**, *101*, 48. (c) Berry, F. J.; Carbuciocchio, M.; Chiari, A.; Johnson, C.; Moore, E. A.; Mortimer, M.; Vetel, F. F. F. *J. Mater. Chem.* **2000**, *10*, 2131.

40.   (a) Rocha, J.; Ferreira, P.; Lin, Z.; Brandao, P.; Ferreira, A.; Pedrosa de Jesus, J. D. *Chem. Comm.* **1997**, 2103. (b) Rocha, J.; Ferreira, P.; Lin, Z.; Brandao, P.; Ferreira, A.; Pedrosa de Jesus, J. D. *J. Phys. Chem. B* **1998**, *102*, 4739. (c) Ananias, D.; Ferreira, A.; Rocha, J.; Ferreira, P.; Rainho, J. P.; Morais, C.; Carlos, L. D. *J. Am. Chem. Soc.* **2001**, *123*, 5735.

41.   Huppertz, H.; Schnick, W. *Angew. Chem.* **1997**, *109*, 2765; *Angew. Chem. Int. Ed. Engl.* 1997, 36, 2651.

42.   Yaghi, O. M.; Li, H.; Davis, C.; Richardson, D.; Groy, T. L. *Acc. Chem. Res.* **1998**, *31*, 474.

43.   (a) Bataoille, T.; Auffrédic, J.-P.; Louer, D. *J. Mater. Chem.* **2000**, *10*, 1707. (b) Vaidhyanathan, R.; Natarajan, S.; Rao, C. N. R. *Chem. Mater.* **2001**, *13*, 185.

44.   (a) Kiritsis, V.; Michaelides, A.; Skoulika, S.; Golhen, S.; Ouahab, L. *Inorg. Chem.* **1998**, *37*, 3407. (b) Serpaggi, F.; Férey, G. *J. Mater. Chem.* **1998**, *8*, 2737. (c) Reineke, T. M.; Eddaoudi, M.; Fehr, M.; Kelley, D.; Yaghi, O. M. *J. Am. Chem. Soc.* **1999**, *121*, 1651. (d) Reineke, T. M.; Eddaoudi, M.; O'Keeffe, M.; Yaghi, O. M. *Angew. Chem.* **1999**, *111*, 2712; *Angew. Chem. Int. Ed. Engl.* **1999**, *38*, 2590. (e) Reineke, T. M.; Eddaoudi, M.; Moler, D.; O'Keeffe, M.; Yaghi, O. M. *J. Am. Chem. Soc.* **2000**, *122*, 4843.

45.   Serpaggi, F.; Férey, G. *J. Mater. Chem.* **1998**, *8*, 2749.

46.   For a review, see: Ciesla, U.; Schüth, F. *Micropor. Mesopor. Mater.* **1999**, *27*, 131.

47.   For examples, see: (a) Antonelli, D. M.; Ying, J. Y. *Angew. Chem.* **1995**, *107*, 2202; *Angew. Chem. Int. Ed.* **1995**, *34*, 2014. (b) Ciesla, U.; Schacht, S.; Stucky, G. D.; Unger, K. K.; Schüth, F. *Angew. Chem.* **1996**, *108*, 597; *Angew. Chem. Int. Ed.* **1996**, *35*, 541. (c) Bagshaw, S. A.; Pinnavaia, T. J. *Angew. Chem.* **1996**, *108*, 1180; *Angew. Chem. Int. Ed.* **1996**, *35*, 1102.

48.   Yada, M.; Kitamura, H.; Machida, M.; Kijima, T. *Inorg. Chem.* **1998**, *37*, 6470.

49.   Yada, M.; Kitamura, H.; Ichinose, A.; Machida, M.; Kijima, T. *Angew. Chem.* **1999**, *111*, 3716; *Angew. Chem. Int. Ed.* **1999**, *38*, 3506.

50.   (a) Yada, M.; Ohya, M.; Machida, M.; Kijima, T. *Chem. Commun.* **1998**, 1941. (b) Yada, M.; Ohya, M.; Ohe, K.; Machida, M.; Kijima, T. *Langmuir* **2000**, *16*, 1535.

51.   Yada, M.; Ohya, M.; Machida, M.; Kijima, T. *Langmuir* **2000**, *16*, 4752.

52.   Wang, Y.; Yin, S.; Palchik, O.; Hacohen, Y. R.; Koltypin, Y.; Gedanken, A. *Chem. Mater.* **2001**, *13*, 1248.

53. (a) Weing, W.; Yang, J.; Ding, Z. *J. Non-Cryst. Solids* **1994**, *169*, 177. (b) Verweij, H. *Adv. Mater.* **1998**, *10*, 1483.
54. Mamak, M.; Coombs, N.; Ozin, G. *Adv. Mater.* **2000**, *12*, 198.
55. Zhang, W.; Pinnavaia, T. J. *Chem. Commun.* **1998**, 1185.
56. Terribile, D.; Trovarelli, A.; Llorca, J.; de Leitenburg, C.; Dolcetti, G. *J. Catal.* **1998**, *175*, 299.
57. Kim, J. M.; Kwak, J. H.; Jun, S.; Ryoo, R. *J. Phys. Chem.* **1995**, *99*, 16742.
58. (a) He, N; Bao, S.; Xu, Q. *Stud. Surf. Sci. Catal.* **1997**, *105*, 85. (b) He, N.; Lu, Z.; Yuan, C.; Hong, J.; Yang, C.; Bao, S.; Xu, Q. *Supramol. Sci.* **1998**, *5*, 553. (c) Araujo, A. S.; Jaroniec, M. *J. Coll. Interface Sci.* **1999**, *218*, 462. (d) Araujo, A. S.; Jaroniec, M. *Stud. Surf. Sci. Catal.* **2000**, *129*, 187.
59. Kloestra, K. R.; van Laren, M.; van Bekkum, H. *J. Chem. Soc., Faraday Trans.* **1997**, *93*, 2111.
60. Kloestra, K. R.; van den Broek, J.; van Bekkum, H. *Catal. Lett.* **1997**, *47*, 235.
61. Clark, D. L.; Hollis, R. V.; Scott, B. L.; Watkin, J. G. *Inorg. Chem.* **1996**, *35*, 667.
62. (a) González, F.; Pesquera, C.; Benito, I.; Mendioroz, S.; Poncelet, G. *J. Chem. Soc., Chem. Comm.* **1992**, 491. (b) Schoonheydt, R. A.; van den Eynde, J.; Tubbax, H.; Leeman, H.; Stuyckens, M.; Lenotte, L.; Stone, W. E. E. *Clays & Clays Minerals* **1993**, *41*, 598.
63. Pires, J.; Machado, M.; de Carvalho, M. B. *J. Mater. Chem.* **1998**, *8*, 1465.
64. For examples, see: (a) Planeix, J. M.; Coustel, N.; Coq, B.; Brotons, V.; Kumbhar, P. S.; Dutartre, R.; Geneste, P.; Bernier, P.; Ajayan, P. M. *J. Am. Chem. Soc.* **1994**, *116*, 7935. (b) Stevens, M. G.; Foley, H. C. *Chem. Commun.* **1997**, 519. (c) Pham-Huu, C.; Neller, N.; Charbonniere, L. J.; Ziessel, R.; Ledoux, M. J. *Chem. Commun.* **2000**, 1871.
65. (a) Lago, R. M.; Tsang, S. C.; Lu, K. L.; Chen, Y. K.; Green, M. L. H. *J. Chem. Soc., Chem. Comm.* **1995**, 1355. (b) Chen, Y. K.; Chu, A.; Cook, J.; Green, M. L. H.; Harris, P. J. F.; Heesom, R., Humphries, M.; Sloan, J.; Tsang, S. C.; Turner, J. F. C. *J. Mater. Chem.* **1997**, *7*, 545.
66. Xu, C.; Sloan, J.; Brown, G.; Bailey, S.; Williams, V. C.; Friedrichs, S.; Coleman, K. S.; Flahaut, E.; Hutchison, J. L.; Dunin-Borkowski, R. E.; Green, M. L. H. *Chem. Commun.* **2000**, 2427.
67. Tamai, H.; Ikeuchi, M.; Kojima, S.; Yasuda, H. *Adv. Mater.* **1997**, *9*, 55.
68. For examples, see: Krause, K. R.; Schabes-Retchkiman, P.; Schmidt, L. D. *J. Catal.* **1992**, *134*, 204. (b) Krause, K. R.; Schmidt, L. D. *J. Catal.* **1993**, *140*, 424.
69. (a) O'Hare, D. in *Inorganic Materials*, Bruce, D. W.; O'Hare, D., Eds. Wiley: Chichester 1992. (b) Morikawa, Y. *Adv. Catal.* **1993**, *39*, 303. (c) O'Hare, D. *New J. Chem.* **1994**, *18*, 989. (d) Lefebvre, F.; de Mallmann, A.; Basset, J.-M. *Eur. J. Inorg. Chem.* **1999**, 361 and references therein.
70. (a) Baba, T.; Koide, R.; Ono, Y. *J. Chem. Soc., Chem. Comm.* **1991**, 691. (b) Baba, T.; Kim, G. J.; Ono, Y. *J. Chem. Soc., Faraday Trans.* **1992**, *88*, 891. (c) Baba, T.; Hikita, S.; Koide, R.; Ono, Y.; Hanada, T.; Tanaka, T.; Yoshida, S. *J. Chem. Soc., Faraday Trans.* **1993**, *89*, 3177. (d) Baba, T.; Hikita, S.; Ono, Y.; Yoshida, T.; Tanaka, T.; Yoshida, S. *J. Mol. Catal.* **1995**, *98*, 49.
71. For review articles, see: (a) Schüth, F. *Ber. Bunsengew. Phys. Chem.* **1995**, *99*, 1306. (b) Zakharov, V. A.; Yermakov, Y. I. *Catal. Rev. Sci. Eng.* **1979**, *19*, 67. (c) Sayari, A. *Stud. Surf. Sci. Catal.* **1996**, *102*, 1. (d) Zhao, X. S.; Lu, G. Q.; Millar, G. J. *Ind. Eng. Chem. Res.* **1996**, *35*, 2075. (e) Corma, A. *Chem. Rev.* **1997**, *97*, 2373. (f) Biz, S.; Occelli, M. L. *Catal. Rev.–Sci. Eng.* **1998**, *40*, 329. (g) Corma, A.; Kumar, D. *Stud. Surf. Sci. Catal.* **1998**, *117*, 201. (h) Guizard, C. G.; Julbe, A. C.; Ayral, A. *J. Mater. Chem.* **1999**, *9*, 55.
72. (a) Moore, P. B.; Shen, J. *Nature* **1983**, *306*, 356. (b) Smith, J. V.; Dytrych, W. J. *Nature* **1984**, *309*, 607. (c) Smith, J. V. *Chem. Rev.* **1988**, *88*, 149. (d) Davis, M. E.; *Acc. Chem. Res.* **1993**, *26*, 111.

73.   Vansant, E. F.; Van Der Voort, P.; Vrancken, K. C., Eds. *Characterization and Chemical Modification of the Silica Surface, Stud. Surf. Sci. Catal.*, **93** Elsevier: Amsterdam, 1995.
74.   (a) Kresge, C. T.; Leonowicz, M. E.; Roth, W. J.; Vartuli, J. C.; Beck, J. S. *Nature* **1992**, *359*, 710. (b) Beck, J. S.; Vartuli, J. C.; Roth, W. J.; Leonowicz, M. E.; Kresge, C. T.; Schmitt, K. D.; Chu, C. T.-W.; Olson, D. H.; Sheppard, E. W.; McCullen, S. B.; Higgins, J. B.; Schlenker, J. L. *J. Am. Chem. Soc.* **1992**, *114*, 10834.
75.   Morey, M. S.; Davidson, A.; Stucky, G. D. *J. Porous Mater.* **1998**, *5*, 195.
76.   (a) Zhao, D.; Feng, J.; Huo, Q.; Melosh, N.; Fredrickson, G. H.; Chmelka, B. F.; Stucky, G. D. *Science* **1998**, *279*, 548. (b) Zhao, D.; Huo, Q.; Feng, J.; Chmelka, B. F.; Stucky, G. D. *J. Am. Chem. Soc.* **1998**, *120*, 6024.
77.   (a) Ishikawa, T.; Matsuda, M.; Yasukawa, A.; Kandori, K.; Inagaki, S.; Fukushima, T.; Kondo, S. *J. Chem. Soc., Faraday Trans.* **1996**, *92*, 1985. (b) Jentys, A.; Pham, N. H.; Vinek, H. *J. Chem. Soc., Faraday Trans.* **1996**, *92*, 3287.
78.   Morrow, B. A.; Cody, I. A. *J. Phys. Chem.* **1976**, *80*, 1995.
79.   Sing, K.S.W.; Everett, D.H.; Haul, R.A.W.; Moscou, L.; Pierotti, R.A.; Rouquérol, J.; Siemieniewska, T. *Pure Appl. Chem.* **1985**, *57*, 603.
80.   Gerstberger, G.; Palm, C.; Anwander, R. *Chem. Eur. J.* **1999**, *5*, 997.
81.   Anwander, R.; *Top. Organomet. Chem.* **1999**, 2, 1.
82.   (a) Parks, G. A. *Chem. Rev.* **1965**, *65*, 177. (b) Treyakov, N. E.; Filimonov, K. N. *Kinet. Katal.* **1972**, *13*, 815. (c) Iler, R. K. *The Chemistry of Silica*, Wiley-Interscience: New York, 1979. (d) Unger, K. K. *Porous Silica*, Elsevier: Amsterdam, 1979.
83.   Anwander, R.; Palm, C. *Stud. Surf. Sci. Catal.* **1998**, *117*, 413.
84.   See also: Morey, M. S.; Stucky, G. D.; Schwarz, S.; Fröba, M. *J. Phys. Chem. B* **1999**, *103*, 2037.
85.   Bouh, A. O.; Rice, G. L.; Scott, S. L. *J. Am. Chem. Soc.* **1999**, *121*, 7201.
86.   Anwander, R.; Roesky, R. *J. Chem. Soc., Dalton Trans.* **1997**, 137.
87.   See also: Morey, M.; Davidson, A.; Eckert, H.; Stucky, G. D. *Chem. Mater.* **1996**, *8*, 486.
88.   Anwander, R.; Palm, C.; Groeger, O.; Engelhardt, G. *Organometallics* **1998**, *17*, 2027.
89.   Eppinger, J. *Ph.D. Thesis*, Technische Universität München, 1999.
90.   Köhler, K.; Engweiler, J.; Viebrock, H.; Baiker, A. *Langmuir* **1995**, *11*, 3423.
91.   Widenmeyer, M.; Grasser, S.; Köhler, K.; Anwander, R. *Micropor. Mesopor. Mater.* **2001**, *44-45*, 327.
92.   (a) Zhou, X.; Ma, H.; Huang, X.; You, X. *J. Chem. Soc., Chem. Comm.* **1995**, 2483. (b) Zhou, X.; Huang, Z.; Cai, R.; Zhang, L.; Zhang, L.; Huang, X. *Organometallics* **1999**, *18*, 4128. (c) Kraut, S.; Magull, J.; Schaller, U.; Karl, M.; Harms, K.; Dehnicke, K. *Z. Anorg. Allg. Chem.* **1998**, *624*, 1193.
93.   Anwander, R. *Top. Curr. Chem.* **1996**, *179*, 33.
94.   Anwander, R.; Runte, O.; Eppinger, J.; Gerstberger, G.; Herdtweck, E.; Spiegler, M. *J. Chem. Soc., Dalton Trans.* **1998**, 847.
95.   (a) Harris, D. H.; Lappert, M. F. in *Organometallic Chemistry Reviews: Organosilicon Reviews*, Elsevier: Amsterdam, 1976, p. 13. (b) Bradley, D. C.; Chisholm, M. H. *Acc. Chem. Res.* **1976**, *9*, 273. (c) Lappert, M. F.; Power, P. P.; Sanger, A. R.; Srivastava, R. C. *Metal And Metalloid Amides*, Ellis Horwood: Chichester, **1980**.
96.   Anwander, R.; Görlitzer, H.; Palm, C.; Runte, O.; Spiegler, M. *J. Chem. Soc., Dalton Trans.* **1999**, 3611.
97.   Nagl, I.; Widenmeyer, M.; Grasser, S.; Köhler, K.; Anwander, R. *J. Am. Chem. Soc.* **2000**, *122*, 1544.
98.   Corker, J.; Lefebvre, F.; Lécuyer, C.; Dufaud, V.; Quignard, F.; Choplin, A.; Evans, J.; Basset, J.-M. *Science* **1996**, *271*, 966.
99.   Adachi, M.; Corker, J.; Kessler, H.; Lefebvre, F.; Basset, J.-M. *Micropor. Mesopor. Mater.* **1998**, *21*, 81.

100. Kagan, H. B.; Namy, J.-L. *Top. Organomet. Chem.* **1999**, *2*, 155.
101. Hou, Z.; Fujita, A.; Zhang, Y.; Miyano, T.; Yamazaki, H.; Wakatsuki, Y. *J. Am. Chem. Soc.* **1998**, *120*, 754 and references therein.
102. Hou, Z.; Fujita, A.; Zhang, Y.; Miyano, T.; Yamazaki, H.; Wakatsuki, Y. *J. Am. Chem. Soc.* **1998**, *120*, 754 and references therein.
103. Cornils, B.; Herrmann, W. A. Eds. *Applied Homogeneous Catalysis by Organometallic Complexes* , VCH: Weinheim, 1996.
104. Tajima, K.; Aida T. *Chem. Commun.* **2000**, 2399.
105. (a) Klein, S.; Maier, W. F. *Angew. Chem.* **1996**, *108*, 2376; *Angew. Chem. Int. Ed.* **1996**, *35*, 2230. (b) Kochkar, H.; Figueras, F. *J. Catal.* **1997**, *171*, 420. (c) Clark, J. H.; Macquarrie, D. J. *Chem. Commun.* **1998**, 853. (d) Müller, C. A.; Maciejewski, M.; Mallat, T.; Baiker, A. *J. Catal.* **1999**, *184*, 280. (e) Price, P. M.; Clark, J. H.; Macquarrie, D. J. *J. Chem. Soc., Dalton Trans.* **2000**, 101.
106. (a) Maschmeyer, T.; Rey, F.; Sankar, G.; Thomas, J. M. *Nature* **1995**, *378*, 159. (b) Oldroyd, R. D.; Thomas, J. M.; Maschmeyer, T.; MacFaul, P. A.; Snelgrove, D. W.; Ingold, K. U.; Wayner, D. D. M. *Angew. Chem.* **1996**, *108*, 2966; *Angew. Chem. Int. Ed.* **1996**, *35*, 2787. (c) Chen, L. Y.; Chuah, G. K.; Jaenicke, S. *Catal Lett.* **1998**, *50*, 107. (d) Davies, L. J.; McCorn, P.; Bethell, D.; Page, P. C. B.; King, F.; Hancock, F. E.; Hutchings, G. J. *J. Mol. Catal. A: Chem.* **2001**, *165*, 243.
107. For an example, see: Oldroyd, R. D.; Sankar, G.; Thomas, J. M.; Özkaya, D. *J. Phys. Chem. B* **1998**, *102*, 1849.
108. (a) Hertl, W.; Hair, M. L. *J. Phys. Chem.* **1971**, *75*, 2181. (b) Sindorf, D. W.; Maciel, G. E. *J. Phys. Chem.* **1982**, *86*, 5208. (c) Haukka, S.; Root, A. *J. Phys. Chem.* **1994**, *98*, 1695.
109. (a) Anwander, R.; Palm, C.; Stelzer, J.; Groeger, O.; Engelhardt, G. *Stud. Surf. Sci. Catal.* **1998**, *117*, 135. (b) Anwander, R.; Nagl, I.; Widenmeyer, M.; Engelhardt, G.; Groeger, O.; Palm, C.; Röser, T. *J. Phys. Chem. B* **2000**, *104*, 3532.
110. For an example, see: Leyrit, P.; McGill, C.; Quignard, F.; Choplin, A. *J. Mol. Catal. A* **1996**, *112*, 395.
111. Anwander, R. *Top. Curr. Chem.* **1996**, *179*, 149.
112. Nagl, I.; Widenmeyer, M.; Herdtweck, E.; Raudaschl-Sieber, G.; Anwander, R. *Micropor. Mesopor. Mater.* **2001**, *44-45*, 311.
113. See also: Toscano, P. J.; Marks, T. J. *J. Am. Chem. Soc.* **1985**, *107*, 653.
114. (a) Lappert, M. F.; Pearce, R. *J. Chem. Soc., Chem. Commun.* **1973**, 126. (b) Schumann, H.; Müller, J. *J. Organomet. Chem.* **1978**, *146*, C5.
115. Watson, P. L.; Parshall, G. W. *Acc. Chem. Res.* **1985**, *18*, 51.
116. Gerstberger, G.; Anwander, R. *Micropor. Mesopor. Mater.* **2001**, *44-45*, 303.
117. Nagl, I. *Ph.D. Thesis*, Technische Universität München, **2001**.
118. For examples, see: (a) Rodger, C.; Smith, W. E.; Dent, G.; Edmondson, M. *J. Chem. Soc., Dalton Trans.* **1996**, 791. (b) Evans, J. *Chem. Soc. Rev.* **1997**, 11.
119. (a) Struchkov, Y. T.; Lindeman, S. V. *J. Organomet. Chem.* **1995**, *488*, 9. (b) Murugavel, R.; Voigt, A.; Walawalkar, M. G.; Roesky, H. W. *Chem. Rev.* **1996**, *96*, 2205. (c) Lorenz, V.; Fischer, A.; Gießmann, S.; Gilje, J. W.; Gun'ko, Y.; Jacob, K.; Edelmann, F. T. *Coord. Chem. Rev.* **2000**, *206-207*, 321.
120. (a) Herrmann, W. A.; Stumpf, A. W.; Priermeier, T.; Bogdanovic, S.; Dufaud, V.; Basset, J.-M. *Angew. Chem.*, **1996**, **108**, 2978; *Angew. Chem. Int. Ed. Engl.*, **1996**, **35**, 2803. (b) Piquemal, J.-Y.; Halut, S.; Brégeault, J.-M. *Angew. Chem.*, **1998**, **110**, 1149; *Angew. Chem. Int. Ed. Engl.*, **1998**, **37**, 1146.
121. (a) Feher, F. J.; Budzichowski, T. A. *Polyhedron* **1995**, *14*, 3239. (b) Abbenhuis, H. C. L. *Chem. Eur. J.* **2000**, *6*, 25 and references therein.
122. Floriani, C. *Chem. Eur. J.* **1999**, *5*, 19 and references therein.

123.   (a) Day, V. W.; Earley, C. W.; Klemperer, W. G., Maltbie, D. J. *J. Am. Chem. Soc.* **1985**, *107*, 8261. (b) Pohl, M.; Lyon, D. K.; Mizuno, N.; Nomiya, K.; Finke, R. G. *Inorg. Chem.* **1995**, *34*, 1413.

124.   Brown, Jr., J. F.; Vogt, Jr., L. H. *J. Am. Chem. Soc.* **1965**, *87*, 4313.

125.   McGeary, M. J.; Coan, P. S.; Folting, K.; Streib, W. E.; Caulton, K. G. *Inorg. Chem.*, 1991, **30**, 1723.

126.   Anwander, R.; Eppinger, J.; Nagl, I.; Scherer, W.; Tafipolsky, M.; Sirsch, P. *Inorg. Chem.* **2000**, *39*, 4713.

127.   Herrmann, W. A.; Anwander, R.; Dufaud, V.; Scherer, W. *Angew. Chem.* **1994**, *106*, 1338; *Angew. Chem. Int. Ed.* **1994**, *33*, 1285.

128.   Feher, F. J.; Newman, D. A. *J. Am. Chem. Soc.* **1990**, *112*, 1931.

129.   Arnold, P. L.; Blake, A. L.; Hall, S. N.; Ward, B. D.; Wilson, C. *J. Chem. Soc., Dalton Trans.* **2001**, 488.

130.   (a) Annand, J.; Aspinall, H. C.; Steiner, A. *Inorg. Chem.* **1999**, *38*, 3941. (b) Annand, J.; Aspinall, H. C. *J. Chem. Soc., Dalton Trans.* **2000**, 1867.

131.   Herron, N. *CHEMTECH* **1989**, 542.

Chapter 3

SILSESQUIOXANES: ADVANCED MODEL SUPPORTS IN DEVELOPING SILICA-IMMOBILIZED POLYMERIZATION CATALYSTS

R. Duchateau
Dutch Polymer Institute, Department of Inorganic Chemistry and Catalysis, Eindhoven University of Technology, P.O.Box 513, 5600 MB Eindhoven, The Netherlands.

Keywords: silsesquioxane, silanol acidity, metallasilesquioxane, polymerization, tethered catalysts

Abstract: To gain more insight, at a molecular level, into processes taking place at silica surfaces, suitable model systems for silica surface silanol sites are of great importance. The major advantage of such homogeneous model systems is that their structural properties and reactivity can be studied in detail using a wide range of powerful techniques such as single crystal X-ray diffraction and multinuclear solution NMR spectroscopy. Several model supports for silica have been studied, ranging from mono-, di- and polysilanols to silsesquioxanes. The latter are probably the most realistic homogeneous model systems for silica surface silanol sites known to date and provide an excellent opportunity to fine-tune synthetic procedures to modify silica surfaces or to anchor metal complexes to the support. In this chapter, the usefulness of silsesquioxanes to provide information about synthetic strategies, stability and reactivity of silica-supported (co)catalysts will be outlined. Physisorption, grafting and tethering are immobilization techniques that will be discussed. As will be shown, metallasilsesquioxanes not only contribute to a molecular level understanding of known heterogeneous catalyst systems; with activities sometimes even exceeding that of commercial silica-supported catalysts, they are interesting catalysts themselves.

1.   SILSESQUIOXANES

Silsesquioxane is the general name for organosiloxide species with the empirical formula $(RSiO_{3/2})_n$ (R = H, hydrocarbon) and closely related compounds.[1] For these compounds, irregular, ladder and cage structures are known, of which the latter are the most familiar. These cage structures can be regarded as small three-dimensional pieces of silica since their degree of oligomerization is sufficient to result in rigid structures that resemble, for example, crystalline forms of silica such as β-cristobalite or β-tridymite.[2]

Fully-condensed silsesquioxanes exist in a wide range of cage structures, such as trigonal prisms $R_6Si_6O_9$, cubes $R_8Si_8O_{12}$ and drum-shaped frameworks like $R_{10}Si_{10}O_{15}$ and $R_{12}Si_{12}O_{18}$.[3] These closed silsesquioxanes have been used mainly as (co-)monomers in polymers and resins to yield organic-inorganic hybrid materials.[4] Incompletely-condensed silsesquioxanes can be regarded as prescursors of the fully-condensed silsesquioxanes. These compounds are especially interesting since they have proven to be realistic homogeneous models for a variety of surface silanol sites. The leading representative of these incompletely-condensed silsesquioxanes is the trisilanol $R_7Si_7O_9(OH)_3$ (**I**, Scheme 1). Although first reported by Brown and Vogt,[5] the names of Feher and co-workers are synonymous with this silsesquioxane ligand.[2b,6,7]

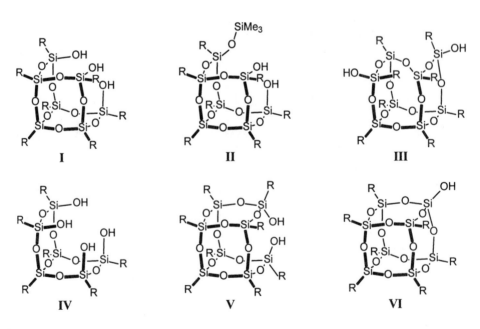

Scheme 1. Examples of incompletely-condensed silsesquioxanes.

1.1 Synthesis

Synthetic methodologies for silsesquioxane preparation are well-documented. The general route consists of hydrolytic condensation of $RSiCl_3$ or $RSi(OR')_3$ precursors (eq. 1). For the synthesis of fully-condensed silsesquioxanes, $RSiX_3$ (X = Cl, OR') with a wide variety of hydrocarbon substituents R can be used.[1,3] Fully-condensed silsesquioxanes can also be prepared by corner-capping reactions of the incompletely-condensed $R_7Si_7O_9(OH)_3$ or by hydrosilylation of $H_8Si_8O_{12}$ (Figure 1) or $R_7Si_8O_{12}H$.[1,4c]

For the preparation of partly-condensed silsesquioxanes (Scheme 1), alkyl trichlorosilanes $RSiCl_3$ are generally required and, interestingly, the choice of hydrocarbon substituent is restricted to cyclopentyl, cyclohexyl or cycloheptyl groups (eq. 1).[2,6] Incompletely-condensed silsesquioxanes with phenyl substituents have also been reported,[8] however, the tetrameric hydroxyl-containing silsesquioxane being the largest well-defined agglomerate (besides polymers), the oligomerization grade of these compounds is considerably lower.

The hydrolytic condensations are straightforward reactions and yield $R_7Si_7O_9$ (R = c-C_5H_9, c-C_6H_{11}, c-C_7H_{13}) as the main product. For R = c-C_6H_{11}, minor amounts of the completely-condensed silsesquioxane (c-$C_6H_{11})_6Si_6O_9$ and the incompletely-condensed (c-$C_6H_{11})_7Si_8O_{11}(OH)_2$ (**III**, Scheme 1) are formed.[2b] For R = c-C_7H_{13},

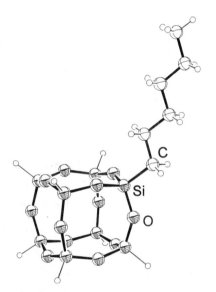

Figure 1. Molecular structure of $H_7Si_8O_{12}(CH_2)_5CH_3$ formed by hydrosilylation of 1-hexene by $H_8Si_8O_{12}$.

significant amounts of (c-$C_7H_{13})_6Si_6O_7(OH)_4$ (**IV**, Scheme 1) are obtained as well.[6b] A drawback of the synthesis of incompletely- condensed silsesquioxanes is that often gestation periods of several months are required. Most attempts to speed up these reactions have failed. For the hydrolysis of $RSiCl_3$ to

$$\xleftarrow[\text{X = OR}]{H_2O} \quad R-Si{\overset{\displaystyle X}{\underset{\displaystyle X}{\diagdown}}}{}^{\!\!\!\!X} \quad \xrightarrow[\text{X = Cl}]{H_2O} \tag{1}$$

$R_7Si_7O_9(OH)_3$ where R= c-C_5H_9 or c-C_7H_{13}, the gestation time can be shortened considerably by heating the reaction mixture to reflux. However, the yield of products is lower since variable amounts of resins are also formed, which require additional purification steps.

Feher and co-workers have carried out impressive work on both improving existing, as well as developing new, silsesquioxane syntheses. Recently, they reported an interesting route to incompletely condensed silsesquioxanes by either acid-[9] or base-mediated[10] splitting of one of the Si-O-Si edges of the fully-condensed silsesquioxanes $R_6Si_6O_9$ and $R_8Si_8O_{12}$ (eq. 2). In principle, incompletely-condensed silsesquioxanes with a wide variety of silicon substituents are available by this route.

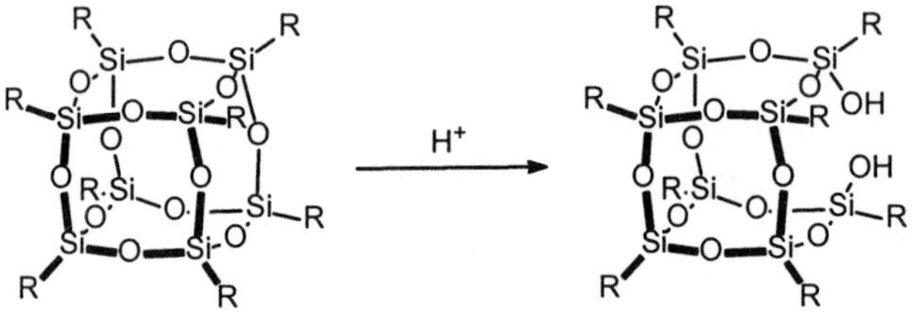

$$\text{(2)}$$

As mentioned above, the leading representative of the incompletely-condensed silsesquioxanes is the trisilanol $R_7Si_7O_9(OH)_3$ (**I**, Scheme 1). It is a suitable model for slightly dehydroxylated silica and can be used to model, for example, silica-grafted species and the reaction of silica with $RSi(OMe)_3$ tethering groups. This tris(silanol) can easily be modified by silylation or stannylation reactions to afford (vicinal) disilanols (**II**, Scheme 1) or isolated monosilanols (**III**, **VI**, Scheme 1), similar to surface sites that are formed upon dehydroxylation of silica (Scheme 2).[2,11]

In the search for simple yet realistic model compounds for isolated silanol functionalities as found in partially dehydroxylated silicas, a promising new approach has recently been published.[12] Instead of double silylation of the trisilanol $R_7Si_7O_9(OH)_3$, the latter compound was allowed to react with $SiCl_4$ in

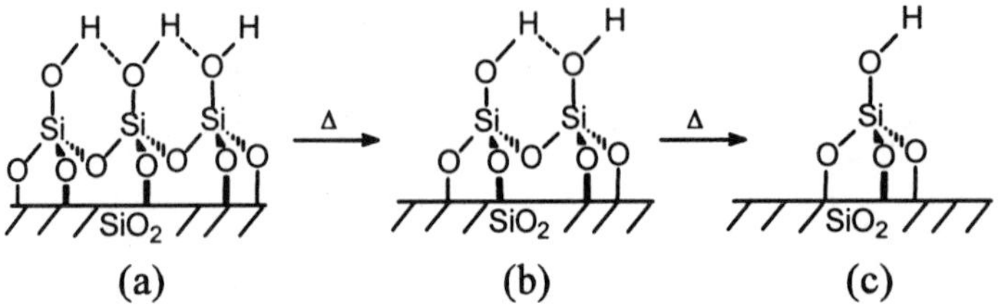

Scheme 2. Silanol sites present at silica surfaces calcined at (**a**) 200°C; (**b**) 500°C; and (**c**) above 700°C .

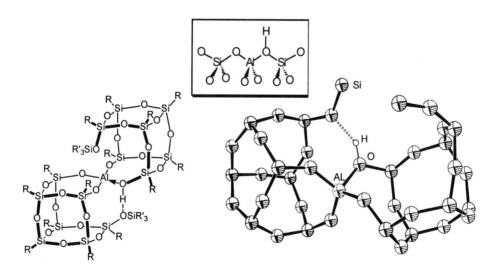

the presence of triethylamine to afford $R_7Si_8O_{12}Cl$, which was subsequently hydrolyzed to give the robust, C_3-symmetric silanol, $R_7Si_8O_{12}(OH)$ (eq. 3, **VI** Scheme 1).

Besides all-silicas, aluminum-doped silicates, aluminosilicates and zeolites also play an important role as catalyst supports. In recent years, several soluble aluminum complexes based on incompletely-condensed silsesquioxanes have been prepared and used as homogeneous models for such aluminosilicates and zeolites.[13]   The aluminosilsesquioxanes $\{[(c\text{-}C_5H_9)_7Si_7O_{11}(OSiR_3)]_2Al\}^-H^+$ ($SiR_3 = SiMe_3$, $SiMePh_2$) are fairly realistic homogeneous models for a Brønsted-acidic aluminosilicate site (Figure 2).[13e,f]   Examples of Lewis acidic aluminosilsesquioxanes are not known as such. The mobility of these homogeneous compounds allows them to form less electronically-deficient dimeric or oligomeric complexes. Typical examples of such latent Lewis acidic complexes are $[(c\text{-}C_6H_{11})_7Si_7O_{12}Al]_2$[13a] and $[c\text{-}C_5H_9)_8Si_8O_{13}]_3Al_2$.[13f]

Figure 2. Schematic and molecular structure of the Brønsted-acidic aluminosilsesquioxane $\{[(c\text{-}C_5H_9)_7Si_7O_{11}(OSiR_3)]_2Al\}^-H^+$.[24]

1.2   Characterization

The homogeneous nature and well-defined structure of silsesquioxanes gives us the opportunity to study the reactivity of the various silica silanol sites independently. Standard characterization techniques such as single crystal X-ray diffraction studies and multinuclear solution NMR spectroscopy provide crucial information that cannot be obtained from the corresponding heterogeneous systems. ^{13}C and ^{29}Si NMR spectroscopy proved to be very informative with respect to the symmetry and purity of silsesquioxanes. The methyne-CH and framework silicon atoms show very characteristic and distinct resonances.[2b,7]

Incompletely-condensed silsesquioxanes not only sterically resemble various silica surface silanol sites (Figure 1); their electronic properties also proved to be compatible with those of silicas since they were found to be considerably more electron-withdrawing than simple silanols and alcohols.[14,15b] Using Hammett correlations, the Si_8O_{12} cage in aryl-substituted spherosilicates Si_8O_{12}-p-C_6H_4X was found to be as electron-withdrawing as a CF_3 group. Based on acidity measurements, with a $pK_{a(THF)}$ of 10.8, Ph_3SiOH is clearly less Brønsted-acidic than the silsesquioxanes (c-$C_5H_9)_7Si_8O_{12}$(OH) ($pK_{a(THF)} = 8.9$) and (c-$C_5H_9)_7Si_7O_9$(OH)$_3$ ($pK_{a(THF)} = 7.6$).[12b] This is in line with the calculated deprotonation energies at 0 K in vacuum for $H_7Si_7O_9$(OH)$_3$ ($\Delta E = 1319$ kJ/mol), $H_7Si_8O_{12}$(OH) ($\Delta E = 1425$ kJ/mol) and Ph_3SiOH ($\Delta E = 1462$ kJ/mol).[12b] Hence, silsesquioxanes are definitely more electron-withdrawing than simple silanols. It is interesting to note the difference in acidity between the monosilanol (c-$C_5H_9)_7Si_8O_{12}$(OH) (**VI**, Scheme 1) and the trisilanol (c-$C_5H_9)_7Si_7O_9$(OH)$_3$ (**I**, Scheme 1; $\Delta pK_{a(I-VI)} = 1.3$). This value is in line with the calculated ($\Delta v = 300$ cm^{-1}) and measured ($\Delta v = 250$ cm^{-1}) difference in the OH-stretching vibrations and agrees with the observed difference in acidity of isolated silanols and silanol nests in zeolitic materials.[12b,16]

2.   METALLASILSESQUIOXANES

During the last decade, a variety of transition metal-containing silsesquioxanes, mainly based on $R_7Si_7O_9$(OH)$_3$ and $R_7Si_7O_9$(OH)$_2$OSiMe$_3$ (**I** and **II**, Scheme 1), have been reported.[7,15] Feher and coworkers clearly are the pioneers in this field. Not only did they report many metallasilsesquioxanes, they also demonstrated that some of these complexes could be used as catalysts. These discoveries have inspired more research groups to study the applicability of metallasilsesquioxanes as models for heterogeneous catalyst systems in detail. In a number of review articles, synthesis and structural characterization of a wide range of silsesquioxanes and metallasilsesquioxanes have been documented.[17]

It is now evident that these complexes are capable of mimicking essential features of heterogeneous silica-supported transition metal catalysts. Important features of silica-supported catalysts that can be observed in the corresponding metal-containing silsesquioxane complexes include a rigid, electron-withdrawing siloxide structure and, for example, the possibility of adjacent oxygen donors to interacting with the metal centre.[18] Silsesquioxanes provide an excellent opportunity to fine-tune synthetic procedures to modify silica surfaces or to anchor metal complexes to the support. However, like all homogeneous model systems for solid supports, (metalla)silsesquioxane compounds are subject to one major problem. Since precursors, reaction intermediates and products are generally all soluble and therefore very mobile, unstable intermediates often react further, affording thermodynamically stable product(s); strongly Lewis acidic metallasilsesquioxane species tend to form dimeric or oligomeric structures to reduce their electron deficiency.

Various heterogeneous catalytic processes such as Diels-Alder,[19] alkene metathesis[20] and alkene epoxidation[21] have been mimicked using metallasilsesquioxane complexes. The focus of this chapter will be on catalytic olefin polymerization, the most broadly studied catalytic process. On the basis of several examples of modelling both classic heterogeneous and immobilized homogeneous olefin polymerization catalysts, both the advantages and drawbacks of silsesquioxanes as silica surface models will be discussed.

2.1 Olefin Polymerization

For the industrial production of polyolefins, heterogeneous catalysts are generally preferred over homogeneous ones. For example, the polymer formed in gas phase, slurry or bulk-monomer processes is usually of high density and insoluble in most solvents. For these processes, only heterogeneous catalysts allow polymer morphology control, narrow particle size distribution, high bulk density and avoid reactor fouling.[22] Besides traditional heterogeneous olefin polymerization catalysts (Ziegler-Natta, Phillips), immobilization of well-defined homogeneous metallocene catalysts is receiving increasing interest.[23] Studying immobilization strategies, stability of the supported catalyst and electronic/steric effects of the support on the catalyst's performance using homogeneous models can significantly contribute to the molecular level understanding of supported catalysts.

2.1.1 Classical Heterogeneous Catalysts

Classical heterogeneous olefin polymerization catalysts such as the Ziegler-Natta and Phillips systems were among the first to be modelled using silsesquioxanes. In 1990, Feher reported the synthesis of $[(c\text{-}C_6H_{11})_7\text{-}$

Si$_7$O$_{11}$(OSiMe$_3$)]CrO$_2$ (Figure 3),[24] the first silsesquioxane-based olefin polymerization catalyst precursor and probably the best homogeneous model for the Phillips catalyst known to date.[15a] Unlike the Phillips catalyst, this complex was not active with ethylene alone and trimethylaluminium was needed to obtain an active catalyst. At room temperature and 1 atm of ethylene, polymerization typically proceeded for several hours before gelation of the solution prevented further uptake of ethylene. The activity of the catalyst was sensitive to the amount of AlMe$_3$ used to generate the active species and was highest when two equivalents of AlMe$_3$ per silsesquioxane chromate were employed. The molecular weights of the polymer were in the range M$_w$ = 61,000-377,000 with M$_w$/M$_n$ = 3.5-6.1, suggesting that more than one active catalyst is present in this system. Not much is known about the nature of the catalytically active species, nor its presumed reduction during activation or possible side-reactions such as leaching induced by the alkylaluminum activator.

Another olefin polymerization pre-catalyst that is related to the silsesquioxane chromate is the vanadyl silsesquioxane complex {[(c-C$_6$H$_{11}$)$_7$Si$_7$O$_{12}$]V=O}$_n$ (n = 1, 2; Figure 4), which exists in both monomeric and dimeric forms.[7d,25] The complex was used in ethylene polymerization after activation with trialkylaluminum species. The monomeric complex was assigned as the catalyst precursor, the dimer being a precursor to the monomer. At 25°C, 1 atm of ethylene and 1-5 equivalents of AlMe$_3$, ethylene polymerization proceeded for 1000-1500 turnovers until gelation of the reaction mixture prevented further uptake of ethylene. Polyethylene with M$_w$ of 48,000 and a narrow M$_w$/M$_n$ of 2.3 was obtained. Hence, the catalyst system can be described as truly single-sited. Similar to the silsesquioxane chromate, the activity of the vanadium catalyst is sensitive to the amount of AlMe$_3$, and the highest activity was observed when three equivalents of AlMe$_3$ were used as cocatalyst.

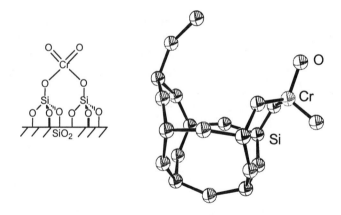

Figure 3. Molecular structure of [(c-C$_6$H$_{11}$)$_7$Si$_7$O$_{11}$(OSiMe$_3$)]CrO$_2$.[24]

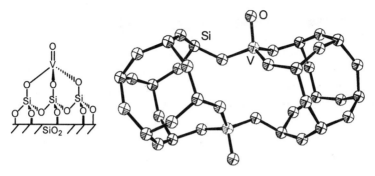

Figure 4. Molecular structure of $\{[(c\text{-}C_6H_{11})_7Si_7O_{12}]V=O\}_2$.[24]

Mechanistic studies of the interaction of $Al(CH_2SiMe_3)_3$ with the vanadyl silsesquioxane indicated a stepwise alkylation of the vanadium by the aluminum alkyl, and the monomeric $[(c\text{-}C_6H_{11})_7Si_7O_9(O_2AlCH_2SiMe_3)OV(=O)(CH_2SiMe_3)_2$ was assigned as the possible catalytically active species (eq. 4).[25b] A survey of the catalyst's reactivity indicated that other olefins could also be polymerized or copolymerized. The activity for propene was rather low (25°C, 3 bar, 3 hrs) and gave low molecular weight atactic polypropene ($M_w < 10^4$). Copolymerization of propene (neat) and ethylene (1 %) gave small amounts of polymer containing 5-10 % propene units. On the other hand, 1,3-butadiene was readily polymerized to mainly *trans*-polybutadiene.

$$ (4) $$

In contrast to the activity exhibited by the silsesquioxane chromate and silsesquioxane vanadyl complexes in the presence of $AlMe_3$, simple siloxy analogues such as $(Ph_3SiO)_2CrO_2$, $\{[O(OSiPh_2)_2]_2Cr(=O)_2\}_2$ and $(Ph_3SiO)_3V=O$ show little or no activity towards olefins under identical conditions.[26] Although the interaction of $(Ph_3SiO)_n(Me_3SiCH_2)_{3-n}V=O$ with $Al(CH_2SiMe_3)_3$ was found to be very similar to that of $[(c\text{-}C_6H_{11})_7Si_7O_{12}]V=O$, none of the AlR_3-activated triphenylsiloxy complexes gave particularly active catalysts, and rapid deactivation was observed as well. This clearly indicates the difference between silsesquioxanes and simple (di)silanols.

Recently, a silsesquioxane-based catalyst system was reported that can be regarded as a homogeneous model for a Ziegler-Natta type catalyst.[27] The catalyst support was synthesized by reacting the trisilanol $(c\text{-}C_6H_{11})_7Si_7O_9(OH)_3$ with an equimolar amount of Mg(Bu)Et, yielding a mixture of compounds with the general formula $[(c\text{-}C_6H_{11})_7Si_7O_{11}(OH)Mg]_n$ (n = 1, 2). Treating this mixture with TiCl$_4$ afforded the pre-catalyst $[(c\text{-}C_6H_{11})_7Si_7O_{12}MgTiCl_3]_n$ (n = 1, 2; Figure 5), which can be activated with AlEt$_3$ to give an active ethylene polymerization catalyst. This system was reported to be more active in ethylene polymerization than its heterogeneous congener. The polydispersity of the polyethylene (GPC: $M_w = 140,000$, $M_w/M_n = 5.5$) was quite high and indicates the presence of multiple active sites. No further information on the nature of the catalytically active species was given.

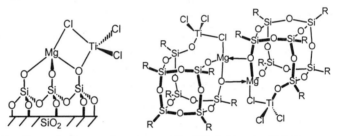

Figure 5. Proposed structure of $[(c\text{-}C_6H_{11})_7Si_7O_{12}MgTiCl_3]_2$.

2.1.2 Surface Organometallic Complexes

Closely related to the classical heterogeneous catalysts are the so-called surface organometallic complexes. The groups of Basset and Scott have developed sophisticated set-ups to prepare and characterize silica grafted organometallic complexes such as silica-grafted zirconium and chromium species.[28] Although recent studies show that oxide-grafted organometallic complexes can be quite uniform in structure, reactivity and distribution, their exact structures often remain elusive.

A good example is the ongoing debate about the structure of the Union Carbide catalyst, formed by grafting Cp$_2$Cr onto silica.[29] Despite advances in spectroscopic, physical and chemical techniques, the nature of the active species remains questionable. Modelling of such species could be very beneficial for more detailed understanding of possible surface structures and reactivity of the active sites. Homogeneous chromium complexes of the type Cp*CrR(THF)$^+$ have been extensively studied by Theopold *et al.* as models for the Union Carbide catalyst.[30] Recent attempts to model the Union Carbide catalyst by reacting silsesquioxanes with Cp$_2$Cr have resulted in a novel trimetallic species, $\{[(c\text{-}C_5H_9)_7Si_7O_{11}(OSiMePh_2)]_2Cr(CrCp)_2\}$ (Figure 6).[31]

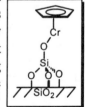

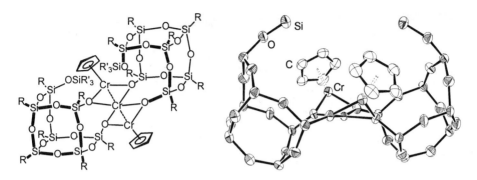

Figure 6. Schematic and molecular structure of $\{[(c\text{-}C_5H_9)_7Si_7O_{11}(OSiMePh_2)]_2Cr(CrCp)_2\}$.[24]

The complex contains one planar Cr(II) atom surrounded by four silsesquioxane siloxy groups and two half-sandwich CpCr(II) fragments, each coordinated to two siloxy groups. One chromium has lost both cyclopentadienyl groups, two have lost one Cp while one intact chromocene co-crystallized with the product. Hence, protonolysis of the Cp groups is clearly facile and not selective. Oxidative addition of a silanol group, proposed as being one of the possible reactions of chromocene with surface silanol sites, was not observed and hence this reaction is probably also unlikely to happen when silica is treated with Cp₂Cr. The trimetallic complex did not show any catalytic activity in ethylene polymerization (50°C, 6 bar, 1 hr). Addition of AlMe₃ in the presence of ethylene (6 bar) gave only traces of polymer, probably formed by decomposition products of the reaction of the trimetallic chromium silsesquioxane with AlMe₃.

A second class of surface organometallic compound and olefin polymerization catalyst that has been modelled using silsesquioxanes includes the oxide-supported zirconium alkyl and hydride species such as those prepared by Basset *et al.*[28a-d] and the ZrR₄/Al₂O₃ system developed by Ittel *et al.* at DuPont.[32] Treatment of the cyclopentyl-substituted trisilanol $(c\text{-}C_5H_9)_7Si_7O_9(OH)_3$ with an equimolar amount of $Zr(CH_2Ph)_4$ yielded a dimeric zirconium benzyl complex, $\{[(c\text{-}C_5H_9)_7Si_7O_{12}]ZrCH_2Ph\}_2$ (Figure 7).[33] Like the oxide-grafted zirconium species that serve as an example for this zirconium silsesquioxane, this dimeric zirconium complex is an active olefin polymerization catalyst without need for additional cocatalyst. The activity of this neutral complex can be explained by the peculiar dimeric structure of the zirconium silsesquioxane complex, which suggests that one of the zirconium atoms acts as an adjacent Lewis acid and activates the other zirconium. This agrees well with the fact that zirconium hydride supported on silica is a more active polymerization catalyst

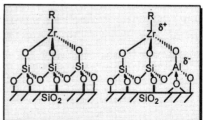

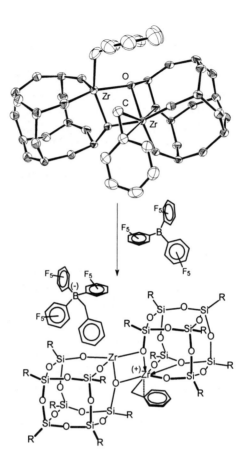

Figure 7. Molecular structure of {[(c-C₅H₉)₇Si₇O₁₂]ZrCH₂Ph}₂, and the cation formed by reaction with B(C₆F₅)₃.[24]

when the surface is doped with Lewis acidic aluminum sites.[28d]

The synergistic effect of the two zirconium centres is intensified when one of the benzyl groups of the dimer is abstracted by a strongly Lewis acidic borane, B(C₆F₅)₃ (Figure 7). For the thus-formed cationic mono(benzyl) complex, an increase in ethylene polymerization activity of over two orders of magnitude was observed (2,400 g PE/mmol·hr) compared to the neutral dimer (10 g PE/mmol·hr). With such activity, this complex is not only a suitable model for silica-supported zirconium species, it is also an interesting catalyst itself. Although the molecular weight ($M_w = 6,600$) is low, the M_w/M_n of 2.3 indicates that the catalyst is single-sited.

2.1.3   Supported Homogeneous Single-site Catalysts

Although metallocenes are among the most versatile olefin polymerization catalysts known to date, their commercial use is limited by their homogeneous nature.[23] Heterogenization of such well-defined homogeneous olefin polymerization catalysts offers opportunities to combine the advantages of both heterogeneous and homogeneous catalysts.[23] In the ideal case, morphology control by replication leads to uniform polymer particles with narrow particle size distribution and high bulk density, whereas the single-site nature of the catalyst provides polymers with desired polymer structures and narrow molecular weight distributions.[34a] Additionally, immobilization of the catalyst can reduce the formation of dormant sites that, in principle, makes it possible to decrease the necessary amount of cocatalyst.[34b]

The most common method to immobilize homogeneous olefin polymerization catalysts consists of treating the silica support with the catalyst precursor and the cocatalyst, often methylalumoxane (MAO).[23] The order of addition of catalyst precursor and MAO is crucial and three main routes have been used. The oldest method consists of grafting the metal precursor that is subsequently activated with an MAO solution. The most common methods in use today are physisorption of metallocenes on a support that is pre-treated with MAO, or treating a silica support with a solution of MAO and pre-activated catalyst. Recently, immobilized boranes and borates have also been reported as suitable cocatalysts.[35,36]

Up to now, catalyst leaching remains a major problem, since it results in the formation of polymer fines and sheets and reactor fouling. Improvement of immobilization techniques remains therefore a topic of great importance. Here we will discuss the three most-used methods to immobilize homogeneous olefin polymerization catalysts, namely grafting and tethering of the catalyst precursor, and immobilization of the catalyst by electrostatic interaction with the supported cocatalyst.

2.1.3.1 Grafting of catalyst precursors

The earliest immobilization methods consisted of grafting metallocenes and half-sandwich complexes onto partly dehydroxylated silica.[23] To activate these systems, they were treated with MAO solutions. Activities were low and leaching remained a serious problem. In view of this immobilization technique, the interaction of silica surface silanols with metallocenes and half-sandwich compounds is of great importance 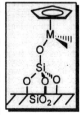 and silica-grafted species of the type [SiO$_2$]-O-M(Cp)X$_2$ (X = chloride, alkyl), containing two chloride or alkyl substituents, are of interest. Preparation of such heterogeneous catalysts can be improved considerably using synthetic strategies developed for homogeneous model complexes.

Maschmeyer and co-workers investigated the interaction of the trisilanol silsesquioxane (c-C_6H_{11})$_7$Si$_7$O$_9$(OH)$_3$ with the group 4 metal complexes Cp$_2$MCl$_2$ and CpMCl$_3$ (M = Ti, Zr, Hf).[37] It appeared that not only the chloride but also the cyclopentadienyl substituents were protonolysed by the silsesquioxane silanols. For titanium, two chlorides and one of the cyclopentadienyls is protonolysed, yielding both monomeric and dimeric complexes with general formula {[(c-C_6H_{11})$_7$Si$_7$O$_{12}$]TiCp}$_n$ (n = 1, 2). For the heavier congeners Zr and Hf, both cyclopentadienyl groups are substituted and complicated structures containing [((c-C_6H_{11})$_7$Si$_7$O$_{12}$)$_4$M$_3$Cl]$^-$ fragments have been proposed.

Recently, Edelmann reported the reaction of (η^5-C_5Me_5)(η^5,η^1-$C_5Me_4CH_2$)Ti$^{(III)}$ with (c-C_6H_{11})$_7$Si$_7$O$_9$(OH)$_2$OSiMe$_3$ that resulted in loss of one pentamethylcyclopentadienyl ligand and oxidation to the tetravalent complex, [(c- C_6H_{11})$_7$Si$_7$O$_{11}$OSiMe$_3$][(c-C_6H_{11})$_7$Si$_7$O$_{10}$(OH)-OSiMe$_3$]TiCp* (Figure 8).[38] These reactions clearly demonstrate the reactivity of surface silanols and show that even the robust cyclopentadienyl ligands of Cp$_2$MCl$_2$ can be displaced by silanols. The latter reaction also explains the low activity of the thus-immobilized catalysts, since grafting might result in protonolysis-mediated decomposition of the catalyst.

Another important criterion for silica-grafted olefin polymerization catalysts is the stability of the metal-siloxy bonds. Early transition metal-oxygen bonds are strong and therefore considered to be fairly stable. However, most of these catalyst precursors are activated with strongly Lewis acidic aluminum-based cocatalysts, for which the Al-O bonds are competitively strong compared to the early transition metal-oxide bonds. Feher *et al.* demonstrated that Al(CH$_2$SiMe$_3$)$_3$ readily splits two of the three vanadium siloxy bonds in the vanadyl

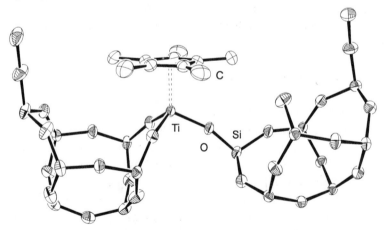

Figure 8. Molecular structure of [(c-C_6H_{11})$_7$Si$_7$O$_{11}$OSiMe$_3$][(c-C_6H_{11})$_7$Si$_7$O$_{10}$(OH)OSiMe$_3$]TiCp*.[24]

silsesquioxane, $(c\text{-}C_6H_{11})_7Si_7O_{12}V=O$, forming $[(c\text{-}C_6H_{11})_7Si_7O_9\text{-}(O_2AlCH_2SiMe_3)OV(=O)(CH_2SiMe_3)_2]$ (eq. 4).[25b] To study the stability of group 4 metal-siloxy bonds, metallasilsesquioxane species of the type $[(c\text{-}C_5H_9)_7Si_7O_{12}]MCp$, $[(c\text{-}C_5H_9)_7Si_7O_{11}(OSiMe_3)]MCp_2$ and $[(c\text{-}C_5H_9)_7Si_8O_{13}]_nM(Cp)X_{3\text{-}n}$ (X = Cl, alkyl; n = 1, 2) were subjected to solutions of MAO in the presence of ethylene (Figure 9).[12] Although virtually all silsesquioxane complexes showed catalytic activity, in all cases the active catalyst resulted from MAO-induced cleavage of the metal siloxy bond. Even the terdentate ligand in $[(c\text{-}C_5H_9)_7Si_7O_{12}]MCp'$ (M = Ti, Zr, Hf; Cp' = C_5H_5, $C_5H_3(SiMe_3)_2$) was readily displaced by MAO to yield an active catalyst. Since silsesquioxanes were used that represent a variety of silanol sites present on silicasurfaces, particularly with regard to the denticity of the siloxy chelates, it might be concluded that many silica-grafted early transition metal polymerization catalysts will leach upon activation with MAO or other aluminum alkyls.

It is noteworthy that the mono(cyclopentadienyl)titanasilsesquioxane dialkyl species $Cp''[(c\text{-}C_5H_9)_7Si_8O_{13}]TiR_2$ (Cp'' = $\eta^5\text{-}1,3\text{-}C_5H_3(SiMe_3)_2$, R = Me, CH$_2$Ph), formed by protonolysis of $Cp''TiR_3$ with $(c\text{-}C_5H_9)_7Si_8O_{12}(OH)$, can be activated with 'non-aluminum' cocatalysts such as $B(C_6F_5)_3$ or $X^+[B(C_6F_5)_4]^-$ (X^+ = Ph_3C^+, $PhN(H)Me_2^+$). The thus-obtained cationic titanasilsesquioxane species, $\{Cp''[(c\text{-}C_5H_9)_7Si_8O_{13}]TiR\}^+\{RB(C_6F_5)_3\}^-$ (eq. 5) proved to be single-site catalysts for the polymerization of ethylene ($M_w = 2.6\times10^5$, $M_w/M_n = 3.3$) and 1-hexene ($M_w = 2.9\times10^3$, $M_w/M_n = 2.0$).[12] Hence, this study suggests that non-leaching silica-grafted half-sandwich catalysts can be formed when aluminum-based cocatalysts are avoided.

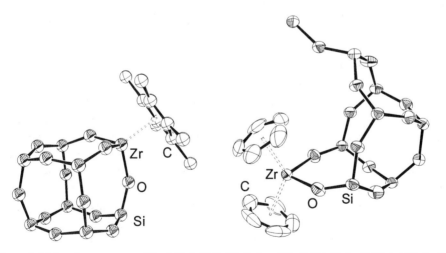

Figure 9. Molecular structures of $[(c\text{-}C_6H_{11})_7Si_7O_{12}]ZrCp*$ and $[(c\text{-}C_5H_9)_7Si_7O_{11}(OSiMe_3)]ZrCp_2$.[24]

$$\text{(5)}$$

2.1.3.2 Tethering of catalyst precursors

Tethering of organometallic compounds to a silica support is, with respect to leaching, probably the best method to anchor a homogeneous catalyst. Unfortunately, it is also the most laborious route. Tethering is a versatile method of anchoring virtually any homogeneous catalyst onto a support and is gaining more and more attention in the recent literature.[39] While several studies on tethering of olefin polymerization catalysts to silicas have been reported, curiously enough hardly any information is available on the effectiveness of the applied immobilization 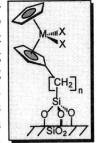 method or the stability, uniform character and the effect of the support on these immobilized catalysts.[40] Two general routes have been applied: building up the ligand and/or metal complex at the surface or immobilizing a pre-synthesized catalyst precursor containing an anchorable functionality. Both routes clearly have their limitations and advantages.[40]

Preliminary results of a study using silsesquioxanes to mimic catalyst tethering have shown that these model supports are also very suitable for optimizing synthetic strategies.[41] The approach used consisted of tethering the (substituted) cyclopentadienyl ligand to the silsesquioxane and subsequent introduction of the metal precursor. To attach the cyclopentadienyl, three general routes can be used that have been applied to tether various organic functionalities to surface supports. They consist of (1) grafting of a silylchloride or silylether-functionalized cyclopentadiene, (2) salt elimination reaction of a cyclopentadienyl-alkali metal salt with a chloropropyl group previously tethered onto the surface by method (1) or sol-gel synthesis, and finally (3) hydrosilylation of an olefin-substituted cyclopentadienyl. Once the cyclopentadienyl ligand is attached to the silsesquioxane, it can be deprotonated and treated with the metal

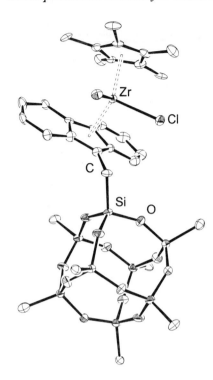

precursor. For silsesquioxanes, these are convenient routes since the silsesquioxane-bonded (substituted) cyclopentadienyl ligand can be purified, and more importantly, there are no adjacent silanol functionalities that can interfere with the intended reaction.[41] Figure 10 shows an example of such a complex, a silsesquioxane-tethered fluorenyl(pentamethylcyclopentadienyl) zirconium dichloride species, $[(c\text{-}C_5H_9)_7Si_7O_{13}CH_2C_9H_8](Cp^*)ZrCl_2$.[41] Keeping in mind that both the Cp and Cl substituents of Cp_2MCl_2 are readily displaced by silanols, this approach will most probably result in multiple surface metal sites and hence for silica supports this approach is probably less attractive.

Figure 10. Molecular structure of $[(c\text{-}C_5H_9)_7SiO_{13}CH_2C_9H_8](Cp^*)ZrCl_2$.[24]

2.1.3.3 Electrostatic interaction with immobilized cocatalyst

Supported alumoxanes

As mentioned at the beginning of this section, the most common route to immobilize metallocenes consists of physisorbing the catalyst precursor onto a MAO-pre-treated silica support or treating silica with a metallocene/MAO mixture.[23] In both cases, the MAO is expected to chemically link to the silica support and the activated metallocene cation will in turn be bonded to the supported MAO by means of electrostatic interactions. The interaction of the MAO with the support is of great importance since this will strongly determine the possibility of leaching. In patent applications, Exxon reported an attractive alternative route to prepare *in situ* supported MAO, consisting of treating hydroxylated silica with trimethylaluminum.[42]

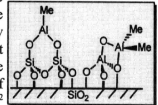

Although many studies have been reported on immobilized MAO, working with amorphous silica and poorly defined MAO is tricky and unpredictable. Since MAO itself is already ill-defined, immobilizing MAO onto model supports will not provide much information either. However, first attempts to mimic the interaction of (wet) silica surfaces with trimethylaluminum (TMA) using silsesquioxane model supports have been reported.[43] A wide variety of methylaluminosilsesquioxanes were obtained and structurally characterized. Silsesquioxane silanols were shown to react in various ratios with TMA. For example, the dimeric silsesquioxane disiloxyaluminum methyl complex $\{[(c\text{-}C_5H_9)_7Si_7O_{11}(OSiMePh_2)]AlMe\}_2$ exists in three different isomeric forms (eq. 6, Figure 11). These dimeric structures are very stable, and prolonged heating (500 hrs, 76°C) is required for these conformers to isomerize to the thermodynamically most stable mixture.

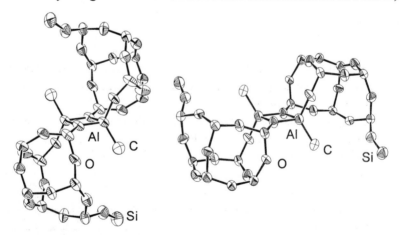

(6)

For the preparation of supported MAO, silicas are used that contain large amounts of silanol groups and even physisorbed water. Reacting such silicas with aluminum alkyls might result in Brønsted acidic aluminosilicate structures, which

Figure 11. Molecular structures of two of the three conformers of $\{[(c\text{-}C_5H_9)_7Si_7O_{11^-}$ $(OSiMePh_3)]AlMe\}_2$.[24]

in turn can react further with additional aluminum alkyls. As was discussed at the beginning of this chapter, such acidic sites were successfully mimicked by Brønsted acids such as $\{[(c\text{-}C_5H_9)_7Si_7O_{11}(OSiMePh_2)]_2Al^-\}[H^+]$.[13e,f] As expected, TMA is readily protonolysed by the acidic proton of this aluminosilsesquioxane, affording the novel C_2 symmetric $[(c\text{-}C_5H_9)_7Si_7O_{11}(OSiMePh_2)]_2Al_2Me_2$ (eq. 7).[24] The structure of this complex is even more robust than its structural isomers $\{[(c\text{-}C_5H_9)_7Si_7O_{11}(OSiMePh_2)]AlMe\}_2$ (Figure 11). Only after prolonged heating (1600 hrs, 76°C) does $[(c\text{-}C_5H_9)_7Si_7O_{11}(OSiMePh_2)]_2Al_2Me_2$ isomerize to the thermodynamically most stable mixture of $\{[(c\text{-}C_5H_9)_7Si_7O_{11}(OSiMePh_2)]AlMe\}_2$.[43]

(7)

Although these methylaluminosilsesquioxane complexes are realistic models for silica surface aluminum sites, they are not necessarily models for supported MAO. Nevertheless, like Barron's *t*-butylalumoxanes, they could possibly be used as olefin polymerization cocatalysts.[44] It was argued that deprotonating the Brønsted acids $\{[(c\text{-}C_5H_9)_7Si_7O_{11}(OSiR_3)]_2Al^-\}[H^+]$ (SiR$_3$ = SiMe$_3$, SiMePh$_2$) might afford weakly coordinating anions supporting cationic metallocene species, in a similar fashion, for example, to that reported by Marks *et al.* for strongly acidic sulfonated zirconia.[45] However, reaction of Cp$_2$ZrMe$_2$ with the Brønsted acidic aluminosilsesquioxane did not afford a cationic zirconocene species but instead resulted in a clean redistribution reaction yielding $[(c\text{-}C_5H_9)_7Si_7O_{11}(OSiMePh_2)]ZrCp_2$, methane and various methylaluminosilsesquioxane species (eq. 8).

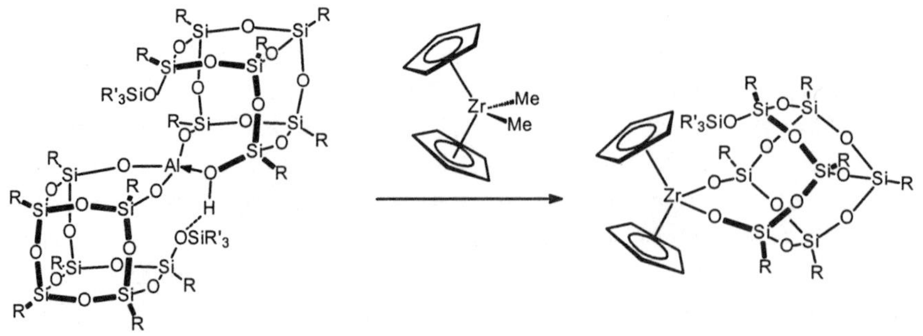

(8)

When Cp$_2$ZrMe$_2$ is treated with the Lewis acidic methyl aluminosilsesquioxanes (eq. 6-7), no active olefin polymerization catalyst is formed either. However, in the presence of Cp$_2$ZrMe$_2$, both $\{[(c\text{-}C_5H_9)_7\text{-}Si_7O_{11}(OSiMePh_2)]AlMe\}_2$ (76°C, 1.5 hrs) and $[(c\text{-}C_5H_9)_7Si_7O_{11}\text{-}(OSiMePh_2)]_2Al_2Me_2$ (76°C, 8 hrs) equilibrate over two orders of magnitude faster than without Cp$_2$ZrMe$_2$. It was proposed that $\{[(c\text{-}C_5H_9)_7Si_7O_{11}\text{-}(OSiMePh_2)]AlMe\}_2$ and $[(c\text{-}C_5H_9)_7Si_7O_{11}(OSiMePh_2)]_2Al_2Me_2$ react as Lewis acids with Cp$_2$ZrMe$_2$, similar to homogeneous Lewis acids such as B(C$_6$F$_5$)$_3$ (eq. 9).[46,47] However, for the Lewis acidic aluminosilsesquioxanes, the interaction with Cp$_2$ZrMe$_2$ is clearly too weak to effectively abstract a methyl to form a cationic zirconocene complex.

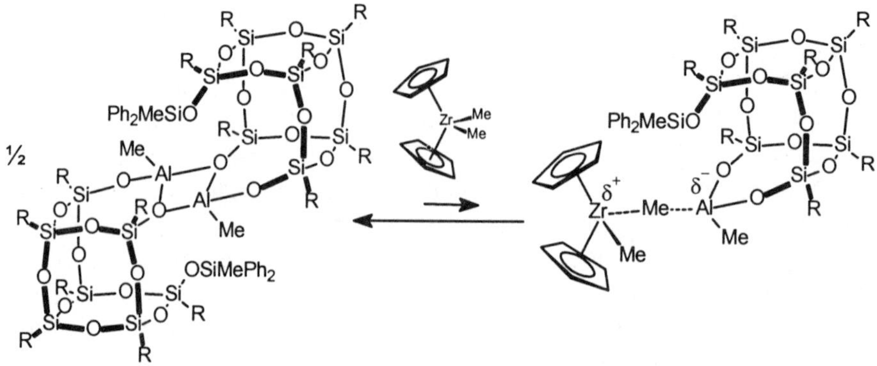

(9)

Supported borates

Recently, immobilized boranes and borates have also been reported to be suitable cocatalysts in olefin polymerization.[35] Not only tethered boron species have been reported, but also physisorbed and grafted ones.[36] Physisorbed or silica-tethered boron species have not yet been modelled. So far, only silica-grafted boranes and borates have been modelled using silsesquioxanes.[48] Representative examples are the neutral silsesquioxane tris(borane), $R_7Si_7O_{12}[B(C_6F_5)_2]_3$ and the silsesquioxane-borato ammonium salt, $\{[(c\text{-}C_5H_9)_7Si_7O_{13}]\,B(C_6F_5)_3\}^-[PhN(H)Me_2]^+$ (Figure 12).

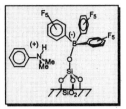

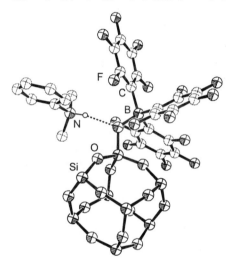

While stable in the absence of Cp_2ZrMe_2, at low temperature the silsesquioxane-borane complex $R_7Si_7O_{12}[B(C_6F_5)_2]_3$ reacts with Cp_2ZrMe_2 to form the inactive $R_7Si_7O_{12}ZrCp_2$ $[B(C_6F_5)_2]$ and $(C_6F_5)_2BMe$ (eq. 10), indicating that the B-O bond is readily split. When reacting zirconocene dibenzyl $Cp_2Zr(CH_2Ph)_2$ with the silsesquioxane-borato ammonium salt $\{[(c\text{-}C_5H_9)_7Si_7O_{13}]B(C_6F_5)_3\}^-$ $[PhN(H)Me_2]^+$ in the presence of ethylene or 1-hexene, an active polymerization catalyst was obtained. However, mechanistic studies showed that the expected $[Cp_2ZrCH_2Ph]^+\{[(c\text{-}C_5H_9)_7\text{-}Si_7O_{13}]B(C_6F_5)_3\}^-$ was not formed, but that as a result of B-O bond splitting

Figure 12. Molecular structure of $\{[(c\text{-}C_5H_9)_7Si_7O_{12}](OB(C_6F_5)_3\}^-$ $\{PhN(H)Me_2\}^{+.24}$

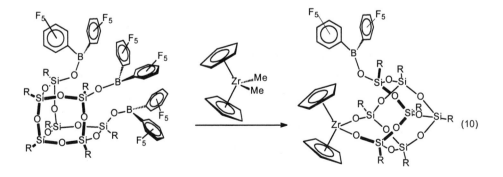

(10)

$$(11)$$

the cationic zirconocene silsesquioxane $\{[(c\text{-}C_5H_9)_7Si_7O_{13}]ZrCp_2\}^+$ $[PhCH_2B(C_6F_5)_3]^-$ was formed instead (eq. 11). The latter is in equilibrium with the neutral $[(c\text{-}C_5H_9)_7Si_7O_{13}]Zr(CH_2Ph)Cp_2$ and free borane $B(C_6F_5)_3$. It was demonstrated that, exclusively in the presence of an excess of $Cp_2Zr(CH_2Ph)_2$, an active catalyst is formed. Hence it was proven that the free borane, formed after splitting of the boron-siloxy bond, appears to be the active cocatalyst that reacts with another equivalent of $Cp_2Zr(CH_2Ph)_2$ to form $[Cp_2ZrCH_2Ph]^+$ $[PhCH_2B(C_6F_5)_3]^-$, the actual active catalyst of this system.[48a]

Hence, facile B-O bond splitting makes the silsesquioxane-borane and borato systems unsuitable as cocatalysts. They are, however, a source of unsupported borane ($MeB(C_6F_5)_2$ or $B(C_6F_5)_3$) that can act as a cocatalyst, which explains the fact that ethylene polymerization activity is still observed. On the basis of these results, it is likely that the B-O bonds in silica-grafted boranes and borates are also labile. Consequently, under the influence of transition metal alkyls, silica-grafted boranes and borates are transformed into unsupported borane, which can be considered the actual cocatalyst.

3.   HETEROGENIZATION OF METALLASILSESQUI-OXANES

As described above, several metallasilsesquioxanes have proven to be useful models for the active sites of the corresponding heterogeneous catalysts. In some cases, these metallasilsesquioxane catalysts are comparably or even more active than their commercially used heterogeneous congeners, which makes them interesting catalysts themselves. If immobilized, these complexes could yield supported catalysts that have the advantage over their heterogeneous analogues of being well-defined and single-sited. Several methods have been described to use metallasilsesquioxanes as building blocks for heterogeneous catalyst systems. These routes range from physisorption and chemical linking to calcination of the metallasilsesquioxanes to give well-defined ceramic materials.[49]

The only attempts at supporting silsesquioxane-based olefin polymerization catalysts to date consist of immobilization by electrostatic interactions. The dimeric metallasilsesquioxane $\{[(c\text{-}C_5H_9)_7Si_7O_{12}]ZrCH_2Ph\}_2$ supported onto MAO pre-treated MCM-41 (activity: 2.6 kg PE/mmol•hr)) and silica (activity: 8.3 kg PE/mmol•hr) gave a catalyst that showed an activity comparable to or even higher than that of the homogeneous complex activated with $B(C_6F_5)_3$. This is quite surprising, since heterogenization of homogeneous metallocene catalysts often goes together with loss of activity. As expected for a heterogeneous catalyst, the polyethylene prepared with the immobilized $\{[(c\text{-}C_5H_9)_7Si_7O_{12}]ZrCH_2Ph\}_2$ showed a much better morphology (spherical particles) and gave considerably less reactor fouling than the homogeneously prepared polymer. Furthermore, the molecular weight of the polymer increased by a factor of 10. However, the rather high polydispersity of the polyethylene indicates that it is probably not formed by a single-site catalyst. This is not surprising keeping in mind that there is a realistic chance that (partial) substitution of the silsesquioxane ligands takes place, since aluminum alkyls (including MAO) are known to be able to split metal siloxy $(M\text{-}OSi\equiv)$ bonds.

4. CONCLUSION

The results presented here clearly demonstrate the success of using silsesquioxanes to mimic silica surface silanol sites. With these model supports, features such as the stability and reactivity of metal-siloxy bonds in the presence of, for example, cocatalysts during catalytic reactions, afforded detailed insight that could never have been obtained with heterogeneous catalyst systems. Besides being useful soluble model systems that can provide significant information about reactions taking place at silica surfaces, several of the metallasilsesquioxane complexes are very interesting catalysts themselves. One of the shortcomings of homogeneous model systems, and hence also silsesquioxane compounds, is that the mobility of the species in solution allow electronically unsaturated species, which are potential active (olefin polymerization) catalysts, to form thermodynamically stable, inactive complexes.

ACKNOWLEDGEMENTS

The research of R.D. in the field of silsequioxane chemistry has been made possible by the financial support of DSM Research B.V. and the Dutch Polymer Institute. R.D. is very grateful for the freedom given by Dr. H.C.L. Abbenhuis and Prof. R.A. van Santen to pursue independent research in their group.

REFERENCES AND NOTES

1. Harrison, P. G. *J. Organomet. Chem.* **1997**, *542*, 141.
2. (a) Voronkov, M. G.; Lavrent'yev, V. L. *Top. Curr. Chem.* **1982**, *102*, 199. (b) Feher, F. J.; Neman, D. A.; Walzer, J. F. *J. Am. Chem. Soc.* **1989**, *111*, 1741.
3. For example see: (a) Feher, F. J.; Budzichowski, T. A. *J. Organomet. Chem.* **1989**, *373*, 153. (b) Feher, F. J.; Schwab, J. J.; Phillips, S. H.; Eklund, A.; Martinez, E. *Organometallics* **1995**, *14*, 4452. (c) Feher, F. J.; Wyndham, K. D. *Chem. Commun.* **1998**, 323. (d) Feher, F. J.; Wyndham, K. D.; Knauer, D. J. *Chem. Commun.* **1998**, 2393.
4. (a) Haddad, T. S.; Farris, A. R.; Lichtenhan, J. D. *ACS Polym. Prep.* **1997**, *38*, 127. (b) Tsuchida, A.; Bolln, C.; Sernetz, F. G.; Frey, H.; Mülhaupt, R. *Macromol.* **1997**, *30*, 2818. (c) Feher, F. J.; Schwab, J. J.; Tellers, D. M.; Burstein, A. *Main Group Chem.* **1998**, *2*, 169.
5. Brown, J. F.; Vogt, L.; H.; Prescott, P. I. *J. Am. Chem. Soc.* **1964**, *86*, 1120.
6. (a) Feher, F. J.; Newman, D. A. *J. Am. Chem. Soc.* **1990**, *112*, 1931. (b) Feher, F. J.; Budzichowski, T. A.; Blanski, R. L.; Weller, K. J.; Ziller, J. W. *Organometallics* **1991**, *10*, 2526. (c) Feher, F. J.; Phillips, S. H.; Ziller, J. W. *Chem. Commun.* **1997**, 829.
7. For example, see: (a) Feher, F. J. *J. Am. Chem. Soc.* **1986**, *108*, 3850. (b) Feher, F. J.; Gonzales, S. L.; Ziller, J. W. *Inorg. Chem.* **1988**, *27*, 3442. (c) Feher, F. J.; Budzichowski, T. A.; Weller, K. J. *J. Am. Chem. Soc.* **1989**, *111*, 7288. (d) Feher, F. J.; Walzer, J. F. *Inorg. Chem.* **1991**, *30*, 1689. (d) Feher, F. J.; Budzichowski, T. A.; Ziller, J. W. *Inorg. Chem.* **1992**, *31*, 5100. (e) Feher, F. J.; Budzichowski, T. A.; Ziller, J. W. *Inorg. Chem.* **1997**, *36*, 4082.
8. (a) Brown, J. F.; *J. Am. Chem. Soc.* **1965**, *87*, 4317. (b) Feher, F. J.; Schwab, J. J.; Soulivong, D.; Ziller, J. W. *Main Group Chem.* **1997**, *2*, 123.
9. (a) Feher, F. J.; Phillips, S. H.; Ziller, J. W. *Chem. Commun.* **1997**, 3397. (b) Feher, F. J.; Soulivong, D.; Lewis, G. T. *J. Am. Chem. Soc.* **1997**, *119*, 11323. (c) Feher, F. J.; Soulivong, D.; Eklund, A. G. *Chem. Commun.* **1998**, 399. (d) Feher, F. J.; Soulivong, D.; Nguyen, F. *Chem. Commun.* **1998**, 1279. (e) Feher, F. J.; Nguyen, F.; Soulivong, D.; Ziller, J. W. *Chem. Commun.* **1999**, 1705.
10. (a) Feher, F, J.; Terroba, R.; Ziller, J. W. *Chem. Commun.* **1999**, 2153. (b) Feher, F, J.; Terroba, R.; Ziller, J. W. *Chem. Commun.* **1999**, 2309. (c) Feher, F. J.; Terroba, R.; Jin, R.-Z. *Chem. Commun.* **1999**, 2513.
11. Feher, F. J.; Budzichowski, T. A.; Rahimian, K.; Ziller, J. W. *J. Am. Chem. Soc.* **1992**, *114*, 3859.

12.   (a) Duchateau, R.; Abbenhuis, H. C. L.; van Santen, R. A.; Thiele, S. K.-H.; van Tol, M.
        F. H. *Organometallics* **1998**, *17*, 5222. (b) Duchateau, R.; Cremer, U.; Harmsen, R. J.;
        Mohamud, S. I.; Abbenhuis, H. C. L.; van Santen, R. A.; Meetsma, A.; Thiele, S. K.-H.;
        van Tol, M. F. H.; Kranenburg, M *Organometallics* **1999**, *18*, 5447.
13.   (a) Feher, F. J.; Budzichowski, T. A.; Weller, K. J. *J. Am. Chem. Soc.* **1989**, *111*, 7288.
        (b) Feher, F. J.; Weller, K. J. *Organometallics* **1990**, *9*, 2638. (c) Feher, F. J.; Weller, K.
        J.; Ziller, J. W. *J. Am. Chem. Soc.* **1992**, *114*, 9686. (d) Edelmann, F. T.; Gun'ko, Y. K.;
        Giessmann, S.; Olbrich, F. *Inorg. Chem.* **1999**, *38*, 210. (e) Duchateau, R.; Harmsen, R.
        J.; Abbenhuis, H. C. L.; van Santen, R. A.; Meetsma, A.; Thiele, S. K.-H.; Kranenburg,
        M. *Chem. Eur. J.* **1999**, *5*, 3130. (f) Skowronska-Ptasinska, M. D.; Duchateau, R.; van
        Santen, R. A.; Yap, G. P. A. *Eur. J. Inorg. Chem.* **2001**, 133.
14.   Feher, F. J.; Budzichowski, T. A. *J. Organomet. Chem.* **1989**, *379*, 33.
15.   (a) Feher, F. J.; Blanski, R. L. *J. Chem. Soc., Chem. Comm.* **1990**, 1614. (b) Feher, F. J.;
        Tajima, T. L. *J. Am. Chem. Soc.* **1994**, *116*, 2145. (c) Feher, F. J.; Rahimian, K.;
        Budzichowski, T. A.; Ziller, J. W. *Organometallics* **1995**, *14*, 3920.
16.   Krijnen, S.; Harmsen, R. J.; Abbenhuis, H. C. L.; van Hooff, J. H. C.; van Santen, R. A.
        Chem. Commun. **1999**, 501.
17.   For example, see: (a) Baney, R. H.; Itoh, M.; Sakakibara, A.; Suzuki, T. *Chem. Rev.* **1995**,
        95, 1409. (b) Feher, F. J.; Budzichowski, T. A. *Polyhedron* **1995**, *14*, 3929. (c)
        Murugavel, R.; Voigt, A.; Walawalkar, M. G.; Roesky, H. W. *Chem. Rev.* **1996**, *96*, 2205.
        (d) Abbenhuis, H. C. L. *Chem. Eur. J.* **2000**, *6*, 25. (e) Lorenz, V.; Fischer, A.;
        Giessmann, S.; Gilje, J. W.; Gun'ko, Y.; Jacob, K.; Edelmann, F. T. *Coord. Chem. Rev.*
        2000, *206-207*, 321.
18.   Herrmann, W. A.; Anwander, R.; Dufaud, V.; Sherer, W. *Angew. Chem., Int. Ed. Eng.*
        1994, *33*, 1285.
19.   Abbenhuis, H. C. L.; van Herwijnen, H. W. G.; van Santen, R. A. *J. Chem. Soc., Chem.*
        Comm. **1996**, 1941.
20.   Feher, F. J.; Tajima, T. L. *J. Am. Chem. Soc.* **1994**, *116*, 2145.
21.   (a) Abbenhuis, H. C. L.; Krijnen, S.; van Santen, R. A. *Chem. Commun.* **1997**, 331.
        Maschmeyer, T.; Klunduk, M. C.; Martin, C. M.; Shephard, D. S.; Thomas, J. M.;
        Johnson, B. F. G. *Chem. Commun.* **1997**, 1847. (b) Crocker, M.; Herold, R. H. M.; Orpen,
        A. G. *Chem. Commun.* **1997**, 2411. (c) Klunduk, M. C.; Maschmeyer, T.; Thomas, J. M.;
        Johnson, B. F. G. *Chem. Eur. J.* **1999**, *5*, 1481. (d) Crocker, M.; Herold, R. H. M. *PCT*
        Int. Appl. 96/05873. (e) Maschmeyer, T.; Klunduk, M. C.; Martin, C. M.; Shephard, D.
        S.; Thomas, J. M.; Johnson, B. F. G. *Inorg. Chem.* **1997**, 1847.
22.   For example, see: Hungenberg, K. H.; Kerth, J.; Langhauser, F.; Marczinke, B.; Schlund,
        R. in *Ziegler Catalysts*, Fink, G.; Mülhaupt, R; Brintzinger, H. H. Eds., Springer-Verlag:
        New York, 1995, Chapter 20.
23.   Hlatky, G. G. *Chem. Rev.* **2000**, *100*, 1347.
24.   Hydrocarbon substituents on silicon are omitted for clarity.
25.   (a) Feher, F. J.; Walzer, J. F.; Blanski, R. L. *J. Am. Chem. Soc.* **1991**, *113*, 3618. (b)
        Feher, F. J.; Blanski, R. L. *J. Am. Chem. Soc.* **1992**, *114*, 5886.
26.   (a) Feher, F. J.; Blanski, R. L. *Makromol. Chem., Macromol. Symp.* **1993**, *66*, 95. (b)
        Abbenhuis, H. C. L.; Vorstenbosch, M. L. W.; van Santen, R. A.; Smeets, W. J. J.; Spek,
        A. L. *Inorg. Chem.* **1997**, *36*, 6431.
27.   Liu, J.-C. *Chem. Commun.* **1996**, 1109.

28. For example, see: (a) Scott, S.; Basset, J.-M.; Niccolai, G. P.; Santini, C. C.; Candy, J.-P.; Lecuyer, C.; Quignard, F.; Choplin, A. *New. J. Chem.* **1994**, *18*, 115. (b) Vidal, V.; Théolier, A.; Thivolle-Cazat, J.; Basset, J.-M.; Corker, J. *J. Am. Chem. Soc.* **1996**, *118*, 4595. (c) Rosier, C.; Niccolai, G. P.; Basset, J.-M. *J. Am Chem. Soc.* **1997**, *119*, 12408. (d) Dufaud, V.; Basset, J.-M. *Angew. Chem.* **1998**, *110*, 848. (e) Amor Nait Ajjou, J.; Scott, S. L. *Organometallics* **1997**, *16*, 86.

29. (a) Karapinka, G. L. US Patent 3,709,853, 1973. (b) Karol, F. J. US Patent 4,015,059, 1977. (c) Karol, F. J.; Karapinka, G. L.; Wu, C.; Dow, A. W.; Johnson, R. N.; Carrick, W. L. *J. Polym. Sci.: Part A1*, **1972**, *10*, 2621. (d) Fu, S.-L.; Lunsford, J. H. *Langmuir*, **1990**, *6*, 1774.

30. For example, see: (a) Thomas, B. J.; Theopold, K. H. *J. Am. Chem. Soc.*, **1988**, *110*, 5902. (b) White, P. A.; Calabrese, J.; Theopold, K. H. *Organometallics*, **1996**, *15*, 5473. (c) Theopold, K. H. *Eur. J. Inorg. Chem.*, **1998**, 15.

31. Duchateau, R. van Santen, R. A. Yap, G. P. A. unpublished results.

32. (a) Ittel, S. D. *J. Macromol. Sci., Chem.* **1990**, *A27*, 1133. (b) Ittel, S. D.; Nelson, L. T. *J. Polym. Prepr.* **1994**, *35*, 665.

33. Duchateau, R.; Abbenhuis, H. C. L.; van Santen, R. A.; Meetsma, A.; Thiele, S. K.-H.; van Tol, M. F. H. *Organometallics* **1998**, *17*, 5663.

34. (a) Hungenberg, K. H.; Kerth, J.; Langhauser, F.; Marczinke, B.; Schlund, R. in *Ziegler Catalysts*, eds. Fink, G.; Mülhaupt, R; Brintzinger, H. H., Springer-Verlag, New York, 1995, Chapter 20. (b) Chien, J. C. W.; He, D. *J. Polym. Sci.: Part A: Pol. Chem.* **1991**, 1603.

35. For example, see: (a) Chen, Y.-X.; Rausch, M. D.; Chien, J. C. W. *J. Polym. Sci., A: Polym. Chem.* **1995**, *33*, 2093. (b) Hlatky, G. G.; Upton, D. J. *Macromol.* **1996**, *29*, 8019. (c) Roscoe, S. B.; Fréchet, J. M. J.; Walzer, J. F.; Dias, A. J. *Science* **1998**, *280*, 270. (d) Bochmann, M.; Pindado, G. J.; Lancaster, S. J. *J. Mol. Catal., A: Chem.* **1999**, *146*, 179. (e) Turner, H. W. *U.S. Pat. Appl.* 5,427,991 (1995). (f) Ono, M.; Hinokuma, S.; Miyake, S.; Inazawa, S. *Eur. Pat. Appl.* 710,663 (1996). (g) Jacobsen, G. B.; Wijkens, P.; Jastrzebski, J.; van Koten, G. *PCT Int. Appl.* 96/28480 (1996). (h) Kaneko, T.; Sato, M. *Eur. Pat. Appl.* 727,433 (1996). (i) Hinokuma, S.; Miyake, S.; Ono, M.; Inazawa, S. *Eur. Pat. Appl.* 775,707 (1997). (j) Carnahan, E. M.; Carney, M. J.; Neithamer, D. R.; Nickias, P. N.; Shih, K.-Y.; Spencer, L. *PCT Int. Appl.* 97/19959. (k) Kaneko, T.; Sato, M. *U.S. Pat. Appl.* 5.807,938 (1998).

36. (a) Hlatky, G. G.; Upton, D. J.; Turner, H. W. *PCT Int. Appl.* 91/09822. (b) Turner, H. W.; *US Pat. Appl.* 5,427,991 (1995). (c) Walzer, J. F. Jr. *PCT Int. Appl.* 95/04319. (d) Ward, D. G.; Carnahan, E. M. *PCT Int. Appl.* 96/23005. (e) Walzer, J. F. Jr. *U.S. Pat. Appl.* 5.643,847 (1997).

37. Buys, I. E.; Hambley, T. W.; Houlton, D. J.; Maschmeyer, T.; Masters, A. F.; Smith, A. K. *J. Mol. Catal.* **1994**, *86*, 309-318.

38. (a) Edelmann, F. T.; Gießmann, S.; Fischer, A. *Chem. Commun.* **2000**, 2153-2154. (b) Edelmann, F. T.; Gießmann, S.; Fischer, A. *J. Organomet. Chem.* **2001**, *620*, 80.

39. For example, see: (a) Gao, H.; Angelici, R. J. *Organometallics* **1999**, *18*, 989. (b) Clark, J. H.; Macquarrie, D. J. *Chem. Commun.* **1998**, 853. (c) Johnson, B. F. G.; Raynor, S. A.; Shephard, D. S.; Maschmeyer, T.; Meurig Thomas, J.; Sankar, G.; Bromley, S.; Oldroyd, R.; Gladden, L.; Mantle, M. D. *Chem. Commun.* **1999**, 1167. (d) de Vos, D. E.; Jacobs, P. A. *Catal. Today* **2000**, *57*, 105. (e) Liu, C.-J.; Yu, W.-Y.; Li, S.-G.; Che, C.-M. *J. Org. Chem.* **1998**, *63*, 7364. (f) Petrucci, M. G. L.; Kakkar, A. K. *Chem. Mater.* **1999**, *11*, 269. (g) Jorna, A. M. J.; Boeldijk, A. E. M.; Hoorn, H. J.; Reedijk, J. *React. Polym.* **1996**, *29*, 101.

40. For example, see: (a) Lee, D.-H.; Yoon, K.-B. *Macromol. Rapid Commun.* **1997**, *18*, 427. (b) Galan-Fereres, M.; Koch, T.; Hey-Hawkins, E.; Eisen, M. S. *J. Organomet. Chem.* **1999**, *580*, 145. (c) Juvaste, H.; Iiskola, E. I.; Pakkanen, T. T. *J. Mol. Catal. A: Chem.* **1999**, *150*, 1. (d) Iiskola, E. I.; Timonen, S.; Pakkanen, T. T.; Kärkki, O.; Seppälä, J. V. *Appl. Surf. Sci.* **1997**, *121/122*, 372. (e) Uusitalo, A.-M.; Pakkanen, T. T.; Iiskola, E. I. *J. Mol. Catal. A: Chem.* **2000**, *156*, 181.

41. Severn, J. R.; Duchateau, R.; van Santen, R. A.; Yap, G. P. A.; Ellis, D. D.; Spek, A. L. *Organometallics* **2002**, *21*, 4.

42. (a) Chang, M. U.S. Pat. 5,008,228, 1991. (b) Tsutsui, T.; Ueda, T. U.S. Patent 5,234,878, 1993. (c) Gürtzgen, S. U.S. Patent 5,446,001, 1995. (d) Herrmann, H. F.; Bachmann, B.; Spaleck, W. Eur. Pat. Appl. 578,838, 1994. (e) Becker, R.-J.; Rieger, R. Eur. Pat. Appl. 763,546, 1997. (f) Becker, R.-J.; Gürtzgen, S.; Kutschera, D. U.S. Patent 5,534,474, 1996. (g) Kutschera, D.; Rieger, R. U.S. Patent 5,789,332, 1998.

43. Skowronska-Ptasinska, M. D.; Duchateau, R.; van Santen, R. A.; Yap, G. P. A. *Organometallics* **2001**, *20*, 3519.

44. Harlan, C. J.; Bott, S. G.; Barron, A. R. *J. Am. Chem. Soc.* **1995**, *117*, 6465.

45. Ahn, H.; Marks, T. J. *J. Am. Chem. Soc.* **1998**, *120*, 13533.

46. (a) Yang, X.; Stern, C. L.; Marks, T. J. *J. Am. Chem. Soc.* **1994**, *116*, 10015. (b) Bochmann, M.; Dawson, D. M. *Angew. Chem., Int. Ed. Engl.* **1996**, *35*, 2226. (c) Coles, M. P.; Jordan, R. F. *J. Am. Chem. Soc.* **1997**, *119*, 8125. (d) Amor, F.; Butt, A.; du Plooy, K. E.; Spaniol, T. P.; Okuda, J. *Organometallics* **1998**, *17*, 5836. (e) Song, X.; Thornton-Pett, M.; Bochmann, M. *Organometallics* **1998**, *17*, 1004. (f) Green, M. L. H.; Saßmannhausen, J. *Chem. Commun.* **1999**, 115. (g) Guérin, F.; Stephan, D. W. *Angew. Chem., Int. Ed.* **2000**, *39*, 1298.

47. See also: (a) Siedle, A. R.; Newmark, R. A.; Lamanna, W. M.; Schroepfer, J. N. *Polyhedron* **1990**, *9*, 301. (b) Harlan, C. J.; Bott, S. G.; Barron, A. R. *J. Am. Chem. Soc.* **1995**, *117*, 6465. (c) Koide, Y.; Bott, S. G.; Barron, A. R. *Organometallics* **1996**, *15*, 2213. (d) Metz, M. V.; Schwartz, D. J.; Stern, C. L.; Nickias, P. N.; Marks, T. J. *Angew. Chem., Int. Ed.* **2000**, *39*, 1312 and references cited therein.

48. (a) Duchateau, R.; van Santen, R. A.; Yap. G. P. A. *Organometallics* **2000**, *19*, 809. (b) Metcalfe, R. A.; Kreller, D. I.; Tian, J.; Kim, H.; Corrigan, J. F.; Taylor, N. J.; Collins, S. *Organometallics* **2002**, *21*, 1719.

49. (a) Krijnen, S.; Abbenhuis, H. C. L.; Hanssen, R. W. J. M.; van Hooff, J. H. C.; van Santen, R. A. *Angew. Chem. Int. Ed. Eng.* **1998**, *37*, 356. (b) Smet, P.; Rioandato, J.; Pauwels, T.; Moens, L.; Verdonck, L. *Inorg. Chem. Commun.* **2000**, *3*, 557. (c) Maxim, N.; Abbenhuis, H. C. L.; Magusin, P. C. M. M.; van Santen, R. A. *Chin. J. Chem.* **2001**, *19*, 30. (d) Wada, K.; Nakashita, M.; Yamamoto, A.; Mitsuda, T. *Chem. Commun.* **1998**, 133.

Chapter 4

THEORETICAL MODELS OF ACTIVE SITES: GENERAL CONSIDERATIONS AND APPLICATION TO THE STUDY OF PHILLIPS-TYPE Cr/SILICA CATALYSTS FOR ETHYLENE POLYMERIZATION

Knut J. Børve, Øystein Espelid
Department of Chemistry, University of Bergen, Norway

Keywords: modelling, *ab initio*, density functional theory, cluster models, Phillips catalysts, chromium, silica, oxidation states, nuclearity, vibrational spectroscopy, electronic transitions, polymerization, initiation, propagation, catalytic activity

Abstract: Quantum chemical cluster-model approximations are briefly reviewed as tools for studying local properties of solid surfaces. In its molecular-cluster version, this technique is applied to Phillips-type Cr/silica catalysts for ethylene polymerization and used to discuss spectroscopic and catalytic properties of low-valent chromium sites. The characterization part includes d-d electronic transitions as well as IR vibrational spectra of adsorbed carbonyl species. Comparison to experimental spectra facilitates identification of surface species. Next, cluster models are used to investigate if and how polymerization may be initiated and sustained at the most topical of the proposed active sites. Two of the chromium sites studied here show promising activity for ethylene polymerization.

1.   INTRODUCTION

The active site is a very powerful concept in heterogeneous catalysis. However, in the case of the Phillips catalyst for the polymerization of ethylene,[1] the active sites have largely eluded identification and are only characterized partly, or indirectly. Based on spectroscopic evidence, and more recently, studies of well-characterized model systems, a number of suggestions have been launched, varying with respect to oxidation state, nuclearity and composition of the coordination sphere.   In this contribution, we apply quantum chemical modelling to cluster models of the various proposed active sites, working in two directions: characterization and catalytic activity.

For the characterization part, we compute spectroscopic signatures of a number of Cr/SiO_2 sites and compare them with experimental d-d diffuse reflectance spectra (DRS) and IR vibrational spectra of adsorbed carbonyl species.[2,3] This facilitates assignment of the spectra as well as identification of surfaces species. However, surface species may well contribute to the spectral properties of the catalyst without providing activity. Due to the short lifetime of the transient species in catalytic reactions, it is often difficult to ascertain the reaction mechanism by experimental means. Modelling offers the possibility to take "snapshots" of the reaction, and thereby to obtain information about the reaction mechanism.   Our second line of approach uses cluster models to investigate if and how polymerization may be initiated and sustained at the most topical of the proposed active sites.[4-6]

2.   MOLECULAR MODELS IN HETEROGENEOUS CATALYSIS

A computational, atomistic model of processes occurring on a solid surface may be discussed in terms of two aspects: (1) the chemical model, *i.e.*, the number of atoms and constraints on their structural arrangement; and (2) the mathematical model used for the forces acting on the nuclei. We will discuss these aspects separately, while recognizing that the choices made in each case are by necessity heavily correlated. While recognizing virtues and limitations in both quantal and classical approaches, only quantum mechanics is considered to be able to describe reliably the progress of a chemical reaction, and the discussion of item (2) will be limited to quantum chemical methods.

When studying large chemical systems, there are many processes that are of mainly local character; of particular interest here are chemisorption and other surface reactions to which the concept *active site* may be applied. One way to approach such systems is to ignore the effects of the extended system on the local interactions, and depict the local defect region as a molecule. This constitutes the molecular cluster model which has three main advantages: (1) it is computationally relatively simple; (2) it opens the door to the use of highly accurate methods for describing the electronic structure; and (3) it establishes a direct connection between molecular chemistry and surface chemistry.[7] However, these advantages come at the cost of neglecting interactions between the cluster and its surrounding, the importance of which depends on the properties of the host system, in particular its metallic or ionic character. The most commonly used support in Phillips catalysis is amorphous silica, which may be classified as a non-metallic, covalent oxide.[8] Since electron mobility is quenched in solids with localized chemical bonds, and since the ionic character of silica is minor, it is likely that molecular clusters can furnish useful models of silica surface sites.

2.1 Molecular Clusters with Saturated Dangling Bonds

The valence electrons in silica are localized in strong, covalent, single bonds between Si and O,[8] and in the process of defining a finite cluster, bonds protruding from the cluster are disrupted. The chemically intuitive way to eliminate these model-induced surface states is to terminate the dangling bonds by hydrogen or some other suitable (pseudo-)atom.[9] Since the electronegativity of hydrogen is intermediate between that of oxygen and silicon, it polarizes the terminal bond in the proper direction, whether substituted for a missing oxygen or silicon atom. A halogen may serve as saturating atom on Si, in place of a missing oxygen.

In order to illustrate how computed properties may be affected by the choice of terminating agent, we consider molecular models of isolated surface silanol groups ($\equiv$SiOH). This functionality dominates the interaction of molecules with dehydrated silica surfaces.[7] A popular model of an isolated silanol group has been the silanol molecule itself, H_3SiOH. While successful at modelling the vibrational properties and the binding of molecules onto isolated surface silanol groups,[7] its gas-phase deprotonation energy of 1502 ± 25 kJ/mol[10] is too high compared to that of an isolated surface hydroxyl group (1390 ± 25 kJ/mol)[11]. With large, cage-like clusters, the experimental value is reproduced by theory,[12,13] but at high computational cost. A more tractable solution is to tune the termination agents used on Si. By replacing an increasing number of hydrogen atoms with fluorine, the deprotonation energy of $H_{3-n}F_nSiOH$ is gradually reduced, and the difluorosilanol molecule is found to have a gas-phase acidity in close agreement with the acidity of a silanol group at the silica surface.

A well-designed cluster model may be expected to provide a good description of the nature of molecule-surface interactions, adsorption geometries, and vibrational and optical transitions.[14] Adsorption energies are more susceptible to the limitations of a cluster model, although one may assume qualitative accuracy.[7] Regarding a molecular cluster as a first approximation, improved models may be obtained by augmenting it with effective potentials that take into account steric and electronic effects of the coupling between the cluster and the extended structure from which it was cut.[15] While representative examples of embedding techniques will be reviewed in the next section, their application to amorphous silica is unfortunately hampered by lack of information about the surface morphology.

2.2   Embedded Cluster Models

Perhaps the most ambitious embedding formalism is the perturbed cluster approach,[16,17] as implemented in the EMBED01 program.[18] The calculations involve iteration of the steps until self-consistency: (1) calculation of the wave function for the molecular cluster in the field of the indented crystal, and (2) correction of the cluster solution to allow for propagation of the wave function into the indented crystal, under the assumption that the density of the states projected onto the indented crystal is that same as in the perfect host crystal. Explicit reference is made to the Hartree-Fock periodic solution for the unperturbed crystal.[19,20] Formally there is no need to saturate dangling bonds in this theory, although strong electronic coupling between the defect region and the rest of the crystal may make it difficult to treat defects in covalent systems. The orthogonality constraints that are introduced may overcompensate for the dangling bonds' density and deform the density in the boundary region.

In order to make the computational problem more tractable, several methods adopt a hybrid approach, in which the cluster is treated quantum mechanically while the surrounding crystal is described in terms of semiclassical potentials. An example of mind is the *QM-Pot* method,[15] in which the cluster environment is described by parameterized interatomic potentials (Pot). An essential aspect is that the cluster is connected to its surroundings through a set of linking (pseudo-)atoms, which are used both to saturate dangling bonds and to place geometric constraints on the cluster. *QM-Pot* provides a working methodology for simultaneous optimization of the structures of both the cluster and the embedding silica matrix, also including transition-state structures.[15] The method has been used with success in studies of zeolites and related ordered systems.

In highly polar crystals, the most important embedding component is the electrostatic (Madelung) potential from the surroundings on the cluster atoms. However, in order to avoid unrealistically large polarization effects within the cluster, it may be necessary to include a short range potential that models Pauli repulsion between the quantum cluster and the immediately surrounding ions.[21] In the *Ab-Initio Model Potential* (AIMP) method,[22,23] a further improved description is used for all atoms in an intermediate region bordering on the cluster. These ions are described in terms of frozen ionic electronic densities computed for the unperturbed crystal, and the computations are made feasible by using model potentials to represent Coulomb, exchange and overlap effects. A particular feature of the *ICECAP*[24,25] program is that polarization effects up to infinite distance are included self-consistently.

3. QUANTUM CHEMICAL METHODS

The time-independent, non-relativistic electronic Schrödinger equation is the starting point of most quantum chemical studies:

$$\hat{H}_e \psi_e = E_e \psi_e$$

The eigenvalue E_e is the energy of the electronic state described by the associated wave function, ψ_e. In principle, the electronic problem has to be solved over and over again for every structural arrangement of the nuclei that may be relevant to the problem at hand. In this process, the electronic energy is mapped out as a function of molecular geometry, *i.e.*, $E_e = E_e(\mathbf{R})$, where $\mathbf{R}$ specifies the relative positions of the nuclei. The potential energy surface of the system is defined as the electronic energy augmented by nuclear-nuclear repulsion terms. It takes on the role of an effective potential for the nuclear motion and is therefore of paramount importance in a theoretical description of chemical reactivity. To *viz.*, reactants, intermediates and products may be identified with local minima on the potential energy surface, while first-order saddle points represent transition states. In this perspective, the main effort in a quantum chemical description of a reaction lies in solving the electronic problem to a sufficiently high degree of accuracy. Excellent textbooks are available[26,27] that describe contemporary methods for this purpose, and we will limit the presentation to a brief summary.

3.1 *Ab initio* Methods

The *Hartree-Fock Self-Consistent Field* (HF-SCF) method is the starting point for most *ab initio* approaches. In this approximation, each electron is taken to move in the mean field of the nuclei and the remaining electrons. This gives rise to one-electron wave functions - molecular spin orbitals - and the full many-electron wave function is obtained by anti-symmetrizing the product of occupied spin orbitals. An *electron configuration* is defined by the orbital chosen to be occupied. Any mean-field model fails if the motion of the constituent particles becomes highly correlated. This aspect limits the accuracy of the HF-SCF method, most notably if two or more electron configurations give rise to states of the same symmetry and almost the same energy. Reaction and activation

energies, as well as electronic excitation energies, are often significantly affected
by electron correlation. A second limiting factor is that the molecular orbitals are
expressed in finite sets of pre-chosen atomic orbitals. The basis sets are chosen
with an eye to computational expense, and limitations in the chosen basis set may
govern the overall accuracy of the model. This aspect applies equally to DFT
methods (see below).

The *Multi-Configurational Self-Consistent Field* (MC-SCF) is an
improvement on Hartree-Fock in that the limitation to a single electron
configuration is lifted. It is well-suited for obtaining zeroth-order wave functions
for excited electronic states,[28] as well as for any molecular system whose
geometry or atomic composition makes it prone to near-degeneracy between
different electronic configurations. However, MC-SCF does not aspire to
quantitative accuracy, since it represents a cost-ineffective way of recovering
dynamical correlation energy. The most popular flavor of MC-SCF is the
Complete Active Space SCF method (CASSCF).[29]

Among *ab initio* models, only methods that take into account very many
electron configurations aspire to high accuracy. *Configuration Interaction* (CI)
and coupled-cluster (CC) expansions[30] are powerful but also very resource-
demanding approaches. *Many-body Perturbation Theory* has become the most
popular *ab initio* approach for estimating the dynamical correlation energy,[31]
starting from either a single configuration (MPn)[32] or a multi-configurational
zeroth-order wave function (CASPT2).[33] Whichever approach is used, *ab initio*
inclusion of electron correlation requires extensive basis sets, and this adds
significantly to the cost. Ways of extrapolating the effects of electron correlation
as computed by means of medium-sized basis sets have therefore been
developed.[34-36]

3.2   Density Functional Theory

The Hohenberg-Kohn theorems assert[37] that the ground-state electronic
energy is completely determined by the electron density, ρ, and may be optimized
variationally with respect to the latter. This forms the basis of *density functional
theory* (DFT). The energy functional is divided into three parts: attraction
between nuclei and electrons $V_{ne}[\rho]$, kinetic energy of electrons $T[\rho]$, and
electron-electron repulsion $V_{ee}[\rho]$. Expressing the electron density in terms of
orbitals opens the door to a powerful reformulation of the energy functional as:

$$E[\rho] = V_{ne}[\rho] + T_s[\rho] + J[\rho] + E_{xc}[\rho]$$
$$E_{xc}[\rho] = (T[\rho] - T_s[\rho]) + (V_{ee}[\rho] - J[\rho])$$

Here, the kinetic energy is split into one part that may be calculated exactly as the
kinetic energy $T_s[\rho]$ of a system of non-interacting electrons, and a small

correction term $(T[\rho] - T_s[\rho])$ which is included in the exchange-correlation functional $E_{xc}[\rho]$. Similarly, $V_{ee}[\rho]$ is split into the classical Coulomb part $J[\rho]$ and a remaining part $(V_{ee}[\rho] - J[\rho])$ which contains both exchange and correlation energy. These considerations lead to the non-linear one-particle Kohn-Sham equations,[38] which are solved iteratively until self-consistency. While there are many similarities between HF and DFT, there is one important difference. If the exact exchange-correlation functional $E_{xc}[\rho]$ were known, DFT would provide the exact total energy, including electron correlation.

The tremendous popularity of approximate DFT methods stems from the fact that modern gradient-corrected functionals provide useful estimates of the exchange-correlation energy at a computational cost comparable to that of Hartree-Fock, which does not include electron correlation. The DFT exchange-correlation functionals consist of local functionals, which are augmented by gradient corrections. A popular gradient correction to the local Slater exchange functional was given by Becke,[39] and is often used in combination with the gradient correction formulated by Perdew[40] for the correlation energy functional. The combination of these forms gives the gradient-corrected exchange-correlation functional termed BP86. A hybrid functional is obtained by including Hartree-Fock exchange in the functional, as exemplified by the B3LYP method.[41]

General problems with present-day DFT formulations are lack of a systematic way of improving the results, the difficulty of extending the theory to electronically excited states, and susceptibility to near-degenerate electron configurations. However, a number of studies show that density functional theory is able to provide accurate estimates of structure, bond dissociation energies and vibrational frequencies of most transition-metal compounds,[42] *i.e.*, systems that by tradition are considered a challenge to quantum chemistry. Moreover, extensive calibration studies show that gradient-corrected DFT is capable of providing accurate energy profiles of the Cossee mechanism[43] for metal-catalyzed olefin polymerization.[44,45]

4.   CLUSTER MODELS OF THE REDUCED Cr/SiO$_2$ SYSTEM

Based on experimental studies, the reduced Cr/silica system is believed to furnish mononuclear and dinuclear Cr(II) sites and mononuclear Cr(III) sites.[46] These di- and trivalent chromium species are expected to be bound by oxygen linkages to the silica surface. The local geometry around chromium at the surface of silica depends partly on the structure and flexibility of the anchoring site and the spacing between neighbouring hydroxyls that may participate in the esterification reaction. In this work, a distribution of mononuclear Cr(II) species is spanned by three cluster models that differ with respect to the $\angle$OCrO bond angle: a pseudo-tetrahedral cluster (**T**, left and top in Figure 1); a pseudo-octahedral cluster (**O**, right and middle); and a cluster with intermediate ∠OCrO bond angle (**I**, right and top). The dinuclear cluster models, also presented in Figure 1, may be regarded as -OCrOCrO- moieties anchored to sites of narrow (**D$_n$**, left and bottom) and intermediate-to-wide "bites" (**D$_w$**, right and bottom), similar to those of the mononuclear **T** and **O** cluster models, respectively. Thermodynamical aspects of the formation of these sites in anchoring reactions are presented in references 4 and 6. Only one mononuclear trivalent cluster model is included in this study, shown as **CrIII** in Figure 1, for the purpose of completeness in the characterization studies. Selected geometry parameters of the cluster models are given in Table 1.

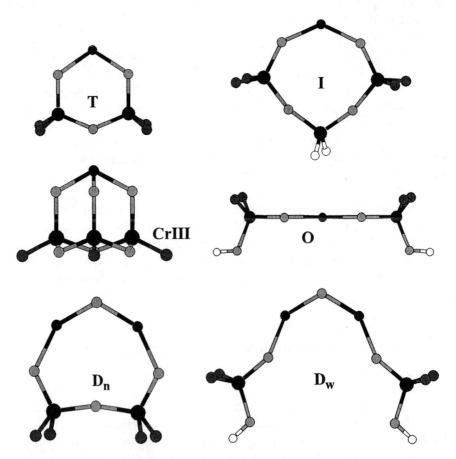

Figure 1. Cluster models of divalent mononuclear chromium sites with the OCrO bond angle in the tetrahedral-to-octahedral range (**T**, **I** and **O**); a trivalent mononuclear chromium site **CrIII**; and divalent dinuclear chromium sites (**D$_n$** and **D$_w$**). The elements are coded on a grey scale according to increasing atomic number: H(white) > C > O > Si > Cr (dark gray).

Table 1. Selected geometry parameters[a] of the cluster models, optimized using the B3LYP method.

	State	*rCrO*	*rSiO*	*∠OCrO*	*∠SiOCr*
T	$^5B_1$	1.84	1.62	108	127
I	$^5B_1$	1.82	1.63	132	141
O	$^5B_1$	1.83	1.62	180	179
D_w[b]	$^9A_2$	1.84	1.59	145	160
D_n[c]	$^9A_2$	1.86	1.61	139	142
CrIII	$^4A_2$	1.81	1.65	104	114

[a] Units: Bond lengths (r) in Å, angles (∠) in degrees. [b] $rCrO_{bridge}$ = 1.77 Å, $∠CrOCr$=123°. [c] $rCrO_{bridge}$ = 1.78 Å, $∠CrOCr$=120°.

The dangling bonds at each silicon atom with an oxygen linkage to chromium are terminated by at most two fluorine atoms. To increase compatibility with cluster models with siloxane oxygen bridges, each silicon atom of the D_w and O cluster models is bound to one oxygen whose dangling bond is terminated by a hydrogen atom. On the basis of the computed gas-phase acidity of the difluorosilanol molecule, this termination scheme is expected to provide cluster models in which the acidic properties of chromium are well described. However, the terminating fluorine atoms may be replaced by hydrogen atoms in cases where higher computation efficiency is desired.

4.1 Computational Details

Most of the calculations presented here were conducted using gradient-corrected density functional theory as implemented in the Amsterdam Density Functional (ADF) set of programs.[47,48] Closed- and open-shell systems were described within spin-restricted and -unrestricted formalisms, respectively. Typically, primitive Slater-type basis sets of triple-zeta (O, F, Cr) and double-zeta (H, C, Si) quality were used, with polarization functions added to all atoms but chromium. Specific information regarding basis sets *etc.*, is detailed in the cited articles. Energy differences refer to electronic degrees of freedom, *i.e.*, without inclusion of zero-point vibrational energies.

5. d-d ELECTRONIC TRANSITIONS IN THE REDUCED CATALYST

Phillips-type catalysts are prepared by impregnating a chromium compound at low loadings onto a support material, most commonly a wide-pore silica.[49] In the subsequent calcination step,[49] chromium is anchored onto the surface of silica with consumption of hydroxyl groups, indicating that chromium becomes attached to the surface through oxygen linkages (Si-O-Cr).[49] Moreover, the oxygen-rich atmosphere in the calcination step causes chromium to become oxidized,[46,49] and the main chromium species have been shown to be hexavalent dichromate and monochromate on sol-gel silica supports, while monochromate dominates on pyrogenic silica supports at low chromium loading.[46] Very recent results from ion-scattering experiments on a surface-science model of the catalyst suggest that only monochromate esters are present at low chromium loading.[50] The lack of pores and internal surfaces in the thin-film amorphous silica support used in the model makes it less than straightforward to generalize this result to Phillips catalysts. Thus, while it certainly topicalizes mononuclear sites, the possibility of dinuclear sites should not be dismissed.

When a calcined catalyst is brought in contact with ethylene, an induction time is observed prior to the onset of polymerization.[49] In this period, chromium is reduced and ethylene is oxidized to formaldehyde.[51] The reduction may also be performed in a separate pretreatment step, *e.g.*, by exposure of the calcined catalyst to carbon monoxide.[52] Such a reduced catalyst polymerizes ethylene without any induction period.[49] It is widely accepted that, after reduction, the dominant chromium species are divalent.[46,49] The deviation from 2.0 in the average oxidation state is attributed to unreactive α-chromia particles[53] and isolated Cr(III) species.[46,54,55]

The d-d region in diffuse reflectance spectra (DRS) of the reduced catalysts is dominated by two broad bands at 7,000-10,000 cm^{-1} and 10,000-13,000 cm^{-1}.[56] These bands have been assigned to pseudo-tetrahedral and pseudo-octahedral Cr(II) species, respectively.[56] Hence, more than identifying oxidation states, DRS studies have been used to differentiate between Cr(II) in different local surroundings. While this is an important step in the direction of identifying the active sites in Phillips catalyst, assignment of the broad d-d bands is hampered by lack of knowledge about the number and positions of the electronic transistions.[53,57]

On the basis of d-d transition energies and intensities computed for cluster models, we have performed a theoretical analysis of d-d transitions in the reduced Cr/silica system.[2] The transition energies were computed by the CASPT2[33] method, whereas transition moments were obtained in the electric dipole approximation from CASSCF wave functions.[29]

In Table 2, the computed d-d transition energies are compared to those observed for the CO-reduced Phillips catalyst, for which we rely on the extensive studies by Zecchina *et al.*,[53,57,58] in addition to more recent work by Weckhuysen and co-workers.[54] The bands observed at 7,500-8,000 cm^{-1} and ~12,500 cm^{-1} match those predicted for the **T** cluster (see Table 2). The observed bands at 15,000-15,500 cm^{-1} and 19,500-21,000 cm^{-1} are in reasonable agreement with those predicted for Cr(II) species. Since no bands are observed at ~11,000 cm^{-1},

Table 2. Comparison between computed and observed d-d bands of the reduced Cr/silica system.

Cluster	$T(cm^{-1})$		Refs.
	Theory	Experiment	
$\mathbf{D_w}$	~4,000	Limit	
$\mathbf{D_n}$, **I**	~5,000	NO BAND	
T	~8,500	7,500-8,000	53, 58, 59
$\mathbf{D_w}$, $\mathbf{D_n}$	~9,000-10,000	10,000	54, 57
CrIII, $\mathbf{D_w}$, **I**	~11,000	NO BAND	
T	~12,500	12,500	54, 57
CrIII	~16,000	15,000-15,500	54, 57
CrIII	~18,000-19,000	19,500-21,000	54

the predicted transitions of Cr(III) species in this region are probably shifted either down to the 10,000 cm^{-1} band, or up to the ~12,500 cm^{-1} band.

The 10,000 cm^{-1} band may be indicative of divalent chromium in a rather relaxed geometry, either bound as a mononuclear species to silica or as part of a dinuclear surface species. Such sites are found to display additional low-energy transitions at 4,000-5,000 cm^{-1} with intensities comparable to those predicted in the region of 10,000 cm^{-1}. However, there are no experimental reports on band maxima in the 4,000-5,000 cm^{-1} region, possibly due to lower-than-computed intensities of the low-energy bands. Alternatively, if the $\angle$OCrO bond angle on the surface is wider than modelled in the $\mathbf{D_w}$ and **I** clusters, this may shift the transition energies below the limit of experimental observation ($\approx$4,000 cm^{-1}).

The good correspondence between our calculations and the observed d-d transition energies supports the widely accepted view of divalent chromium as the main surface species after reduction of the catalyst. Furthermore, a fairly consistent interpretation of the spectra is afforded by a population of pseudo-tetrahedral mononuclear Cr(II) species, dinuclear Cr(II) species analogous to the wide-bite $\mathbf{D_w}$ cluster, as well as a minor population of Cr(III) species.

6. VIBRATIONAL SPECTROSCOPY OF CARBON MONOXIDE ADSORBED ONTO THE REDUCED CATALYST

Vibrational spectroscopy of adsorbed probe molecules is a powerful technique in catalyst characterization,[60] and IR spectroscopy of chromium-carbonyl species holds a special position within characterization of the reduced Phillips system. Carbonyl IR spectra have been used to determine the nuclearity[61,62] and oxidation state[61] of chromium adsorption sites, as well as their relative acidity[63,64] and coordination environment.[61,63] Oxygen donors in the form of silanol and siloxane groups are present at the surface and may coordinate to chromium. In this respect, the divalent chromium species have been classified into three families (A, B, C), differing in coordinative unsaturation (A>B>C) and, consequently, in their ability to chemisorb CO.[61,63]

In order to facilitate a theoretical analysis of CO adsorption on the reduced Cr/silica system, the binding energies and vibrational properties have been computed for oligocarbonyl complexes formed at the cluster models presented above. The computations were conducted using two density functional methods, BP86 and B3LYP (section 3.2), both of which show useful accuracy. While the two DFT methods predict the same trend with respect to the coordination energy of higher carbonyls,[3] improved adsorption energies are obtained in a simple scaling procedure, calibrated by means of a high-quality *ab initio* method, CCSD(T). A full account of these calculations is available in Reference 3.

The calculated vibrational IR transitions appear in three frequency regions, with carbonyl complexes of dinuclear Cr(II), mononuclear Cr(II), and mononuclear Cr(III) making up the low, intermediate and high frequency regions respectively, Figure 2. The experimental spectrum is dominated by two well-separated triplets of bands:[64] the so-called room temperature triplet at 2178-2191 cm^{-1} and the low temperature triplet at 2035-2120 cm^{-1}, in addition to weak bands at high frequencies that are observed at room temperature and high CO pressure.[62]

The **T** cluster shows the highest affinity for carbon monoxide, and even at a low partial pressure of CO, the adsorption equilibrium is expected to be shifted toward the dicarbonyl complex **T-2$_S$** in Figure 2. This species gives rise to two vibrational modes, an asymmetric **T-2(B$_1$)** mode, and a symmetric **T-2(A$_1$)** mode,[3] which we assign to bands observed at 2187 and 2191 cm^{-1}, respectively.[63] Precoordination of an oxygen donor to chromium is likely to modify and possibly also limit the uptake of carbon monoxide at Cr(II)-B sites.[63] This scenario has been explored with silanol and siloxane molecules at different distances and orientations relative to the **T** cluster. With a siloxane molecule at an O···Cr distance of 2.42 Å (see **Cr(II)-B-1** in Figure 2), only one equivalent of CO is adsorbed onto chromium. Based on the computed shift from **T-2** we assign the third component of the room temperature triplet at 2184 cm^{-1} to monocarbonyl complexes of Cr(II)-B sites that are significantly affected by the presence of a non-displaceable oxygen donor.[3]

A displaceable oxygen donor is expected to reduce the adsorption energy of CO without affecting the spectroscopic signature of the dicarbonyl complex once it is formed.[65] However, in the case of a silanol donor, the O-H stretching frequency may be used to monitor the coordination. By comparing our calculated shifts upon changes in the silanol coordination to a Cr(II) site to those observed,[65] we deduce that the coordinating silanol is limited to an O···Cr distance of about 4 Å.[3] The silanol moiety is not displaced until the second CO equivalent is

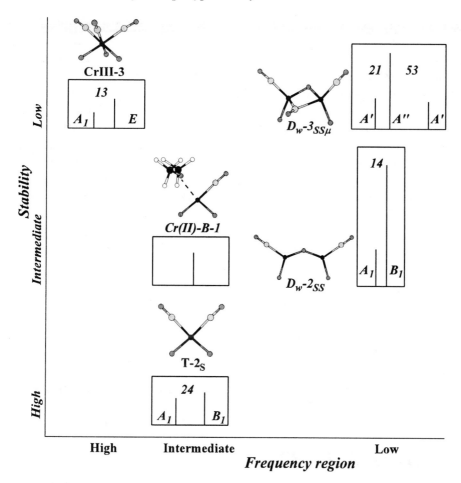

Figure 2. Schematic overview of carbonyl-chromium complexes *vs.* carbonyl stretching frequency and coordination energy. The bar diagrams associated with each carbonyl complex show the splitting pattern of the carbonyl stretching frequencies (in cm^{-1}) and relative infrared intensities, obtained from BP86 density functional calculations. The elements are coded on a grey scale according to increasing atomic number: H(white) > C > O > Si > Cr (dark gray). Only the top section of each carbonyl complex is shown

coordinated, and the monocarbonyl species is found to contribute in the region of the asymmetric **T-2** mode.

New bands develop at lower frequencies in the room temperature spectra with increasing CO pressure.[61,64] However, these bands are best seen in the low temperature spectra, where a band at 2100 cm^{-1} appears first.[64] A dicarbonyl complex with one terminal carbonyl coordination at each chromium center of the dinuclear site (see **D$_w$-2$_{SS}$** in Figure 2) gives rise to an intense asymmetric B$_1$ stretching mode, which is assigned to the band observed at 2100 cm^{-1}.[64] The significantly less intense and blue-shifted symmetric stretching mode is assigned as the band observed at 2120 cm^{-1}.[64] Carbonyl complexes arising from a third terminal coordination at a dinuclear site are unable to account for the observed evolution of the low temperature triplet. However, a tricarbonyl complex may be formed at a dinuclear site in yet another way, namely by coordination of the third CO in a bridging position, as exemplified by D_W-$3_{SS\mu}$ in Figure 2. This structure is only predicted by BP86. Compared to the antisymmetric stretching mode involving the terminal CO ligands, the vibrational frequency of the bridging CO is strongly red-shifted and assigned to the band observed at 2035-2045 cm^{-1}. Moreover, the entropy loss associated with CO adsorption in a bridging position is computed to be 40% larger than that of a terminal position, suggesting an explanation as to why these bands are best studied at low temperatures.

Finally, the high-frequency region was found to be dominated by carbonyl complexes of the **CrIII** cluster. The coordination energies for the first three adsorptions are very similar and considerably lower than for terminal adsorption at Cr(II) sites. This indicates that saturation is reached over a narrow range at high CO pressure.

A consequence of our interpretation of these spectra is that both mononuclear and dinuclear Cr(II) species exist in appreciable amounts in reduced Cr/silica systems, together with small amounts of mononuclear Cr(III) species. Furthermore, the population of mononuclear Cr(II) sites with nondisplaceable oxygen donors most likely exceeds the population of less saturated Cr(II) species, including both unsaturated chromium sites and those with displaceable oxygen coordination.

7.   CATALYTIC ACTIVITY

A number of proposals have appeared in the literature concerning possible polymerization-active sites on Phillips catalysts. Since these catalysts do not require an alkylating agent to be activated, it has been suggested that hydrogen supplied from a surface silanol group may serve to start the polymer chain.[66,67] However, the observed inverse correlation between hydroxyl population and activity suggest that surface hydroxyls may inhibit polymerization by detrimental coordination to active sites.[68] The fact that excellent catalysts may be obtained with catalysts dehydroxylated by chemical means, *e.g.*, by fluorination, led to the suggestion that the starting structures for polymerization evolve from a reaction between ethylene and the divalent chromium species.[49] Of these, the coordinatively unsaturated mononuclear Cr(II)-A type sites are claimed to be the more active for polymerization.[49,63,69] However, dinuclear Cr(II) sites have also been proposed to be active.[61]

This section presents a theoretical investigation of how polymerization may be initiated and sustained upon contact with ethylene at divalent chromium species on reduced Cr/silica Phillips-type catalysts. The density functional method BP86 was used exclusively.

In the selection of which chromium sites to examine, we rely on conclusions presented in the preceding sections concerning the population of chromium sites in the reduced Cr/silica system. Hence, reactive interactions between ethylene and chromium at pseudo-tetrahedral mononuclear Cr(II)-A sites are investigated first.[4] Next, the possibility of active sites arising through reactive interactions involving silanols at mononuclear Cr(II)-B sites is examined.[5] Finally, the catalytic properties of an unsaturated dinuclear Cr(II) site are explored.[6] Our choice of focus is supported by a comparison made by Rebenstorf[55] of the carbonyl region of IR spectra recorded from catalyst samples exposed to CO before and after short-time polymerization. He observed changes in the bands that we ascribe to Cr(II)-A sites and Cr(II)-B sites with a displaceable donor, and dinuclear Cr(II) sites, but none in the band at 2184 cm^{-1} which we assign to CO at Cr(II)-B sites with a nondisplaceable donor.

7.1 Reactions between Mononuclear Cr(II)-A Sites and Ethylene

Mononuclear Cr(II)-A species are commonly held to be active for polymerization,[49,63,69] and form the natural starting point for an investigation of polymerization activity in the Phillips system. The energy data presented here are from calculations on the tetrahedral cluster model terminated by hydrogen atoms.[4]

Mononuclear Cr(II)-A species with an $\angle$OCrO bond angle in the tetrahedral range may adsorb one or two equivalents of ethylene. In each case, ethylene coordinates *trans* to an ester oxygen ligand, as in structures **1a** and **1e** in Figure 3, giving rise in the latter to a square planar diethylene complex. The coordination is characterized by donation from ethylene to chromium. Taking the loss of entropy into account, it appears that both mono- and diethylene complexes are present in appreciable amounts. Only the monoethylene complex is found to undergo transformation to a covalently bound complex, structure **1b** in Figure 3, in which the donor bond is supplemented by back-donation from Cr 3d to the π^* orbital of the olefin. The transformation is close to isoenergetic and implies that chromium formally becomes oxidized to Cr(IV). **1b** is best described as a trigonal planar complex. A second ethylene may attack the exposed axial position to give a diethylene complex with a skew tetrahedral configuration, structure **1c** in Figure 3. Chromacyclopentane, shown as structure **1d** in Figure 3, may arise through the formation of a carbon-carbon bond between the two ethylene units. The transition state (see **TS[1c-1d]** in Figure 3) is positioned at a forming C-C bond length of 2.08 Å, and constitutes a very modest barrier. Structure **1d** is stable by 186 kJ/mol relative to free ethylene and the tetrahedral Cr(II) cluster,

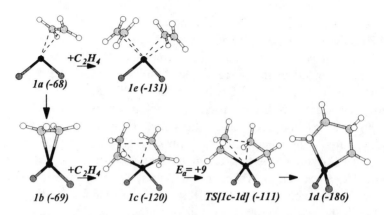

Figure 3. Initation reactions at a mononuclear Cr(II)-A site. Electronic energies relative to the separated cluster and ethylene molecule(s) are given in parentheses (kJ/mol). E_a denotes an activation energy. The elements are coded on a grey scale according to increasing atomic number: H(white) > C > O > Si > Cr (dark gray). Only the top section of each cluster is shown.

and displays a distorted tetrahedral configuration around chromium, with $\angle$ CCrC and $\angle$ OCrO bond angles of 88° and 112°, respectively.

Despite the acute $\angle$ CCrC bond angle, **1d** displays no affinity for ethylene, and the energy barrier for direct insertion of ethylene into one of the Cr-C bonds in **1d** was computed to be 119 kJ/mol. Rearrangement of **1d** to a covalently coordinated 1-butene may be achieved through intramolecular β-H transfer to the α-carbon at the far end of the carbon chain. The transition state takes the form of a four-membered pseudo-ring structure, whose energy is 192 kJ/mol above that of **1d**. Rearrangement to a linear carbene species, butylidenechromium, through α-H transfer reaction between the two carbon atoms bonded to chromium, is another possibility that meets with a prohibitively high energy barrier.

Rather than forming the chromacyclopentane structure, ethylene may add oxidatively to the Cr(II)-A site to give an ethenylhydrido species with chromium in oxidation state IV.[4] Insertion of the first ethylene into the chromium-hydride bond is predicted to be facile, and the next insertion presumably takes place in the chromium-ethenyl bond, resulting in a but-3-enylethylchromium(IV) product. There is no tendency for the alkenyl to act as a bidentate ligand, and this structure is therefore representative of a tetrahedral dialkylchromium(IV) species. There is no driving force for coordinating ethylene to this structure, and direct insertion of ethylene into the chromium-ethyl bond requires an activation energy of 113 kJ/mol. This is considered inconsistent with high catalytic activity towards polymerization, and is in agreement with the lack of activity recently reported by Amor Nait Ajjou *et al.* for a dialkylchromium species anchored to silica by two oxygen ester linkages.[70,71] After heating at 70°C for a significant period of time

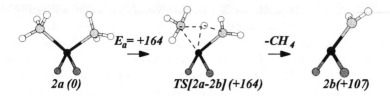

2a (0)   TS[2a-2b] (+164)   2b(+107)

Figure 4. Conversion of dimethylchromium to methylidenechromium by concerted intramolecular α-H elimination. Electronic energies relative to the dimethylchromium species are given in parentheses (kJ/mol). E_a denotes an activation energy. The elements are coded on a grey scale according to increasing atomic number: H(white) > C > O > Si > Cr (dark gray). Only the top section of each cluster is shown.

(~4 hours), the bis(neopentyl)chromium structure was converted with a stoichiometry in accordance with transformation to a carbenechromium surface species.[71,72] The reaction is expected to proceed by intramolecular α-H elimination.[71,72]

We compute a high reaction barrier of 164 kJ/mol for intramolecular α-H elimination in a model dimethylchromium structure,[4] Figure 4. The suggestion by Amor Nait Ajjou *et al.* that coordinating oxygen atoms from nearby siloxane bridges may aid the reaction[72] gains support in simple test calculations with water as the donor.[4] Taking into account the entropy gain associated with liberation of gaseous alkanes at 70°C, the proposed conversion to a carbenechromium structure appears viable. Rather than generating a carbenechromium(IV) structure from a dialkylchromium intermediate, it is conceivable that ethylene reacts directly with the Cr(II)-A site to give an ethylidenechromium species.[73] The overall reaction is slightly exothermic.[4]

Chain propagation at a carbenechromium surface species has been proposed to proceed by alternating alkylidene/chromacyclobutane intermediates.[73] Starting from an ethylidenechromium species, we find ethylene to participate in a non-activated and strongly exothermic [2+2] cycloaddition reaction, resulting in a chromacyclo(methyl-)butane product. In order to regenerate the alkylidene functionality, intramolecular α-H transfer is proposed to occur, either as a concerted step or involving a hydridochromium intermediate. Either way, the activation energy turns out to be prohibitively high (>200 kJ/mol), and regeneration of the alkylidene species appears unlikely.

7.2 Reactions between Mononuclear Cr(II)-B Sites and Ethylene

In conventionally prepared catalysts, the calcination step causes dehydroxylation of the silica surface.[68] However, a residual population of hydroxyls is present at the surface even at the highest temperatures used in the activation procedure.[68] In principle, a surface silanol may act to either protonate or to hydrogenate a nearby chromium site. Insertion of ethylene into the Cr-H bond of a cationic Cr(IV) cluster model has been computed to proceed with a low activation energy barrier.[74] However, deprotonation of a silanol group requires some 1400 kJ/mol,[7] compared to the proton affinity of 664 kJ/mol computed for the **T** cluster model. In order for proton transfer to become energetically feasible, it appears that significant interaction within the ion pair is required.

With this background, we explore if and how a coordinating silanol moiety may interact reactively at a mononuclear Cr(II)-B site, to activate chromium for polymerization. The pseudo-tetrahedral mononuclear Cr(II) cluster model was extended to include a silanol moiety close to the chromium centre, structure **3a** in Figure 5. Starting from the monoethylene complex, proton transfer to ethylene from the coordinating silanol moiety occurs with a modest activation energy of 54 kJ/mol. The oxygen anion coordinates closely to chromium to give a four-coordinate alkylchromium(IV) product, implying that the reaction is effectively one of hydrogen transfer rather than proton transfer. Thus, in contrast to what was found in the previous section for Cr(II)-A sites, at monochromium B-type sites, the formation of inert chromacycloalkane species may be avoided by a competitive reaction with the coordinating silanol group. While hydrogen transfer to a coordinated ethylene appears viable in conventionally prepared catalysts,[49] in Amor Nait Ajjou *et al.*'s model catalyst,[71,72] hydrogen transfer to a carbenechromium species appears more likely. We find hydrogen transfer to an alkylidenechromium structure to proceed with a similar barrier and considerably higher exothermicity[5] than the reaction shown in Figure 5.

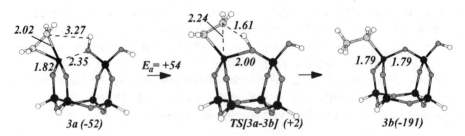

Figure 5. Fully relaxed reaction paths for hydrogen transfer to a chromium(IV)-ethylene complex of C_s symmetry. Electronic energies relative to the separated cluster and ethylene molecule(s) are given in parentheses (kJ/mol), and energy barriers are denoted by E_a. Bond lengths are included in Å. The elements are coded on a gray scale according to increasing atomic number: H(white) > C > O > Si > Cr (dark gray). Adapted with permission from *J. Catal.* **2002**, *205*, 366. Copyright 2002 Elsevier Science (USA).

Figure 6. Models of a difluorosilanol molecule coordinating to a pseudo-tetrahedral cluster with chromium in the divalent (left) (**T**) or hexavalent (right) state (**T**(oxo)$_2$). The elements are coded on a gray scale according to increasing atomic number: H (white) < O < F < Si < Cr (dark gray). Reproduced wither permission from *J. Catal.* **2002**, *205*, 177. Copyright 2002 Elsevier Science (USA).

Fully relaxed, the resulting monoalkyl structure **3b** has a tetrahedral ligand arrangement and does not show affinity toward ethylene. However, with three of the ligands bound to the surface, it is likely that surface strain keeps the ligand arrangement at the monoalkychromium(IV) site from relaxing freely upon hydrogenation. Further studies between silanols and the pseudo-tetrahedral monochromium site were perfomed by means of the cluster models shown in Figure 6, subject to the following geometric constraints: (1) the bottom O(SiF$_2$)$_2$ part of the cluster was frozen; and (2) the SiF$_2$ part of the difluorosilanol molecule was oriented parallel to the SiOSi moiety at the bottom of the cluster, at distances ranging from 4.5 to 6.0 Å. These constraints effectively reduce the flexibility of both the silanol molecule and the OCrO moiety. At a distance of 5.5 Å, the resulting model reproduces the observed frequency shift of -42 cm^{-1}, assigned to hydroxyls coordinating to reduced chromium surface species by Nishimura and Thomas.[65]

Taking into account geometric relaxation following hydrogenation by decreasing the distance constraint by 0.5 Å, the hydrogen transfer reaction at a triplet-coupled ethylene-Cr(IV) complex turns out to be close to isoenergetic. For comparison, hydrogenation of the bare Cr(II)-B site to give a hydridochromium(IV) species is endothermic by 100 kJ/mol, while hydrogenation of an ethylidenechromium(IV) reactant turns out to be exothermic by ~50 kJ/mol under these conditions. A modest distortion of the resulting ethylchromium(IV) structure toward a square-pyramidal geometry with a vacant site is necessary in order to introduce affinity to ethylene, as shown by the values of ∠OCrO listed for **4a** in Table 3.

Insertion of ethylene into the Cr-ethyl bond in **4a** is presented in Figure 7, with selected geometry parameters given in Table 3. The insertion is assisted by an α-agostic interaction at the transition state. Along the path of insertion, the weakly-bound oxygen ligand (O$_1$) is temporarily displaced by the incoming ethylene molecule, Figure 7 and Table 3. The strength of the ethylene-chromium

Table 3. Geometry parameters of stationary points along the reaction path of ethylene insertion into the Cr-ethyl bond in the ethylchromium(IV) species **4a**. **4b**: π-complex; **TS[4b-4c]**: transition state; **4c**: butylchromium.[a]

	4a	4b	TS[4b-4c]	4c
$rCrC_1$	-	2.81	2.11	2.00
$rCrC_2$	-	2.80	2.34	3.01
$rCrC_3$	2.00	2.02	2.14	4.38
rC_1C_2	$(1.33)^b$	1.34	1.41	1.51
rC_2C_3	-	3.38	2.26	1.54
$rCrO_1$	1.99	2.03	2.21	1.98
$rCrO_2$	1.80	1.82	1.82	1.80
$rCrO_3$	1.91	1.94	1.91	1.90
$\angle O_1CrO_3$	153	161	169	154
$\angle O_2CrO_3$	109	105	106	108
$\angle O_1CrO_2$	93	91	85	93

[a] Units: bond lengths (r) in Å, angles (∠) in degrees. [b] Free monomer.
Reprinted with permission from *J. Catal.* **2002**, *205*, 366. Copyright 2002, Elsevier Science (USA).

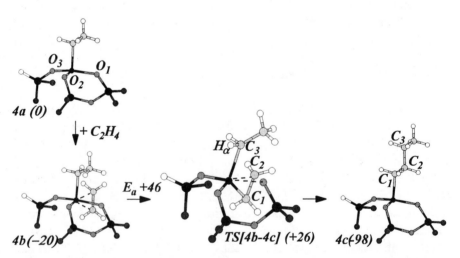

Figure 7. Stationary points along the reaction path of ethylene insertion into a Cr-ethyl bond in an ethylchromium(IV) species. Electronic energies relative to the separated cluster and ethylene molecule(s) are given in parentheses (kJ/mol), and the energy barrier is denoted by E_a. The elements are coded on a gray scale according to increasing atomic number, H(white) < O < Si < Cr(dark gray). Adapted with permission from *J. Catal.* **2002**, *205*, 366. Copyright 2002, Elsevier Science (USA).

coordination bond is too low to make the ethylene complex **4b** a realistic resting state in this system. The contributions from coordination and insertion, $\Delta H_\pi = -20$ kJ/mol and $\Delta H^\ddagger = +46$ kJ/mol, respectively, amount to an apparent activation energy of $\Delta H_{app} = 26$ kJ/mol, which agrees nicely with that reported for Scott and Amor Nait Ajjou's model system ($\Delta H_{obs} = 30.2 \pm 0.9$ kJ/mol).[69] Moreover, our model also contains mononuclear chromium in oxidation state IV also after polymerization, and it supports a propagation mechanism that proceeds by insertion of the monomer *via* an α-agostic transition state. These features agree with properties deduced from the experimental study by Scott and Amor Nait Ajjou.[69,75] The low stability of the hydridochromium(IV) structure implies that chain termination is not likely to occur through transfer of hydrogen to the metal, in agreement with the observed lack of influence of ethylene pressure on the chain length of the produced polymer.[76]

How does the concept of an active site presented here agree with the observed properties of Phillips-type catalysts? First, in light of the rather low complexation energy of ethylene at the activated ethylchromium(IV), the presence of additional silanol groups close to the vacant site may be enough to prevent ethylene coordination and thus also chain propagation. Consistent with this perspective, it is important to reduce the silanol population, as observed during heating.[68] Second, dehydroxylation increases the surface strain[77] and possibly also the acidity of the remaining silanol groups.[78] This may be beneficial, since we have seen that the catalytic activity of the Cr(II)-B sites depends on geometric constraints that affect the orientation of the silanoic group at the hydrogenated site, as well as the strength of the chromium-oxygen bonds, to allow displacement of the weakest-bound oxygen ligand during insertion. Third, dehydroxylation results in strained siloxane defects which are capable of dissociative chemisorption of molecules.[77] The presence of H_2 in the feed increases activity, while there is no evidence of hydrogenation of the polymer. Thus the strained siloxane defects may dissociatively chemisorb H_2 to give $\equiv$SiOH and $\equiv$SiH,[79] and subsequently activate suitably positioned Cr(II) sites. Alternatively, hydrogen may react with inactive chromacyclobutane and -pentane species, to reduce the metal with release of alkanes and consequent reactivation.

7.3 Dinuclear Cr(II) Sites

Complete removal of surface hydroxyls, *e.g.*, by fluorination, still allows for highly active Phillips-type catalysts.[68] Thus, it appears that a starting structure for polymerization may evolve in a reaction between ethylene and chromium of the reduced catalyst.[49] This view gains support in a recent study by Ruddick and Badyal, where trimerization of ethylene was found to proceed without incorporation of hydrogens external to the feed.[80] Rebenstorf suggested that the growing hydrocarbon chain may be suspended between two chromium centres at a dinuclear site,[81] leaving each chromium in a monoalkyl configuration while alleviating the need for additional hydrogen atoms. This suggestion is investigated in the following section and, in more detail, in Ref. 6. Due to limitations implicit in a single-configurational description of the electronic structure, for some of the steps postulated for the initiation phase we have not been able to determine the activation energy.

The original suggestion by Rebenstorf [81] involves a μ-ethano-μ-oxo-dichromium(III) starting structure for polymerization. Our calculations suggest that this species has a drive toward adsorption of an additional equivalent of ethylene, and to form a chromacyclopentane species. However, contrary to what is the case at the monochromium A-type sites, dinuclear sites may undergo subsequent conversion to a μ-butano-μ-oxo-dichromium(III) structure. Subsequent insertion of ethylene into the adjacent Cr-C bond takes place with a low energy barrier of 55 kJ/mol, to give a μ-hexano product structure in an exothermic reaction.

Formation of 1-hexene from the μ-hexano structure is examined along the path of β-hydrogen transfer, starting with the reactant in a pronounced β-agostic conformation. This structure is included as **5a** in Figure 8 and is only +21 kJ/mol less stable than the global minimum conformation. The reaction path corresponds to concerted β-hydrogen transfer, where H_β is transferred from C_β to $C_{\alpha2}$, Figure 8. As the C_β-H_β bond is stretched, C_β coordinates to $Cr^{(2)}$ and a double bond starts to develop between C_α and C_β. Prior to the transition state, the budding vinyl moiety binds covalently to both chromium atoms. In the transition state **TS[5a-5b]**, a six-membered pseudo-ring is formed by the catalytic chromium atom

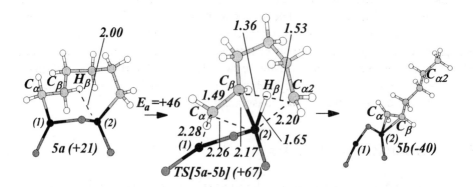

Figure 8. Concerted β-H transfer reaction in a μ-hexano ligand suspended between the chromium atoms in the D_w cluster model, with reactant **5a**, transition state **TS[5a-5b]**, and product **5b**. The elements are coded on a gray scale according to increasing atomic number, H(white) < O < Si < Cr(dark gray). Only the top section of each cluster is shown. Adapted with permission from *J. Catal.* **2002**, *206*, 331, Copyright 2002, Elsevier Science (USA).

together with most of the carbon backbone, leaving out only the $OCr^{(1)}C_\alpha$ moiety. From the **5b** product, 1-hexene may be displaced to the gas phase by an incoming ethylene monomer. While the transition state, **TS[5a-5b]**, constitutes an energy barrier of only 46 kJ/mol relative to **5a**, it is important to realize that this is due to a favorable difference in ring strain between that of the nine-membered ring in the reactant structure, and a six-membered pseudo-ring formed at the transition state By extension, based on ring strain in cycloalkanes,[82] 1-hexene seems to be the only oligomerization product that may form at this dinuclear site. Moreover, the 1-hexene thus produced does not contain hydrogen from sources other than ethylene. These predictions are in line with the observations made by Ruddick and Badyal.[80]

Returning to the issue of chain propagation, the energetics of the insertion reaction depend on the length of the μ-alkano ligand, again with ring strain as a main factor. In the case of a μ-hexano ligand, the activation energy for monomer insertion into the Cr-C bond is 39 kJ/mol, computed relative to the corresponding ethylene-chromium complex. Since the apparent barrier for β-hydrogen transfer in this structure is computed to be 67 kJ/mol, chain growth is favored over the formation of 1-hexene.

As a limiting case, we consider ethylene insertion into a Cr-C bond in a μ-oxo-di(ethylchromium) structure, **6a** in Figure 9. The formation of a molecular ethylene-chromium π-complex (**6b** in Figure 9) is exothermic by 45 kJ/mol, with ethylene coordinating at chromium-carbon bond distances of 2.25 and 2.51 Å and with the ethylene double bond stretched to 1.42 Å. The insertion is assisted by an α-agostic interaction. Relative to the π-complex, the activation energy is computed to 55 kJ/mol, which is considered to be consistent with catalytic activity.

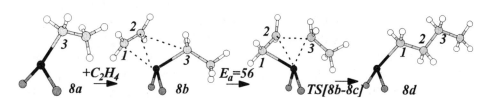

Figure 9. Insertion of ethylene into a Cr-C single bond in di(ethylchromium) structure **6a**, with π complex **6b**, transition state **TS[6b-6c]**, and product **6c**. The elements are coded on a gray scale according to increasing atomic number, H(white) < O < Si < Cr(dark gray). The reaction was modelled at the D_w cluster, of which only the top section is shown. Adapted with permission from *J. Catal.* **2002**, *206*, 331, Copyright 2002, Elsevier Science (USA).

8.   CONCLUSION

The aim in this work has been to provide molecular-level insight into Cr/silica Phillips-type catalysts for ethylene polymerization. To this end, cluster models were constructed to represent mononuclear and dinuclear chromium(II) and mononuclear chromium(III) sites in the reduced Cr/silica system. In combination with state-of-the-art quantum mechanical methods, these cluster models support a consistent interpretation of (1) the observed d-d transitions for the reduced Cr/silica system; and (2) the observed carbonyl IR spectra of the reduced Cr/silica system exposed to CO.

While abundantly present at the reduced Cr/silica surface, coordinatively unsaturated, mononuclear chromium(II)-A sites lead to stable and catalytically inactive structures when brought into contact with ethylene. However, two of the other chromium sites investigated were found to support catalytic polymerization of ethylene. These are monochromium(II) sites with a coordinating silanol moiety, and dichromium(II) sites. This result is in accordance with McDaniel and Welch's conclusion[68] that the Phillips-type Cr/silica catalyst possesses two types of active sites, producing high and low molecular weight (MW) polymers.[68] Since dinuclear sites may be activated without surface silanol groups, we expect these sites to be responsible for the observed activity in catalysts where all hydroxyl groups have been removed, such as in fluorinated catalysts.[68] Since the latter system produces high MW polymer, it appears that dinuclear sites may be responsible for the high MW fraction from Phillips catalysts prepared in the conventional manner.[68] The Cr(II)-B sites investigated here depend upon a hydrogen source to become activated, and may be responsible for the production of low MW polymer.[68]

Since the monoalkylchromium(IV) site carries a linear polymer chain, termination by β-hydrogen transfer to monomer leads to linear, even-numbered α-olefins. Such a prolonged series of 1-alkenes is observed for the Phillips catalyst[83] as well as for the model system of Scott and Amor Nait Ajjou, which displays only mononuclear sites.[69] The dinuclear sites are predicted to form 1-hexene selectively during the early stages of catalysis. This is in line with the observations made by Ruddick and Badyal,[80] and may be understood in light of the present results if only dinuclear sites were activated in their catalyst.

Our assignment of the carbonyl region of the IR spectrum of the reduced Cr/silica catalyst and the identification of active sties find mutual support in carbonyl IR spectra recorded before and after short-time polymerization.[55] The peaks that suffer a decrease in intensity after short-time polymerization are indeed those that we assign to the chromium sites which are also proposed to be active toward polymerization.

Based on the chromium/silica surface sites identified herein, it may be possible in future work to use structural information from silicalite and cristobalite to improve the cluster models by taking into account embedding effects. By combining theoretical investigations and ambitious, well-defined experiments, one may get closer to defining the chemical species that make the Phillips catalyst highly active for polymerization. With this information at hand, one may eventually realize the strengths of Phillips catalysts within the concept of single-site catalysts.

ACKNOWLEDGEMENTS

The Research Council of Norway supported this research through grants of computing time (Programme for Supercomputing). Para//ab High Performance Computing Centre, Bergen, Norway, is thanked for extensive amounts of computer time.

REFERENCES

1. Hogan, J. P.; Banks, R. L. (Phillips Petroleum Company, Belgian Pat. 530,617(1995); U.S. Patents 2,825,721 (1958); 2,846,425(1958); 2,951,816(1960).
2. Espelid, Ø.; Børve, K. J. *Catal. Lett.* **2001**, *75*, 49.
3. Espelid, Ø.; Børve, K. J. *J. Catal.* **2002**, *205*, 366.
4. Espelid, Ø.; Børve, K. J. *J. Catal.* **2001**, *195*, 125.
5. Espelid, Ø.; Børve, K. J. *J. Catal.* **2002**, *205*, 177.
6. Espelid, Ø.; Børve, K. J. *J. Catal.* **2002**, *206*, 331.
7. Sauer, J.; Ugliengo, P.; Garrone, E.; Saunders, V. R. *Chem. Rev.* **1994**, *94*, 2095.
8. Duffy, J. A. *Bonding, Energy Levels and Bands in Inorganic Solids*, Longman: New York, 1990.
9. Sauer, J. *Chem. Rev.* **1989**, *89*, 199.
10. Damrauer, R.; Simon, R.; Krempp, M. *J. Am. Chem. Soc.* **1991**, *113*, 4432.
11. Paukshtis, E. A.; Soltanov, R. I.; Yurchenko, E. N. *React. Kinet. Catal. Lett.* **1981**, *16*, 93.
12. Sauer, J.; Hill, J.-R. *Chem. Phys. Lett.* **1994**, *218*, 333.
13. Civalleri, B.; Garrone, E.; Ugliengo, P. *Chem. Phys. Lett.* **1998**, *294*, 103.
14. Pacchioni, G. *Heterogeneous Chem. Rev.* **1995**, *2*, 213.
15. Sauer, J.; Sierka, M. *Comput. Chem.* **2000**, *21*, 1470.
16. Pisani, C.; Dovesi, R.; Nada, R.; Kantorovich, L. N. *J. Chem. Phys.* **1990**, *92*, 7448.
17. Pisani, C.; Birkenheuer, U. *Comput. Phys. Commun.* **1996**, *96*, 221.
18. Pisani, C.; Birkenheuer, U.; Casassa, S.; Corà, F. *EMBED01 Users Manual*, University of Torino: Torino, 1999.
19. Pisani, C.; Dovesi, R.; Roetti, C. "Hartree-Fock ab-initio Treatment of Cystalline Systems". In *Lecture Notes in Chemistry, Vol. 48*, Springer Verlag: Heidelberg, 1998.
20. Saunders, V. R.; Dovesi, R.; Roetti, C.; Causá, M.; Harrison, N.; Orlando, R.; Zicovich-Wilson, C. *CRYSTAL 98 User's Manual*, University of Torino: Torino, 1988.
21. Børve, K. J. *J. Chem. Phys.* **1992**, *96*, 6281.
22. Huzinaga, S.; Seijo, L.; Barandiaran, Z.; Klobukowski, M. *J. Chem. Phys.* **1987**, *86*, 2132.
23. Barandiaran, Z.; Seijo, L. *J. Chem. Phys.* **1988**, *89*, 5739.

24.   Harding, J.; Harker, A.; Keegstra, P.; Pandey, R.; Vail, J.; Woodward, C. *Physica* **1985**, *131B*, 151.
25.   Vail, J.; Pandey, R.; Kunz, A. *Rev. Solid State Sci.* **1991**, *5*, 241.
26.   Veszprémi, T.; Fehér, M. *Quantum Chemistry*, Kluwer: Dordrecht, 1999.
27.   Jensen F. *Introduction to Computational Chemistry*, Wiley: Chichester, 1999.
28.   Malmqvist, P. -Å. *"Molecules in the Stellar Environment"* in *Lecture Notes in Physics*, Vol. 428, Jørgensen, U. G., Ed., Springer-Verlag: New York, 1994.
29.   Roos, B. O. In *Ab initio Methods in Quantum Chemistry*, Lawley K. P., Ed. Wiley: New York, 1987.
30.   Cizek, J. *J. Chem. Phys.* **1975**, *62*, 3258.
31.   Bartlett, R. J.; Silver, D. M. *J. Chem. Phys.* **1975**, *62*, 3258.
32.   Møller, C.; Plesset, M. S. *Phys. Rev.* **1934**, *45*, 618.
33.   Roos, B. O.; Andersson, K.; Fülscher, M. P.; Malmqvist, P.-Å.; Serrano-Andrés, L.; Pierloot, K.; Merchán, M. *Adv. Chem. Phys.* **1996**, *93*, 216.
34.   Siegbajn, P. E. M.; Bloomberg, M. R.; Svensson, M. *Chem. Phys. Lett.* **1994**, *223*, 35.
35.   Curtiss, L. A.; Raghavachari, K.; Pople, J. A. *J. Chem. Phys.* **1993**, *98*, 1293.
36.   Ochterski, J. W.; Petersson, G. A.; Montgomery, J. A. *J. Chem Phys.* **1996**, *104*, 2598.
37.   Hohenberg, P.; Kohn, W. *Phys. Rev. B* **1964**, *136*, 864.
38.   Kohn, W.; Sham, L. *Phys. Rev. A* **1965**, *140*, 1133.
39.   Becke, A. D. *Phys. Rev. A* **1988**, *38*, 3098.
40.   Perdew, J. P. *Phys. Rev. B* **1986**, *33*, 8822.
41.   Stevens, P. J.; Devlin, F. J.; Chablowski, C. F.; Frisch, M. J. *J. Phys. Chem. A* **1994**, *98*, 11623.
42.   Espelid, Ø.; Børve, K. J.; Jensen, V. R. *J. Phys. Chem. A* **1998**, *102*, 10414.
43.   Cossee, P. *J. Catal.* **1964**, *3*, 80.
44.   Jensen, V. R.; Børve, K. J. *Organometallics* **1997**, *16*, 2514.
45.   Jensen, V. R.; Børve, K. J. *J. Comput. Chem.* **1998**, *19*, 947.
46.   Weckhuysen, B. M.; Wachs, I. E.; Schoonheydt, R. A. *Chem. Rev.* **1996**, *96*, 3327.
47.   Baerends, E. J. *et al. ADF 2000.02 Computer Code*, 2000.
48.   Guerra, C. F.; Snijders, J. G.; te Velde, G.; Baerends, E. J. *Theor. Chem. Acc.* **1998**, *99*, 381.
49.   McDaniel, M. P. *Adv. Catal.* **1985**, *33*, 47.
50.   Thüne, P. C.; Linke, R.; van Gennip, W. J. H.; de Jong, A. M.; Niemantsverdriet, J. W.; *J. Phys. Chem. B.* **2001**, *105*, 3073.
51.   Baker, L. M.; Carrick, W. L. *J. Org. Chem.* **1968**, *33*, 616.
52.   Krauss, H.-L.; Stach, H. Z. *Anorg. Allg. Chem.* **1969**, *366*, 280.
53.   Fubini, B.; Ghiotti, G.; Stradella, L.; Garrone, E.; Morterra, C. *J. Catal.* **1980**, *66*, 200.
54.   Weckjuysen, B. M.; De Ridder, L. M.; Schoonheydt, R. A. *J. Phys. Chem.* **1993**, *97*, 4756.
55.   Rebenstorf, B. Z. *Anorg. Allg. Chem.* **1989**, *571*, 148.
56.   Weckhuysen, B. M.; Schoonheydt, R. A.; Jehng, J.l Wachs, I. E.; Cho, J.; Ryoo, R.; Kijlstra, S.; Poels, E. *J. Chem. Soc., Faraday Trans.* **1995**, *91*, 3245.
57.   Zecchina, A.; Garrone, E.; Ghiotti, G.; Morterra, C.; Borello, E. *J. Phys. Chem.* **1975**, *79*, 996.
58.   Ghiotti, G.; Garrone, E.; Gatta, G. D.; Fubini, B.; Giamello, E. *J. Catal.* **1983**, *80*, 249.
59.   Krauss, H.-L.; Rebenstorf, B.; Westphal, U. *Z. Anorg. Allg. Chem.* **1975**, *414*, 97.
60.   Niemantsverdriet, J. W. *Spectroscopy in Catalysis*. VCH: Weinheim, 1995.
61.   Rebenstorf, B. *J. Polym. Sci., Polym. Chem.* **1991**, *29*, 1949.
62.   Koher, S. D.; Ekerdt, J. G. *J. Phys. Chem.* **1994**, *98*, 4336.
63.   Ghiotti, G.; Garrone, E.; Zecchina, A. *J. Mol. Catal.* **1988**, *46*, 61.
64.   Zecchina, A.; Spoto, G.; Ghiotti, G.; Garrone, E. *J. Mol. Catal.* **1994**, *86*, 423.

65. Thomas, J. M.; Thomas, W. J. *Principle and Practice of Heterogenous Catalysis*, VCH: Weinheim, 1997.
66. Grovenevald. G.; Wittgen, P. P. M.; Swinnen, H. P. M.; Wernsen, A.; Schuit, G. C. A. *J. Catal.* **1983**, *83*, 346.
67. Jozwiak, W. K.; Dalla Lana, I. G.; Fiederow, R. *J. Catal.* **1990**, *121*, 182.
68. McDaniel, M. P.; Welch, M. B. *J. Catal.* **1983**, *82*, 53.
69. Scott, S. L.; Amor Nait Ajjou, J. *Chem Eng. Sci.* **2001**, *56*, 4155.
70. Amor Nait Ajjou, J.; Scott, S. L. *Organometallics* **1997**, *16*, 86.
71. Amor Nait Ajjou, J.; Scott, S. L.; Paquet, V. *J. Am. Chem. Soc.* **1998**, *120*, 415.
72. Amor Nait Ajjou, J.; Rice, G. L.; Scott, S. L. *J. Am. Chem Soc.* **1998**, *120*, 13468.
73. Kantcheva, M.; Dalla Lana, I. G.; Szymura, J. A. *J. Catal.* **1995**, *154*, 329.
74. Schmid, R.; Ziegler, T. *Can. J. Chem.* **2000**, *78*, 265.
75. Amor Nait Ajjou, J.; Scott, S. L. *J. Am. Chem. Soc.* **2000**, *122*, 8968.
76. Blom, R.; Follestad, A.; Noel, O. *J. Mol. Catal.* **1994**, *91*, 237.
77. Chuang, I.; Maciel, G. E. *J. Phys. Chem. B* **1997**, *101*, 3052.
78. Morrow, B. A.; McFarlen, A. *J. Non-Cryst. Solids* **1990**, *120*, 61.
79. Ferrari, A. M.; Garrone, E.; Spotto, G.; Ugliengo, P.; Zecchina, A. *Surf. Sci.* **1995**, *323*, 151.
80. Ruddick, V. J.; Badyal, J. P. S. *J. Phys. Chem. B* **1998**, *102*, 2991.
81. Rebenstorf, B.; Larsson, R. *J. Mol. Catal.* **1981**, *11*, 247.
82. Isaacs, N. S. *Physical Organic Chemistry*, Longman: Essex, 1995.
83. Krauss, H.-L.; Hums, E. *Z. Naturforsch.* **1979**, *34B*, 1628.
84. Spoto, G.; Bordiga, S.; Garrone, E.; Ghiotti, G.; Zecchina, A. *J. Mol. Catal.* **1992**, *74*, 175.

Chapter 5

LATE TRANSITION METAL COMPLEXES IMMOBILIZED ON STRUCTURED SURFACES AS CATALYSTS FOR HYDROGENATION AND OXIDATION REACTIONS

Cathleen M. Crudden,* Daryl P. Allen, Irina Motorina, Meredith Fairgrieve
Department of Chemistry, University of New Brunswick, Fredericton, New Brunswick, E3B 6E2, Canada. Current address: Queen's University, Department of Chemistry, Kingston, Ontario K7L 3N6, Canada

Keywords: Asymmetric dihydroxylation, cinchona alkaloids, enantioselective catalysis, mesoporous, nanostructured, supported, reduction

Abstract: New advances in the preparation of catalysts on well-defined active sites are reviewed, including molecular imprinting, organic zeolites, chiral polymers and supported dendrimers. The effect of the support, the grafting technique and the choice of ligand are discussed with particular emphasis on hydrogenation and dihydroxylation catalysts. The use of highly ordered mesoporous materials such as MCM-41,[1] SBA-15,[2] or FSM-16[3] in catalysis is of particular importance.[4,5] Phosphine[6,7,8] and amine[9] ligands have been immobilized onto the surfaces of mesoporous silicates. The grafting of phosphines (chiral and achiral) for use in hydrogenation reactions is described in detail.[6,7] Similarly, chiral cinchona alkaloids grafted onto SBA-15 are shown to be very effective ligands for the osmium-catalyzed dihydroxylation of olefins.[10,11]

Nanostructured Catalysts, edited by S. Scott *et al.*
Kluwer Academic/Plenum Publishers, 2003

1.   INTRODUCTION

The development of homogeneous transition metal complexes as reagents for organic synthesis has altered the landscape of modern organic chemistry. Coupling reactions, hydrogenations, epoxidations, metathesis reactions and dihydroxylations are now commonplace in synthetic schemes.[12] In addition to their ability to carry out reactions that would require multiple steps using "traditional" reagents, transition metals can be modified by chiral ligands and used in enantioselective transformations. The great advantage of this approach is that only catalytic amounts of the chiral reagent are required. Using catalytic asymmetric hydrogenation and oxidation reactions, a wide variety of organic molecules can be produced with exquisite control of stereochemistry. Noyori, Sharpless and Knowles received the 2001 Nobel Prize in chemistry for their seminal contributions to this area of chemistry.

Despite the utility of asymmetric catalysis and the widespread application of transition metals in organic synthesis, asymmetric catalytic methods are not widely employed in industry. This is predominantly because viable methods for the recovery of the catalyst and its chiral ligands from the reaction medium are yet to be developed.[13] The cost of ligands for asymmetric catalysis is often extreme, and so the ability to reuse them is critical if reactions are to be performed on an industrial scale.

Asymmetric *heterogeneous* catalysis is an obvious alternative if catalyst recovery is an issue. Unfortunately, few heterogeneous catalysts provide high enough enantioselectivity to warrant their use. A notable exception is the hydrogenation of alpha keto-esters using cinchona-modified Pt on alumina.[14] As shown in eq. 1, this reaction gives ethyl lactate in high enantiomeric excess (ee).

Hybrid systems, in which molecular transition metal complexes are heterogenized by tethering them to an insoluble support, have been explored since the early days of asymmetric catalysis. In 1973, Kagan reported the first example of an effective asymmetric catalyst that was recoverable by virtue of being tethered to a polymeric support (eq. 2).[15] Bailar,[16] Grubbs,[17] Collman[18] and

$$Ph\text{-}C(=O)\text{-}CH_3 \xrightarrow[\text{HSiR}_3]{\text{PS-DIOP•[Rh(COD)Cl]}_2} Ph\text{-}CH(OSiR_3)\text{-}CH_3 \quad (2)$$

58% ee

PS-DIOP = (PS)—O—⟨CH₂PPh₂ / ''CH₂PPh₂⟩   (PS) = polystyrene

Capka[19] also reported heterogenized achiral transition metal catalysts in the early 1970's.[20]

Although supported asymmetric catalysts can be effective, these early studies showed that finding a technique to separate the catalyst and the ligand from the reaction mixture was only part of designing a recoverable catalyst. It was evident even in Kagan's study that a homogeneous transition metal complex tethered to a polymeric support was not necessarily an accurate representation of the same complex in solution. The effect of the micro-environment of the catalyst *outside* the immediate coordination sphere of the metal is significant. In the case of chiral catalysts employed in asymmetric transformations, the effect can be severe. Although decreases in activity and selectivity are the norm,[21] there are cases where the support can actually improve the activity of the tethered transition metal complex.[22] Hydrogenation and oxidation reactions are often better on supports since the catalysts benefit from site isolation,[23] although the nature of the support is critical to observe this effect.[24] The method of anchoring the catalyst is also an integral part of the microenvironment, and can affect the behavior of the catalyst. After discussing the support and the method of anchoring, specific examples of catalysts supported on mesoporous materials and used in reduction and dihydroxylation reactions will be discussed.

2. THE SUPPORT

2.1 Amorphous Organic Polymers

2.1.1 Partially Cross-Linked Polymers

Among polymeric organic supports, partially cross-linked polystyrene is the most common.[25] This support was used by Merrifield in his Nobel prize-winning work on the solid phase synthesis of peptides.[26] Soon after this seminal report, the attachment of transition metal complexes to polystyrene was investigated. In 1971, Grubbs reported the preparation of a polymeric version of Wilkinson's catalyst (eq. 3).[17] This catalyst was effective for the hydrogenation of simple alkenes at ambient temperature and pressure, but large cyclic olefins were significantly less reactive with the polymeric catalyst than with $ClRh(PPh_3)_3$

$$\text{(PS)}-CH_2Cl \xrightarrow[\text{24 hrs}]{\text{LiPPh}_2} \text{(PS)}-CH_2PPh_2 \xrightarrow[\text{2-4 weeks}]{\text{ClRh(PPh}_3)_3} \text{(PS)}-CH_2\overset{Ph_2}{P}-RhCl(PPh_3)_2 \quad (3)$$

itself. Grubbs and Kroll attributed this to the fact that the reactive sites were on the inside of the polymer bead, so diffusion of the olefin inside the polymeric bead was required. They observed a linear relationship between the rate of reduction and the size of the substrate.[17]

After this report, a considerable amount of research into polystyrene-supported catalysts ensued. It was found that, in general, the supported systems suffered from decreased activity, leaching, and were only active in certain solvents. The polymer backbone itself is thermally and mechanically unstable, and loss of the metal due to lability of the Rh-phosphine bond is always a concern.[27] However, the major drawback of this support is that the large majority of functionalized sites are on the inside of the polymer beads, necessitating swelling with an appropriate solvent, and therefore restricting the number of solvents that can be used. Research into alternative polymer backbones has led to the development of polymeric supports that are swellable in a variety of solvents.[28]

2.1.2   Non-Cross-Linked Polymers

Grafting transition metal complexes on *soluble* supports is an alternative approach which removes the problems of diffusivity, but not of solvent specificity. After the reaction, recovery of the transition metal complex is generally accomplished by precipitation of the polymer using a second solvent.[29] Two common soluble polymers are non-cross-linked polystyrene and linear polyethylene glycol (PEG). The important advantage offered by this approach is that the catalyst is in the same phase as the substrate during the reaction. In general, soluble polymers are highly effective at mimicking the behavior of non-polymer-bound catalysts. The disadvantage of this method is that it is too solvent specific. More importantly, flow processes are not possible and precipitation may require significant amounts of solvent. The long-term stability of highly functionalized polymers like polyethylene glycol is another potential problem.

Polyethylene itself can be used as a support for transition metal catalysts.[30] As well as being stable and unreactive, polyethylene is valuable because of its "thermomorphic" (miscible/soluble at one temperature and immiscible/insoluble at another) phase properties. Transition metal complexes of PPh$_2$-terminated polyethylene are easily prepared (eq. 4) and display the same phase behavior as linear polyethylene, namely solubility in organic solvents at temperatures above 90 °C, and precipitation below this point. These polymeric complexes are generally good at mimicking the activity of their simple homogeneous precursors.

$$\text{BuLi} + \text{n H}_2\text{C}=\text{CH}_2 \longrightarrow \text{Bu} \left[\!\!\!\underset{n}{\bigwedge}\!\!\!\right]\!\!\overset{\text{Li}}{} \quad \xrightarrow[\text{2. ClRh(PPh}_3)_3]{\text{1. ClPPh}_2} \quad \text{Bu} \left[\!\!\!\underset{n}{\bigwedge}\!\!\!\right]\!\!\overset{\overset{\text{Ph}_2}{\text{P}-\text{RhCl(PPh}_3)_2}}{} \quad (4)$$

The Bergbreiter group has also used solvent mixtures that are thermomorphic for catalyst separation.[31] For example, a mixture of heptane and 90% ethanol is miscible at 70°C, but separates into two phases at room temperature. Catalysts tethered to poly-*N*-isopropylacrylamide are distributed throughout this mixture at 70°C, but upon cooling the polymeric catalyst is dissolved preferentially in the more polar ethanol phase.

2.1.3   Highly Cross-Linked Polymers

Highly cross-linked polymers, or macroreticular resins, have considerable potential as catalyst supports, since their use is not restricted to one type of solvent and they have relatively high surface areas. Nozaki,[32] Gagné,[33] Severin[34] and others have reported examples of catalysts immobilized on macroreticular polymers for applications in hydroformylation, hydrogenation, hydroboration and oxidation reactions. Nozaki and Takaya demonstrated that Binaphos (**2a**) can be modified to include one or more vinyl groups (**2b-d**), and then incorporated into macroreticular polystyrene (eq. 5 and 6).[32]

$$\textbf{2b-d} + \text{Ar}\!\!\diagdown\!\!\!\underset{\textbf{3}}{} + \underset{\textbf{4}}{\text{[divinylbenzene]}} \quad \xrightarrow[\text{2. [Rh(acac)(CO)}_2]]{\text{1. Initiator}} \quad \textbf{5b-d•Rh} \quad (5)$$

$$\textbf{2b-d} \quad \xrightarrow{\text{[Rh(acac)(CO)}_2]} \quad \textbf{2b-d•Rh} \quad \xrightarrow[\text{Initiator}]{\textbf{3 + 4}} \quad \textbf{5b-d•Rh} \quad (6)$$

The presence of significant quantities of divinylbenzene (**4**) causes cross-linking of the polymer chains. Heterogeneous catalysts prepared from **2b** or **2c** give enantioselectivities equal to those reported for Binaphos, but **2d**, which has three vinyl groups, is less selective (68% ee, Table 1). However, if the Rh complex is formed before polymerization (eq. 6), **2d** gives excellent enantioselectivities (Table 1, entry 5). It is likely that pre-forming the Rh complex before polymerization prevents the ligand from being polymerized in less selective conformations. Considering the utility of the products of hydroformylation, this particular catalyst is likely to be of wide applicability.

Table 1. Hydroformylation of styrene with homogeneous catalyst **2** and heterogeneous catalyst **5**.

Entry	Catalyst	B:L	ee (%)
1	**2a•Rh**	89:11	92
2	**5b•Rh**	84:16	89
3	**5c•Rh**	89:11	89
4	**5d•Rh**	88:12	68
5	**5d•Rh***	87:13	85

*Rh added before polymerization

2a R$_1$ = R$_2$ = H
2b R$_1$ = CH=CH$_2$, R$_2$ = H
2c R$_1$ = H, R$_2$ = CH=CH$_2$
2d R$_1$ = R$_2$ = CH=CH$_2$

2.2   Ordered Organic Polymers

2.2.1   Molecularly Imprinted Polymers

One of the advantages of the rigid backbone in macroreticular resins is the ability to engineer specific interactions into the resin. Using a technique called "molecular imprinting", polymeric materials can be prepared that have pores with specifically designed structures, and functional groups located appropriately within the pores. Although primarily employed to prepare polymers for chromatographic applications,[35] molecular imprinting has recently been used to create polymers that contain catalytically active organic groups[36] or transition metal complexes.[37-40] The ultimate aim is to place ligands for the metal in close association with well-defined pockets in the polymer. The material is not meant to swell or gel in the solvent, since shape of the cavities should not be fluxional. The microenvironment is controlled by the use of a template that is attached to the metal catalyst during the polymerization, and is then removed. This template can be either a well-defined chiral species[37,38] or a transition-state mimic for the desired reaction.[39-41]

Gagné and co-workers have described the formation of chiral cavities that have associated phosphine ligands within polyethyleneglycol-dimethylacrylate (poly-EGDMA) as shown in Scheme 1.[37] Pt complex **6**, containing a chiral Binol substituent and a polymerizable phosphine, is mixed with the acrylate monomer in a 3:97 ratio, and polymerization is initiated with AIBN[42] in the presence of a porogen (PhCl). After polymerization, the Binol template is removed by treatment with acid. Depending on the strength and size of the acid, varying amounts of the Binol template are removed. Gagné clearly demonstrates that the chiral pockets are capable of binding one enantiomer of racemic Binol with high selectivity. The only drawback to this elegant method for controlling the

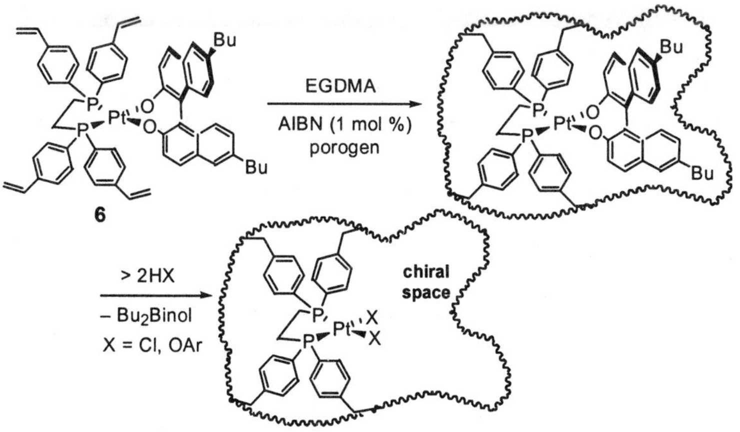

Scheme 1. *Creation of chiral space within porous rigid polymers.*

microenvironment of the metal *outside* the ligand sphere is that several different types of sites are produced, with different selectivities. Those sites that are more accessible are understandably less selective. Gagné estimates that the selectivity of the more accessible sites for rebinding Binol of the same stereochemistry as the template is on the order of 46%. The less accessible sites have selectivities between 89 and 97%. Selective poisoning of the more accessible sites may be able to solve this problem. This technique is used to modify imprinted polymers for chromatographic applications, but with the exception of a recent report from the Gagné lab,[38] it has not been applied to imprinted catalysts.

Severin[39] has examined the asymmetric transfer hydrogenation of acetophenone (PhCOCH$_3$) with imprinted Rh- and Ru-containing polymers. In this case, the removable ligand is designed to be a mimic of the transition state of the reaction. The co-polymerization of Rh complex 7 and ethylene glycol/dimethylacrylate was carried out as shown in eq. 7. The phosphate ligand is expected to imitate the transition state of the hydrogenation. Since the phosphate is configurationally labile, it has no effect on the enantioselectivity of the imprinted catalyst. However, treatment with BnNEt$_3$Cl replaces the phosphonate with a chloride, and provides a product-sized cavity in the polymer, leading to a doubling of catalyst activity. Like the homogeneous version, catalyst **poly-7** gives very high enantioselectivites for the reduction of acetophenone (up to 95% ee).

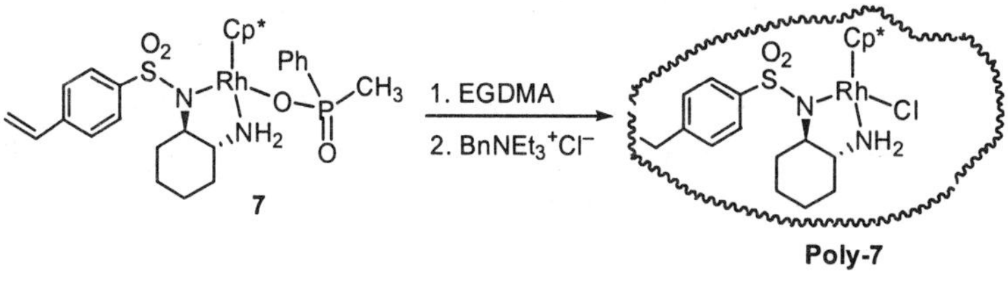

(7)

Poly-7

Mosbach reported the first example of a carbon-carbon bond forming reaction within an imprinted polymer.[40] Cobalt complex **8** was co-polymerized with styrene and divinyl benzene (eq. 8). Removal of the template molecule and recharging the polymer with cobalt (which is lost during template removal) gives a catalyst that is eight times more active than the corresponding homogeneous complex for the aldol reaction between acetophenone and benzaldehyde (eq. 9).

2.2.2 Chiral Polymeric Ligands

Chiral polymers have also been investigated as ligands for transition metal catalysts. For example, poly-L-leucine can be employed as a ligand for the Pd-catalyzed carbonylation of crotyl alcohol into lactone **9** (eq. 10).[43]

One restriction inherent to the use of naturally available chiral polymers is that the ligating sites are nitrogen- and oxygen-based, even though many late transition metals prefer softer ligands such as phosphines. Also the types of chiral structures that can be employed are limited.[44] Synthetic chiral polymers designed on ligand architectures known to be effective in solution are viable alternatives to natural polymers. Moreau has prepared polymers containing 1,2-*trans*-diaminocyclohexane in the backbone and employed them as ligands for the asymmetric transfer hydrogenation of acetophenone.[45] The chiral polymer gives significantly better results than the homogeneous catalyst (58% ee vs. 33% homogeneous). Chan,[46] Pu[48-50] and Lemaire[51-54] have also prepared polymers incorporating binaphthyl-based ligands in the backbone. Chan reported the synthesis of a chiral co-polymer by the condensation of a chiral diol **10**, bisbenzoylchloride **11** and chiral amino-BINAP **12** (eq. 11).[47]

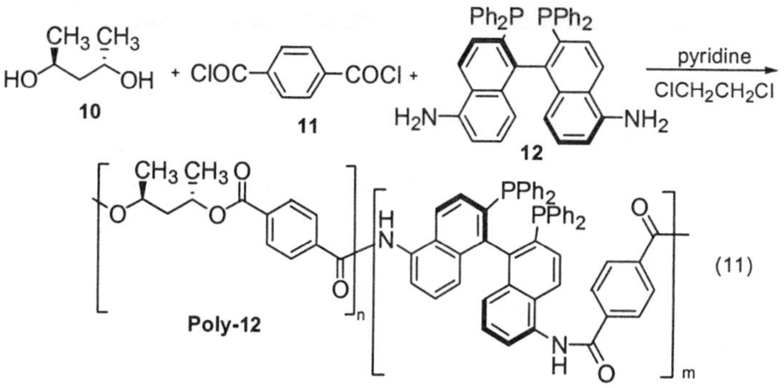

(11)

In the presence of $[Ru(cymene)Cl_2]_2$ and **poly-12**, the hydrogenation of 2-(6'-methoxy-2'-naphthyl)acrylic acid (**13**) yields the non-steroidal anti-inflammatory agent Naproxen™ (**14**) in 88% ee (eq. 12). The solvent employed for the transformation is critical. In the optimum solvent (2:3 methanol:toluene), the polymeric catalyst was actually more active than the homogeneous system (**poly-12**: 96% conversion, 4h; BINAP: 56%, 4h; bis benzoyl BINAP, 64%, 4h). However, in 9:1 methanol:toluene, **poly-12** was significantly less active, giving only 37.5% conversion after 60 hrs.

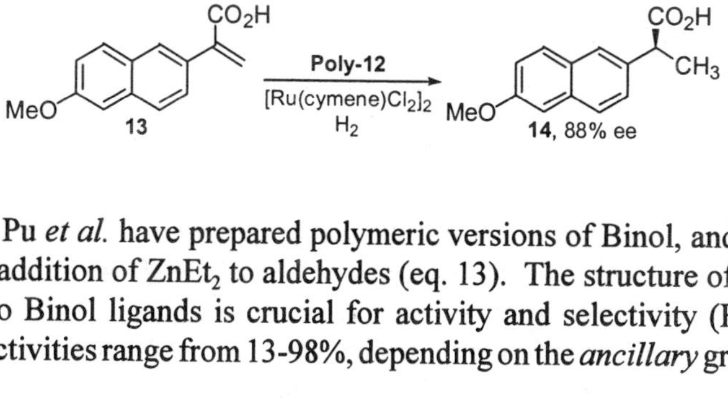

(12)

Lin Pu *et al.* have prepared polymeric versions of Binol, and employed them in the addition of $ZnEt_2$ to aldehydes (eq. 13). The structure of the spacer between two Binol ligands is crucial for activity and selectivity (Figure 1).[47] Enantioselectivities range from 13-98%, depending on the *ancillary* groups on the backbone.

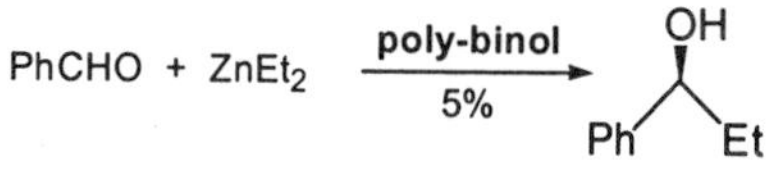

(13)

Poly-BINAP was also prepared by the Pu group.[48] In the asymmetric hydrogenation of dehydroamino acids, Rh-polyBINAP performed as well as the analogous homogeneous BINAP-Rh complex. Ru-polyBINAP was also effective for the asymmetric reduction of ketones.[48] Ingeniously, Pu prepared a co-polymer containing both Binol and BINAP, and used the mixed ligand for a tandem asymmetric diethylzinc addition/asymmetric reduction of *p*-acetylbenzaldehyde (eq. 14).[49] The tandem asymmetric reactions give the desired product (**16**) in 99%

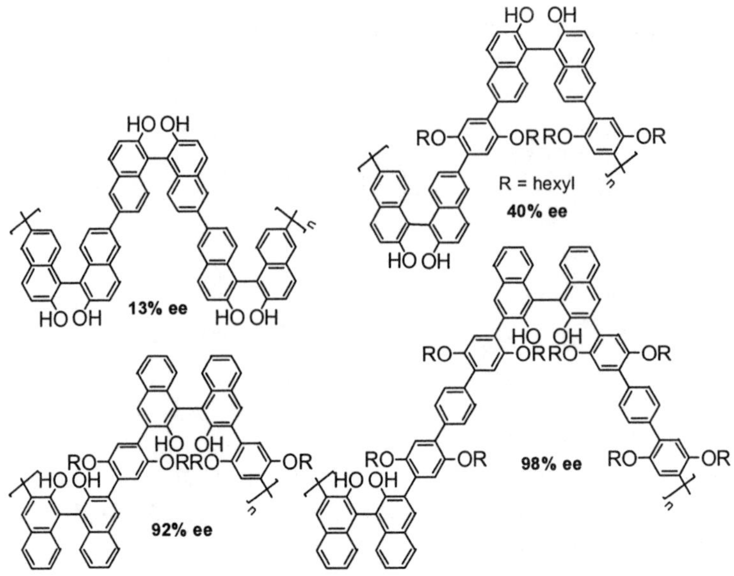

Figure 1. Binol polymers for asymmetric additions to aldehydes.

yield, with an enantioselectivity of 92% and a diastereoselectivity (de) of 89%, similar to the homogeneous reactions.

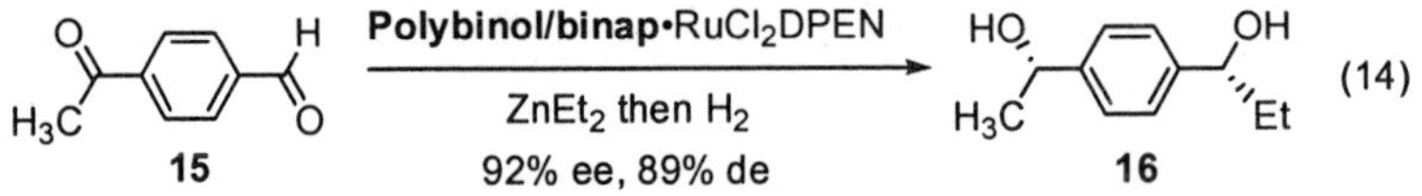

$$(14)$$

Chiral polyamides and polyureas have been prepared and tested as ligands for transition metals by the Lemaire group.[50] *Diam*-BINAP (**17**) can be prepared in five steps from Binol, and can be converted into an oligomer by treatment with toluene diisocyanate (**18**) (eq. 15).[51] A Ru complex of the resulting polymeric ligand **19** catalyzes the hydrogenation of ketoesters with 90-99% ee, compared to 98-99% ee for the homogeneous catalyst.

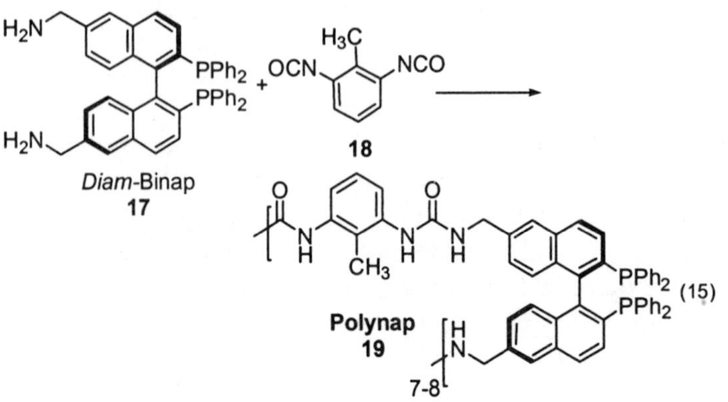

$$(15)$$

Lemaire has combined the concepts of polymeric ligands and molecular imprinting in a study of polyurea-Rh complexes for asymmetric transfer hydrogenation.[52,53] The homogeneous Rh complex of (R,R)-bisdiamine **20** catalyzes the transfer hydrogenation of acetophenone to give the S isomer with 55% enantioselectivity, and a polymeric version (**poly-22**, eq. 16) catalyzes the same reaction with reduced ee (33%).

An imprinted version of this polymer (**S-IMP-poly-22•Rh**) can be prepared by adding (S)-PhCH(CH$_3$)ONa prior to polymerization (eq. 17). After removal of this chiral template, the resulting catalyst displays enhanced selectivity for the reduction of acetophenone (43% ee) vs. the non-imprinted version.

Interestingly, imprinting with the R-isomer leads to only a slight reduction in enantioselectivity (30% ee), with the S-isomer still the major product, indicating that the ligand chirality is more important than imprinting (eq. 18). Higher selectivity is observed for the reduction of propiophenone, but the homogeneous catalyst is still superior (66% ee imprinted, 47% ee non-imprinted, 80% ee homogeneous).

2.2.3 Organic Zeolites

Organic materials with organized, well-defined pores have been prepared by several routes.[54] Wuest and co-workers have described the use of "molecular tectons" or building blocks, to organize and control the pores in organic solids in a rational manner.[55] Douglas Gin and co-workers have prepared organic polymers that have well-defined and ordered pores[56] by taking advantage of the self-assembly behavior of polymerizable lyotropic liquid crystals such as **23**.

When exposed to water, compound **23** phase separates forming an inverse hexagonal micelle. If this self-assembly is carried out in the presence of divinyl benzene (**4**), the resulting inverse micelle can be polymerized with light, leading to an ordered organic polymer (**24**), (eq. 19).

Since the channels of **24** are lined with carboxylate groups, it is an effective catalyst for the Knoevenagel reaction (eq. 20). The activity of **24** was compared with ordered inorganic materials (Na-MCM-41 and zeolite NaY), and with an amorphous organic polymer (2% cross-linked polyacrylate) having the same number of basic sites as **24**.

In refluxing THF, **24** gave approximately 80% conversion after 7 hrs, while all other catalysts tested were inactive. The dramatic difference in activity is attributed to a significant enhancement of basicity that is caused by electrostatic effects from close packing of the carboxylates, and immobilization in a low-dielectric medium. Interestingly, a lamellar form of **24** was significantly less active than the hexagonal form, even though the basicity was estimated to be similar. Thus the ordered structure of the active site is crucial to the high activity observed, since it affects the accessibility of the catalytic sites (carboxylates).

In subsequent work, Gin and co-workers have shown that Sc-doped organic nanostructures (**Sc-26**, eq. 21) are effective catalysts for a variety of carbon-carbon bond forming reactions, including the Mannich reaction and the Mukayama aldol reaction (eq. 22).[57] In both cases, the diastereoselectivity was higher using ordered Sc catalyst **26** than using $Sc(OTf)_3$ itself, Sc-surfactant complexes, or Sc on amorphous polymers, which all gave a 50/50 mixture of *syn* and *anti* isomers.

$$\text{(22)}$$

Sc-26 syn/anti = 76/24
Sc(OTf)$_3$ syn/anti = 50/50

2.3 Inorganic Polymers

Inorganic materials generally have higher thermal, oxidative, and mechanical stability as well as higher rigidity than organic supports. Since these materials are always in a different phase than the substrate, they can be used in almost any solvent including water. Inorganic supports are classified broadly by their elemental composition and include carbon, alumina, clay, silicates and zeolites. The extent and type of porosity the materials exhibit is often critical to activity. Microporous materials have pores between 2 and 20 Å, mesoporous materials possess pores in the 20-500 Å range, and macroporous materials have >500 Å pores.[58]

2.3.1 Amorphous Inorganic Polymers

The use of amorphous inorganic polymers as catalyst supports has been the subject of several reviews to which the reader is referred.[59]

2.3.2 Ordered Inorganic Polymers

Ordered inorganic polymers can be natural, synthetic or semi-synthetic. Clays and zeolites have a natural structure that can be employed to direct reactivity, or can be modified by chemical reactions.[60] Completely synthetic materials with high degrees of order and varying porosity are also readily available.

2.3.2.1 Clays

Transition metals can be incorporated into the interstitial sites of clays either by ion exchange with the unmodified clay,[61] or by transformation of surface silanols into ligands for transition metals. Working primarily with Montmorillonite clay, B.M. Choudhary and co-workers[62] have described the preparation of a variety of clay-supported transition metal catalysts (Scheme 2). The Alper group has also studied Ru-[63] and Pd-clay,[64] of which the latter is used to prepare isocyanates from aryl amines. Recently, Claver and Fernandez have shown that asymmetric metal complexes can be immobilized on clay and used in the hydrogenation of imines,[65] or in the hydroboration of styrene.[66] The latter reaction is challenging due to the sensitivity of the hydroborating reagent, but the

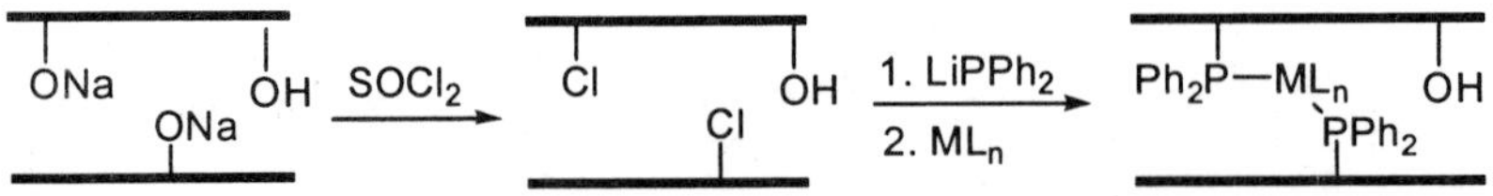

Scheme 2. Preparation of clay-supported metal complexes.

immobilized catalysts perform well, giving the same enantioselectivity as the homogeneous system. The ion-exchanged catalysts are shown to be recyclable over several runs.

2.3.2.2 Zeolites

Zeolites have been used as solid acid catalysts in industry for decades.[67] These naturally occurring aluminosilicates are prized for their ability to differentiate between organic compounds based on seemingly little difference in shape and size. Unfortunately, the small size of the pores prevents the application of most natural zeolites in the synthesis of larger organic molecules, and the encapsulation of large organometallic catalysts.[68]

However, if the pores are artificially widened, spectacular results can be obtained.[69] Using a chiral Rh catalyst immobilized on the surface of zeolite USY (**30b**), Corma achieved extraordinary results in the hydrogenation of dehydroamino acids (**27**, eq. 23).[70] USY is a modified zeolite which has mesopores resulting from de-alumination and destruction of the sodalite unit to permit communication between several α-cages. For comparison, a catalyst supported on amorphous silica (**30a**), and a homogeneous catalyst (**29**) were tested. The zeolite-supported catalyst consistently out-performed both the silica-supported Rh complex *and* the homogeneous catalyst (Table 2). The most impressive example is entry 3 in which the homogeneous catalyst gives only 55% ee, while the zeolite-supported catalyst give the product in 94% ee!

Table 2. Hydrogenation of **27** with Rh catalysts **29** and **30**.

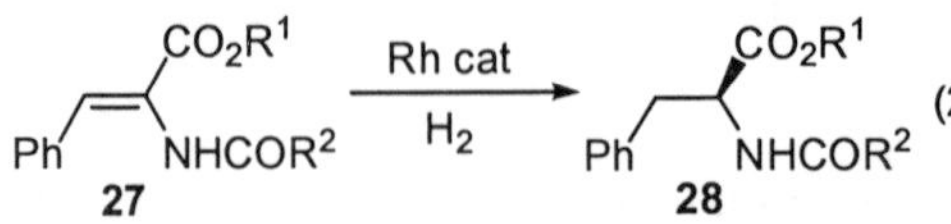

27 **28** (23)

R^1	R^2	Cat 29	Cat 30a	Cat 30b
H	Me	84%ee	88%ee	98%ee
H	Ph	90%ee	94%ee	97%ee
Et	Me	**55%ee**	**58%ee**	**94%ee**

Rh cat =

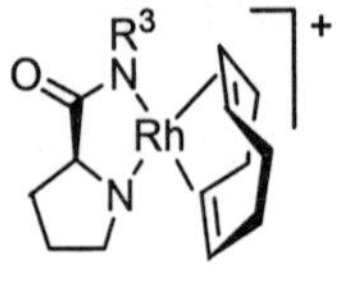

29 R^3 = ^{t}Bu
30a R^3 = $(CH_2)_3$–SiO_2
30b R^3 = $(CH_2)_3$–USY

2.3.2.3 Imprinted silicates

Molecular imprinting has been used extensively to prepare controlled-porosity inorganic materials. In a recent example, Davis described the preparation of microporous silicates in which the pores are created by an organic template (**31**) included during the synthesis of the silicate.[71] The difference between this method and the approach generally employed is that the template is actually polymerized into the bulk silica via covalent bonds, and then removed by a chemical reaction. Thus, the hydrolysis of **31** in the presence of excess $Si(OEt)_4$ (1 : 150) leads to an organic/inorganic composite. The covalently-attached organic template is then removed by treatment with TMSI, followed by washing with $H_2O/MeOH/NaHCO_3$ (eq. 24).

Imprinted silicate

Imprinted silicates were also prepared with 1,4-disubstituted aromatic templates, and the resulting amine-functionalized imprinted silicates were catalysts for the Knoevenagel condensation (eq. 20) using malononitrile ($NC-CH_2-CN$). The imprinted silicates react more slowly than amorphous aminopropyl silica, but they are considerably more selective.

2.3.2.4 Templated mesoporous molecular sieves

Materials with significantly larger pores can be prepared if the silicate hydrolysis is carried out in the presence of a supramolecular array of template molecules, rather than a molecular template. The 1992 report[1] of a reliable method to prepare all silica (or silica/alumina) molecular sieves with porosity in the *mesoscopic* range (20-500 Å) was a landmark in synthetic silicate chemistry. This report by Beck, Kresge, Vartulli *et al.* from Mobil Corporation describes the use of surfactant assemblies as templates for the synthesis of mesoporous molecular sieves of controlled pore sizes.[1] The removal of the template was

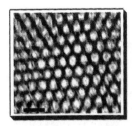

Figure 2. TEM of mesoporous silicate (SBA-15-type); scale bar = 25 nm.

accomplished by calcination, leaving a silicate with highly ordered pores, large pore volumes and surface areas of >1000 m²/g. It is believed that the ordered materials are prepared by a cooperative templating method in which the silicate oligomers assist the organization of the surfactant into a liquid-crystal-type structure.[72,73] Hydrolysis and condensation of the silicate around this supramolecular template leads to the observed structure (Figure 2). For a detailed discussion of methods used to control porosity and structure, see Stein and Schroden's chapter in this book.

Mesoporous molecular sieves have been shown to be superior supports for a wide number of catalysts compared to their amorphous counterparts.[74] Catalysts for the polymerization of alkenes, diethyl zinc addition to aldehydes, allylic amination, hydrogenation and dihydroxylation of alkenes have been reported.[4-11] The latter two reactions will be discussed in detail in section 4 of this chapter, while the others are discussed in chapters by Aida/Tajima and Brunel in this book.

As well as surfactants, organic gels can also act as templates for the synthesis of mesoporous silicates. Shinkai and co-workers have shown that certain organic compounds act as gelators of organic solvents (Figure 3).[75] The gels that are formed have a right- or left-handed helical structure depending on the absolute chirality of the gelator. Although neither **32a** nor **33a** catalyze the hydrolysis and condensation of Si(OEt)₄,charged versions of these compounds (**32b** and **33b**, Figure 3) do.

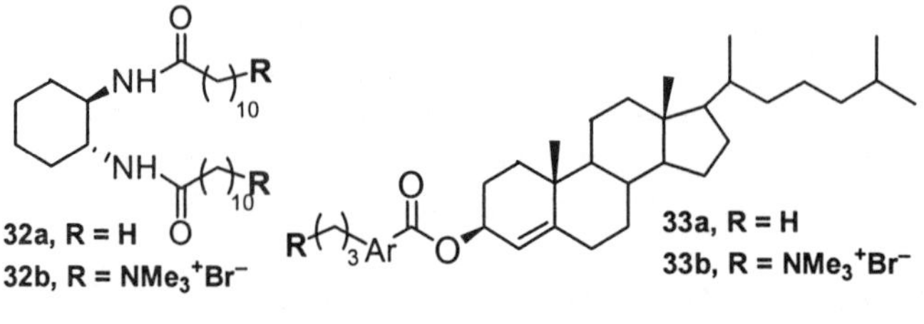

32a, R = H
32b, R = NMe₃⁺Br⁻

33a, R = H
33b, R = NMe₃⁺Br⁻

Figure 3. Organic gelators for the preparation of chiral silicates.

Thus, in the presence of **32b** or **33b**, a silica skeleton of the helical gels can be generated, and retained upon removal of the template. Although the use of the resulting materials in catalysis has not yet been reported, this technique is obviously a very powerful method for the preparation of ordered inorganic materials with chiral morphology.

2.3.2.5 Inorganic/organic composites

An alternative method for transferring chiral information to silica is to *permanently* incorporate chiral organic molecules into the structure of the silicate. These organic-inorganic composites can be prepared from a variety of organic precursors, usually containing an Si(OR)₃ unit.[76] The challenge in this area is retaining sufficient porosity and surface area so that chiral information can be effectively transferred to the substrates during catalysis.

Chiral organic/inorganic composites that also have chiral twist morphology can be prepared by hydrolysis and condensation of siloxane-functionalized organic gelators such as **34** (eq. 25).[78] The resulting helical material has a surface area of 120 m²/g and a broad distribution of pores (5 to 50 nm).

Chiral inorganic/organic composites of similar structure were used for the preparation of Rh catalysts (eq. 26).[78] When the urea structure is omitted, the compound is not a gelator, and chirality is not transmitted to the macroscopic structure. However, polymer **36** is an active and enantioselective catalyst for the transfer hydrogenation of acetophenone. Although it is unclear how the Rh is bound, the polymeric catalyst is more enantioselective than the homogeneous catalyst (58% ee vs. 26% ee). A catalyst prepared by grafting **35** onto preformed silica is less selective (22% ee) showing the importance of the microenvironment of the Rh complex.

3. METHOD OF IMMOBILIZATION

3.1 Immobilization Without Ligand Modification

Several methods are available for grafting metal complexes onto polymeric supports. These can be classified into non-covalent, covalent or ionic attachment. Non-covalent attachment processes such as impregnation or occlusion yield catalysts in which the metal is supported without the benefit of ancillary ligands.[79] Often these processes are followed by high temperature calcination or reduction in order to create a support covered with a highly dispersed metal. An alternative approach to create a "ligandless" metal on silica supports is surface organometallic chemistry.[80] Using this technique, an organometallic species is reacted with the surface of carefully prepared silica to create a well-defined surface species. Scott,[81] Basset,[82] Anwander[83] and others have shown that surface organometallic chemistry is a powerful technique for the creation of well-defined heterogeneous catalysts. For a thorough review of this subject see the chapters by Scott and by Anwander in this book.

3.1.1   Non-covalent Immobilization of Metal and Ligands

Several methods are available for the non-covalent immobilization of the metal *and* its ligands. In the simplest case, a metal complex is physically entrapped in the support by assembling the polymer around the preformed metal complex. Organic polymers can be employed to immobilize catalytically active complexes using technique called microencapsulation.[84] Kobayashi has shown that polystyrene and related polymers can adsorb catalytically active metal complexes such as OsO_4 when both are dissolved in a solvent such as THF at high temperature.[84] Cooling causes phase separation and treatment with methanol hardens the resulting microcapsules, which can then be used, along with a homogeneous ligand, for a variety of reactions. Interestingly, the ligand, substrate and reagents are able to diffuse into and out of the microcapsules, but the metal remains trapped. The drawback of this method is that the chiral ligand is not immobilized along with the metal.

Entrapment of metal complexes can also be accomplished using sol-gel chemistry. Blum, Avnir and co-workers reported that Wilkinson's catalyst, Vaska's complex[85] and RuBINAP[86] could be effectively isolated in sol-gel matrices by admixing them with $Si(OMe)_4$ during the polymerization process. The entrapped complexes were active for several reactions including the isomerization, asymmetric hydrogenation, hydroformylation and disproportionation of olefins, and were reusable without leaching of the metal.[87] Blocking of the pores was occasionally observed, but the catalysts could be reactivated by washing and sonication. In some cases, the entrapped catalysts were *more* active than the homogeneous complexes.[88]

Vankelecom and Jacobs have shown that polydimethylsiloxane membranes are able to physically immobilize an asymmetric Rh catalyst.[89] By mixing RhDuphos (37), a silicone polymer terminated by vinyl groups (38), a cross-linking agent (39) and a hydrosilylation catalyst, Vankelecom and Jacobs are able to prepare an immobilized chiral catalyst in the form of a membrane. The catalyst displayed high enantioselectivity in the hydrogenation of methylacetoacetate (41), although the homogeneous system was still superior (Scheme 3).

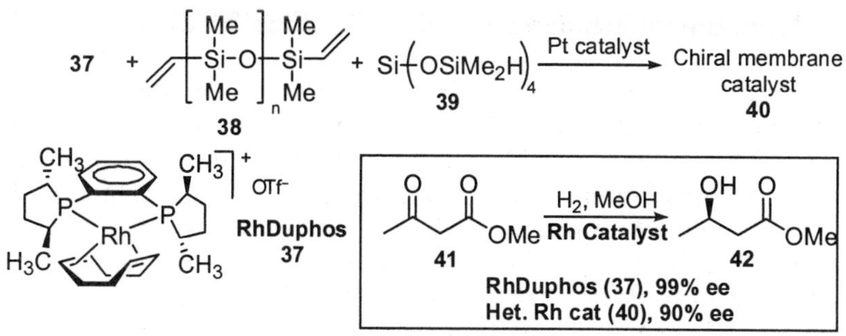

Scheme 3. Asymmetric hydrogenation with chiral membrane cataylsts.

The reactivity of the Rh catalyst also suffered after encapsulation, with 28 turnovers/hr observed for the immobilized complex and 482 turnovers/hr for the homogeneous case. Although the decrease in activity is significant, this method has the benefit that the ligands do not need to be modified in order to immobilize the catalyst. The composites can also be used for asymmetric hydrogenation in water.[90]

Augustine has described what promises to be an extremely powerful method for the immobilization of catalysts.[91] Homogeneous asymmetric catalysts such as RhDuphos, RhDiPamp[92] and RhBPPM[93] can be immobilized on a variety of supports by simply pretreating these supports with heteropolyacids such as phosphotungstic acid (PTA). Many supports can be used including alumina, carbon and Montmorillonite clay, with alumina being the most effective. Although it is unclear how the catalysts are bound to the surface, they are not leached after multiple uses and exhibit enhanced reactivity vs. the homogeneous catalysts. It is likely that some chemical change takes place during hydrogenation, since the first run gives significantly different results from subsequent runs (Table 3, eq. 27).[94] Entry 3 in Table 3 is the most remarkable: the enantioselectivity of the hydrogenation of **43** increases from 21% in the first run to 87% in the third. The catalysts could be *reused 15 times without loss of activity or leaching of Rh.* A variety of Rh complexes can be immobilized using Augustine's technique. These complexes also exhibit improved reactivity upon binding to the support.

Table 3. Asymmetric hydrogenation with PTA-treated catalysts.

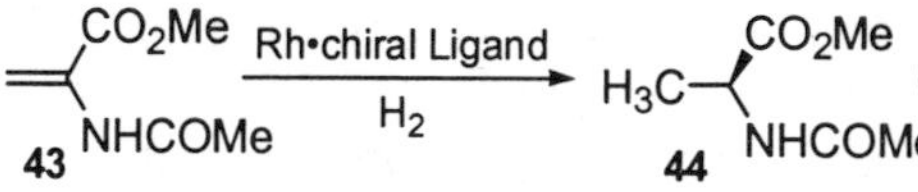

Chiral Ligand	Homogeneous Catalyst	Supported Catalyst	
		1st Run	3rd Run
Dipamp	76%ee	90%ee	93%ee
MeDuphos	96%ee	83%ee	96%ee
BPPM	84%ee	21%ee	87%ee

3.1.2   Ionic Immobilization

A compelling example of immobilization of unfunctionalized chiral complexes by ionic interactions has been reported by Broene, Tumas and co-workers.[95] Through judicious choice of counterion, cationic Rh complexes can be attached to mesoporous silica. A supported version of RhDuphos (**37**) was prepared by simply treating a CH_2Cl_2 solution of the catalyst with MCM-41. The support took up the catalyst from the solution, giving an active, and recyclable catalyst for the hydrogenation of dehydroamino acid **43** (eq. 27). In hexane, 99% ee was obtained, compared with 87% for the homogeneous catalyst in the same solvent.[96] The supported Rh complex can be reused several times without loss of activity or leaching.

It is likely that the catalyst is bound to the surface via hydrogen bonds between the acidic silanols and the triflate counterion, based on the following data. If the highly lipophilic $B(C_6F_5)^-$ anion is used in place of the triflate, the catalyst does not bind to the surface. The addition of $NaB(C_6F_5)_4$ to the supported catalyst causes leaching from the surface. Trimethylsilyl-protected silicates have a significantly lower capacity for Rh, as does amorphous silica gel, which has a lower number of silanols compared with MCM-41. Subsequent to the Tumas/Broene work, the preparation of similar supported complexes was described by Bianchini and by Hölderich (on Al-MCM-41).[97] The only disadvantage of this method is that polar solvents disrupt the hydrogen bonding at the surface and result in leaching.

When ionic complexes are supported on polar surfaces, there is always the possibility that the charged metal will interact with the surface, leading to an inactive catalyst. Milstein has reported an inventive strategy designed to take advantage of this phenomenon. Milstein, Lahav and co-workers have shown that complex **45a** forms a monolayer at the air-water interface that can be transferred to a glass slide using Langmuir-Blodgett technology.[98] Using regular hydrophilic glass, the positively charged Rh species bind to the glass surface, and form a triple layer (Figure 4).

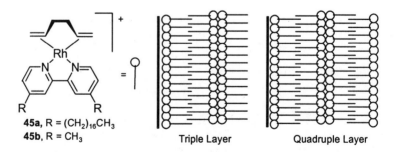

45a, R = $(CH_2)_{16}CH_3$
45b, R = CH_3
Triple Layer   Quadruple Layer

Figure 4. Self-assembled monolayers as hydrogenation catalysts.

The resulting supported catalyst is inactive for the hydrogenation of acetone. However, when a quadruple layer is prepared that exposes the catalytically active Rh center to the solution, a highly active catalyst is obtained that hydrogenates acetone with a turnover frequency of 1460 hr^{-1}, compared with 10.4 hr^{-1} for the homogeneous catalyst (**45b**) in acetone. Interestingly, the ordered nature of the surface has an effect on the activity of the catalyst. As the surface becomes more disordered at higher temperatures, the activity decreases. The high activity is restored by cooling the catalyst.

3.2 Covalent Immobilization Using Modified Ligands

The large majority of immobilized catalysts are prepared by introduction of a new substituent that chemically reacts with the surface to form a covalent bond. In order to do this without disturbing the coordination sphere of the metal, it is necessary to introduce the substituent somewhere on the periphery of the ligand. Therefore, it is important that the ligand-metal bond in question be strong, otherwise leaching of the metal will result.

3.2.1 Sulfonated Phosphines

Sulfonated phosphines, initially developed for applications in aqueous phase catalysis,[99] can also be supported on silica using the interaction between the negatively charged SO_3^- group and the acidic silanols on the surface (Figure 5). Although the phosphine must be modified to introduce the SO_3^- functionality, many sulfonated phosphines are commercially available, and the SO_3^- unit can be introduced easily by electrophilic aromatic sulfonation.

Bianchini has reported that the sulfonated analog of his tridentate phosphine triphos (**46**) can be immobilized on silica surfaces, and the resulting catalyst used in hydrogenation and even hydroformylation reactions.[100] The hydroformylation of olefins is a challenging reaction for a supported catalyst since the presence of CO in the reaction mixture provides a good ligand for Rh, and often causes leaching.[101] Despite this fact, **46•SiO₂** can be used multiple times in the hydroformylation of hexene without any loss in activity, or leaching of Rh into the solution. Presumably the tridentate nature of the ligand is responsible for retaining the Rh on the support. Furthermore, the supported catalyst **46•SiO₂** is more active than **46** in THF or in a biphasic MeOH/*iso*-octane mixture. This is likely a site-isolation effect since there is evidence that the homogeneous species is deactivated by formation of hydrogen-bridged dimers.

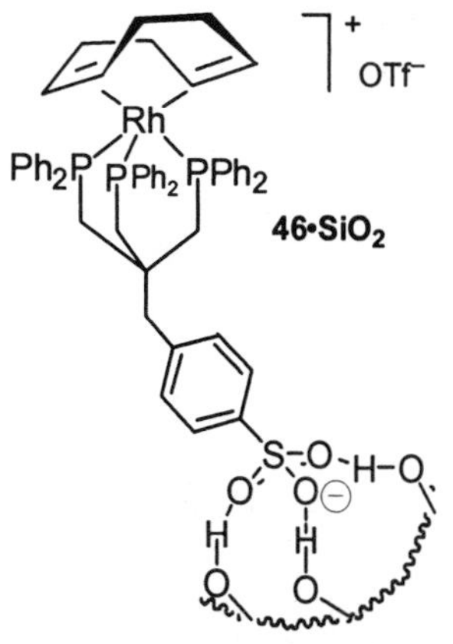

Figure 5. Supported-hydrogen bonded catalyst **46•SiO₂**.

Sulfonated phosphines can also be employed in supported aqueous phase (SAP) systems.[102] In this unique approach developed by Davis *et al.*, the catalyst is dissolved in a thin film of water on the surface of porous glass. Sulfonated BINAP derivative **47** (Scheme 4) supported in this fashion gives high ee and activity for the hydrogenation of olefins.[103] Since it was found that water was detrimental to enantioselectivity, a thin film of ethylene glycol was used in its place.[104] With the substrate (**13**, eq. 12) dissolved in a 1:1 mixture of CHCl$_3$:cyclohexane, the hydrogenation took place with 96% enantioselectivity and no leaching was observed. Remarkably, the supported catalyst is more active (131 turnovers/hr) than the homogeneous catalyst (41 turnovers/hr).

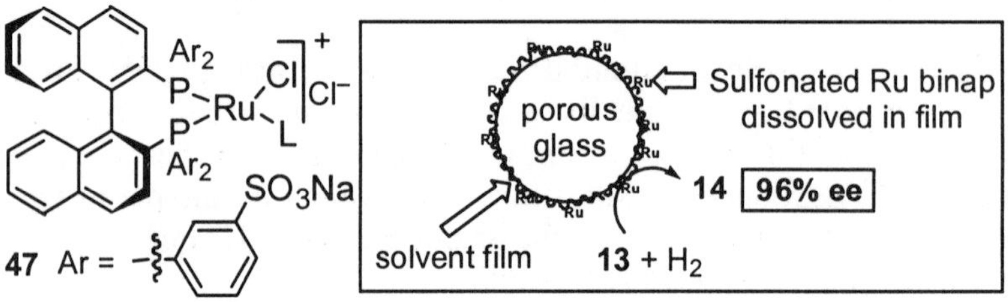

Scheme 4. Asymmetric hydrogenation using supported aqueous phase technology.

3.2.2   Siloxane-modified Phosphines/Amines

Modification of ligands to incorporate a trialkoxysilyl group is the most commonly employed method for the preparation of supported catalysts.[105] The limitation of this method is that several synthetic steps are required to introduce the Si(OEt)$_3$ group or its equivalent. Grafting is accomplished by interaction of the Si(OEt)$_3$ group with water and Si-OH groups on the surface of the silica support.[106] With three alkoxides on silicon, grafting can occur to multiple sites on the surface, or via a linear condensation between two different silanes (coating).[107,108] Another method of grafting has also been observed in which the alkoxysilane adds across an Si–O bond of a strained surface siloxane formed by high temperature dehydroxylation of the surface.[109]

The possibility of forming different linkages means that the catalyst may exist in several different environments. This is especially problematic since site-isolated catalysts often display different activity or selectivity than catalysts in close contact. One method that can increase isolation is to use a monoalkoxy group (such as SiMe$_2$OEt) that cannot undergo cross-condensation.[110] Another method is to use an inert alkylsiloxane, such as EtSi(OEt)$_3$, to separate the catalysts on the surface.[111]

Despite these facts, the immobilization of ligands using Si(OEt)$_3$ groups can be very effective. Alper, Manzer and co-workers have described a hydroformylation catalyst that is supported by a siloxane linker.[112] Unlike many supported hydroformylation catalysts, leaching of Rh is not observed, and the optimized catalysts can be reused multiple times without losses in activity. The uniqueness of the Alper approach is that the catalyst is isolated from the siloxane linkage and from the surface by long chain amide linkages arranged in a dendrimeric fashion. The best catalysts are obtained with the longest linkers (**48c**), which serve not only to isolate the catalyst from the surface, but also from each other (Figure 6). The Alper group has also introduced a more complex peptidic dendrimer supported on polystyrene (**49a,b** Figure 6).[113] Catalyst **49a**, which differs only in the ancillary groups on the *other side* of the support, displays better stability and recyclability than **49b**.

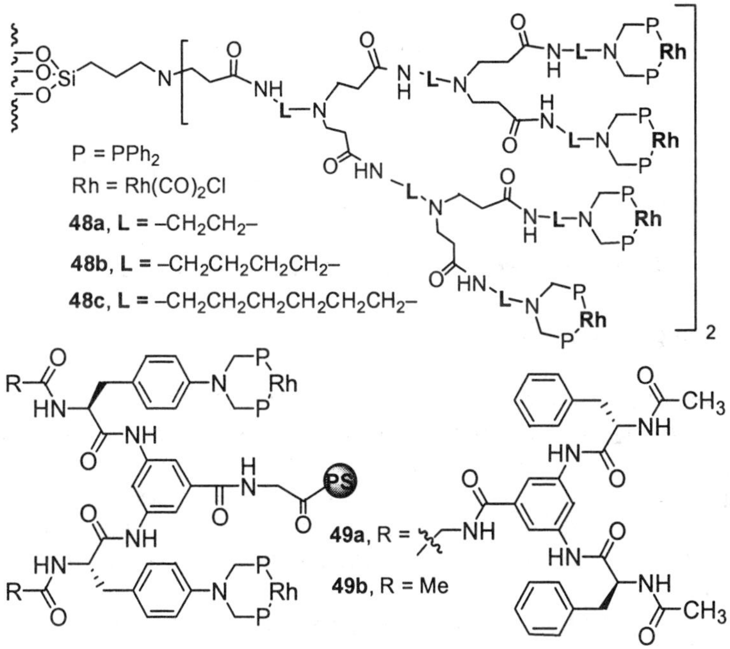

Figure 6. Supported dendrimers for hydroformylation.

An alternative method, which does not require the introduction of an $Si(OR)_3$ unit into the ligand, has been described by Kakkar and co-workers.[114] After washing the surface of a glass slide with piranha reagent, $SiCl_4$ is added to generate a monolayer containing reactive Si-Cl groups. Treatment of this material with $HNEt_2$ generates silylamides. The reaction of silyl amides with alcohol-containing ligands then takes place with elimination of $HNEt_2$, and formation of a surface siloxane bond. This method has the advantage that the catalyst is never exposed to acidic silanols on the surface,[113] yet it is immobilized by a strong Si-O bond.

3.2.3   Thiol Immobilization on Gold

Although the formation of self-assembled monolayers on flat gold surfaces has been extensively studied, the use of gold as a support for catalysts has received surprisingly little attention.[116] Gold nanoclusters can be prepared by reduction of gold salts such as $AuCl_4^-$ in the presence of long chain thiols as stabilizers.[117] Depending on the ratio of gold to thiol, various cluster sizes can be prepared. The resulting "monolayer-protected-clusters", or MPC's, are similar to dendrimers in terms of their physical properties. They are stable in air, can be precipitated, redissolved and purified by gel-permeation chromatography. As with dendrimers, MPC's can be difficult to handle since they form fine powders that can clog filters. Functionalized thiols can be introduced onto the MPC's by displacing the alkyl thiols used in the preparation of the MPC (the so-called "place exchange" method).[118] For an the immobilization of a chiral alkaloid on an MPC, see section 4.2.1.

4.   CATALYSTS IMMOBILIZED ON ORDERED MATERIALS FOR OXIDATION AND REDUCTION

4.1   Catalysts on Ordered Materials: Reduction

4.1.1   Mesoporous Molecular Sieves

Two reports have appeared on the hydrogenation of olefins using Rh complexes supported on mesoporous silicates. Shyu *et al.* have described the preparation and catalytic activity of a neutral Rh catalyst grafted onto MCM-41.[6] Our group has described the preparation and catalytic activity of a bidentate, cationic complex grafted onto SBA-15.[7] In both cases, the supported complexes are approximately three times more active than the corresponding homogeneous complexes.

In Shyu's method, MCM-41 was first treated with water to hydrate the surface, and then with $(EtO)_3SiCH_2CH_2CH_2PPh_2$ in refluxing benzene (eq. 28). Solid state CPMAS NMR spectroscopy showed that there was virtually no phosphorus(V) species resulting from interaction of the phosphine with the surface, as is occasionally observed.[115] The resulting material was treated with $ClRh(PPh_3)_3$ to give a supported catalyst with excess phosphine on the surface. Analysis of this material by CPMAS NMR was consistent with a supported analog of Wilkinson's catalyst, although the new peak was too broad and featureless to permit exact assignment of structure. Importantly, free phosphine remained on the surface.

The catalyst was tested for the hydrogenation of cyclohexene at 150 psi H_2 and 75 °C. Under these conditions, the heterogeneous catalyst was three times more active than Wilkinson's catalyst itself. Increased activity was obtained after the first run, stabilizing at approximately 7200 turnovers/hr. At room temperature, a turnover frequency of 150 turnovers/hr was calculated, after subtracting an induction period required to activate the catalyst.

Leaching of the catalyst was estimated by comparing the UV spectrum of the reaction mixture, filtered through regular filter paper, with that of the reaction mixture filtered through silica gel which is expected to trap any dissociated Rh in the solution. Knowing the extinction coefficient of Wilkinson's catalyst, the authors estimated the leaching of Rh from the surface to be minimal (<0.3%) when tested after the fifth run.[6,119]

Our group prepared a bidentate ligand (52) as shown in Scheme 5.[7] Treatment of diethylmalonate with sodium ethoxide and allyl bromide generated the required carbon skeleton. The ester groups were reduced with $LiAlH_4$ and then converted into chlorides. Hydrosilylation of the olefin took place in the presence of platinic acid, and the chloride groups were converted into phosphines. The phosphine was purified by silica gel chromatography after protection as the bis-BH_3 derivative. After purification, the phosphine groups were revealed by treatment with morpholine.

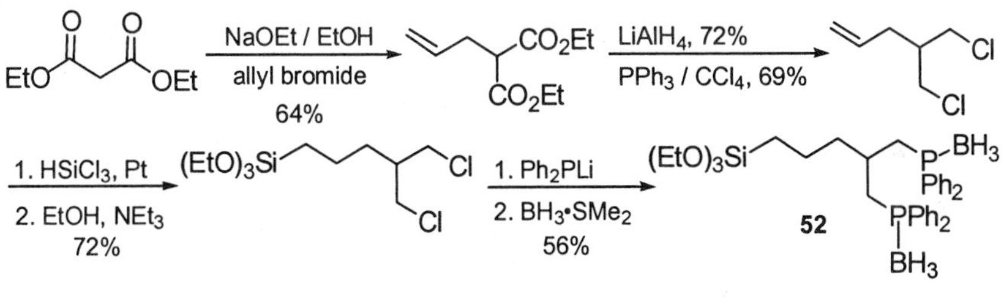

Scheme 5. Preparation of siloxane-modified bisphosphine.

Rh complex **51** was then prepared by reaction of deprotected phosphine **52** with $[Rh(COD)(THF)_2]^+BF_4^-$ (Scheme 6). The cationic complex was chosen to prevent any possible dimerization through chloride bridges, which is known to inactivate the catalyst.[120]

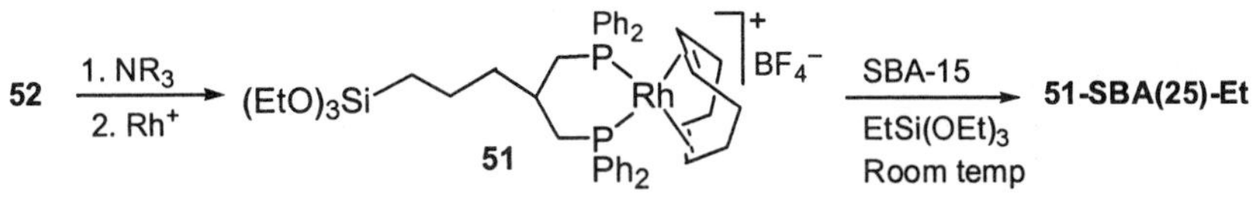

Scheme 6. Preparation of supported hydrogenation catalysts.

The resulting complex was characterized by solution NMR (Figure 8a). In order to prepare a heterogenized catalyst, Rh complex **51** and SBA-15 were stirred in an aromatic hydrocarbon solvent at different temperatures. Varying amounts of P(V) species were detected by ^{31}P CPMAS NMR as a broad signal centered at *ca.* 35 ppm. Blumel, Wasylishen and Fyfe have observed this same species when grafting phosphines onto silica.[115] Blumel attributes this signal to a compound of type **53** (Figure 7).

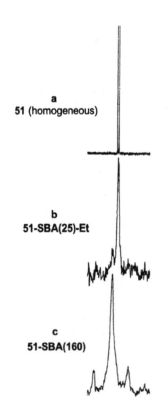

Figure 7. Surface oxidation of phosphines.

The amount of this species increased with the grafting temperature. At 165°C, the P(V) species was the only one observed (Figure 8c), but even at room temperature some of this species was present. Notably, when **51** was grafted at room temperature simultaneously with EtSi(OEt)$_3$ (**51-SBA(25)-Et**),[121] only minimal amounts of **53** were observed (Figure 8b). Of all the catalysts examined, **51-SBA(25)-Et** was the most active for the hydrogenation of isosafrole (**54**) giving 3300 turnovers/hr at room temperature (Table 4). Under the same conditions, a homogeneous catalyst prepared by mixing [Rh(COD)$_2$]$^+$BF$_2^-$ and DPPP in a 1:1 ratio gave only 350 turnovers/hr, and the homogeneous Rh complex itself (**51**) gave 700 turnovers/hr. The activity of the catalyst seems to be inversely proportional to the loading, a fact also observed by Shyu[6,119] and Bianchini.[100] This may be explained by increased site isolation at lower loadings, which is consistent with the increased activity observed when EtSi(OEt)$_3$ is present during grafting. An alternative explanation is that the reaction is diffusion limited and so the less accessible Rh catalysts inside the pores are not contributing to the observed rate of reaction.[22b] This fact is currently being tested by examining catalysts of different particle sizes.

The leaching of Rh from the surface was addressed by filtering the catalyst through a 0.45 µm Nalgene syringe filter. The resulting filtrate was treated with aqueous acid and the Rh content determined by ICP-MS. The amount of Rh that leached from the catalyst depends on which run is examined and also on the loading of catalyst. For example, catalyst **51-SBA(25)-Et**

a
51 (homogeneous)

b
51-SBA(25)-Et

c
51-SBA(160)

Figure 8. NMR spectra of **51** in solution (**a**), on support grafted at 25°C with EtSi(OEt)$_3$ (**b**), and at 160°C without additives (**c**).

Table 4. Hydrogenation of isosafrole (**54**) with Rh catalysts.

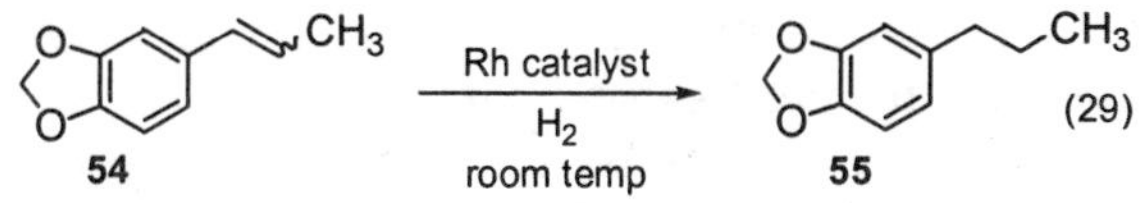

Entry	Catalyst	%Rh (wt%)	TOF (hr^{-1})
1	**Rh$^+$/DPPP**	n.a.	350
2	**51**	n.a.	750
3	**51-SBA(110)**	1.4%	2000
4	**51-SiO$_2$ (110)**	0.98%	1330
5	**51-SBA(25)**	2.0%	200
6	**51-SBA(165)**	1.4%	1030
7	**51-SBA(25)-Et**	0.34%	3300
8	**Rh$^+$/SBA**	3.3%	2000

lost 21-36% of the Rh after the first use, but only 5% in subsequent runs, while catalyst **51-SBA**(110) lost only 2-3% Rh after the first run. It should be noted that in these cases, the catalysts still retained over 90% of their original activity despite the leaching observed.

Our current efforts are directed towards the preparation of a chiral phosphine supported on mesoporous silica. The route being examined is based on Lemaire's synthesis of diaminoBINAP (**17**, eq. 15). In order to permit maximum conformational mobility, we will bind the catalyst to the surface using only one Si(OEt)$_3$ group. Compound **56** can be prepared by reaction of bis-TBDMS protected Binol with one equivalent of bromine, followed by conversion of the TBDMS groups into triflates (Scheme 7). CuCN treatment generates **57**, which is reacted with HPPh$_2$ in the presence of NiCl$_2$dppe. This reaction is capricious,

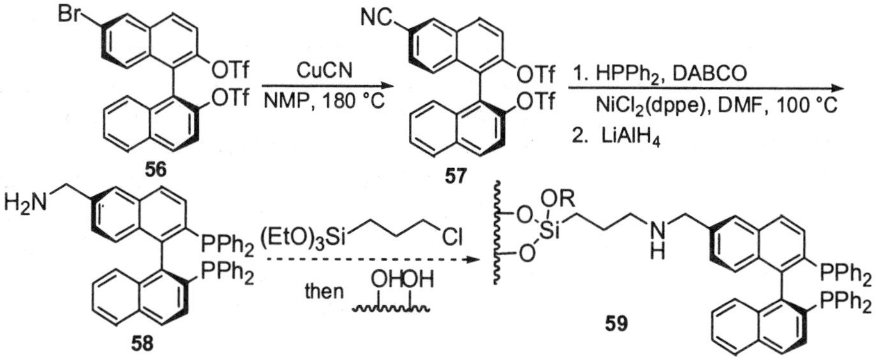

Scheme 7. Preparation of a supported BINAP ligand.

and must be carried out with strict exclusion of air and on a moderately large scale. Reduction of the cyano substituent with LiAlH$_4$ yields the primary amine **58**. Reaction of this species with 3-chloropropyl trimethoxysilane is slow, but catalysis by Bu$_4$NI is being investigated. Compound **59** will be reacted with the SBA surface in the presence of EtSi(OEt)$_3$ to prepare a supported chiral catalyst. The use of this material as a ligand for asymmetric hydrogenation will then be examined.

4.2   Catalysts on Ordered Materials: Asymmetric Di-hydroxylation

The dihydroxylation reaction converts olefins into diols with high enantioselectivity using derivatives of natural products called cinchona alkaloids, that are readily available in enantiomerically pure form (**62**, Scheme 8).[122] The reaction proceeds with >90% enantioselectivity, has excellent functional group tolerance and wide substrate scope.

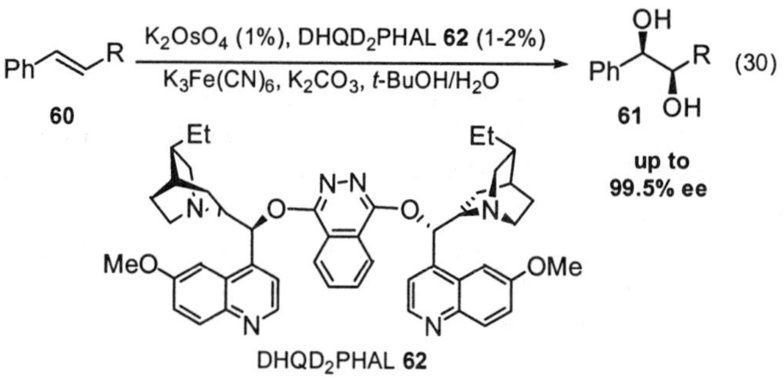

Scheme 8. The asymmetric dihydroxylation (AD) reaction.

4.2.1   Cinchona Alkaloids on Amorphous Supports

Sharpless and Kim reported the first example of a supported cinchona alkaloid (**63**, Figure 9).[123] Polymer **63** was a poor promoter (68% conversion after 7 days, ee not reported), but Salvadori showed that its activity could be improved by decreasing the loading of the alkaloid. At 20% loading there is no reaction, but at 4% loading, the dihydroxylation of *trans*-stilbene (**60**, R = Ph) took place with 27% ee.[124] Higher ee (45%) was obtained by varying the chlorobenzoyl group on the alkaloid. Adding a spacer between the polymer chain and the ligand as in **64** gave an even more effective system (*ca.* 85% ee).[124] Salvadori[125] and others[126] also reported the preparation and testing of polystyrene-based ligands such as **65**.

In more recent work, the Salvadori group showed that optimization of the polymer backbone (for example increasing its polarity) provided

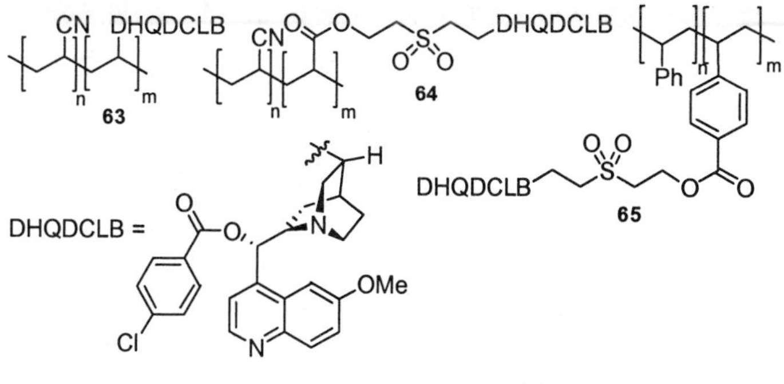

Figure 9. Cinchona alkaloids supported on organic polymers.

enantioselectivities rivaling those of the homogeneous ligand.[127] Careful tuning of the backbone structure[128] was necessary because of the solvent mixture used in this reaction, 1:1 'BuOH/H$_2$O, is challenging for organic polymers.[129]

A further complication of organic polymer-bound ligands is that the chiral ligand can end up merely occluded in, not chemically bound to, the polymer. For example, the free-radical polymerization of unmodified alkaloids such as H$_2$C=CH-DHQDCLB will only occur with highly electron-deficient monomers such as acrylonitrile. Several publications have appeared describing the apparent polymerization of simple alkaloid monomers with acrylate derivatives,[130] which have been subsequently refuted.[131]

Since the second generation dimeric Sharpless ligands of the DHQD$_2$-PHAL type (**62**) give significantly improved results, most recent studies have focused on incorporating models of this ligand. Immobilization on silica gel was first described by Lohray, but the resulting catalysts were less active and less selective than the homogenous catalyst.[132] By using a very rigid tether to restrict interactions of the catalyst with the surface, significantly better results were obtained by Bolm (**66a**, Figure 10).[133] A similar strategy was used by the Bolm

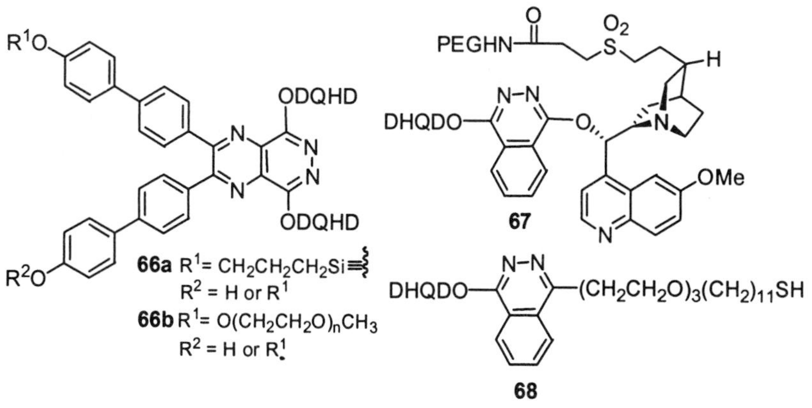

Figure 10. Second generation supported AD ligands.

group in the preparation of soluble polymer-bound ligand **66b**.[134]   Another promising soluble ligand was described by Janda (**67**), who was the first to describe the use of soluble polymers in this area.[135] Finally, Mrksich and co-workers reported the immobilization of a monomeric cinchona alkaloid (**68**) on gold nanoclusters.[116] The monomeric alkaloid (DHQDPHAL) was modified to incorporate a long chain thiol (**68**) that was then immobilized on 2.5 nm gold clusters. The resulting ligands are active in the AD reaction, and show good enantioselectivity and activity. Isolation of the supported ligand is a bit problematic, since the nanoparticles clog regular filter paper and gel permeation chromatography must be employed. A decrease in enantioselectivity is also observed upon reuse. Despite these facts, the use of gold nanoclusters as catalyst supports is an under-developed and intriguing concept.

4.2.2   Cinchona Alkaloids on Mesoporous Molecular Sieves

An alternative approach to distancing the chiral ligand from the amorphous support is to use a support that puts the catalyst in an ordered microenvironment. Two groups have reported the use of mesoporous molecular sieves as supports for cinchona alkaloid ligands.[10,11] Both methods begin with compound **72**, which is prepared as shown in Scheme 9.

Kim *et al.* reacted the olefin **72** with hydroxythiols under radical conditions in a similar manner to the original work of Sharpless and Salvadori (eq. 31). In our approach, **72** was hydrosilylated with HSi(OEt)3 in the presence of catalytic H_2PtCl_6. Compound **73** was then immobilized by reaction with SBA-15 or with amorphous silica gel in CH_2Cl_2 at room temperature in the presence of

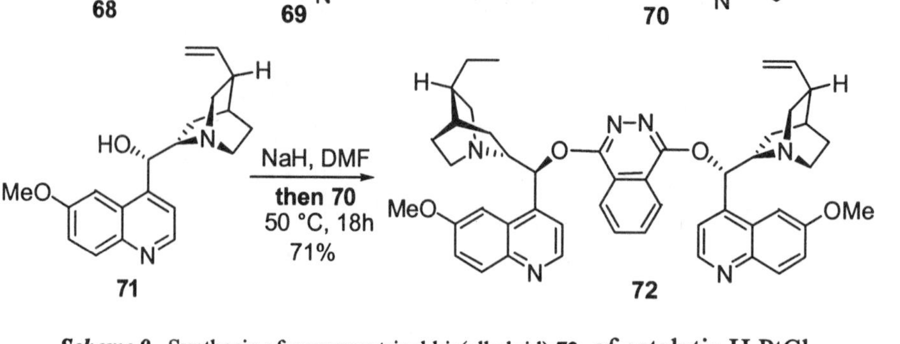

Scheme 9. Synthesis of unsymmetrical bis(alkaloid) **72**. of catalytic H_2PtCl_6.

triethylamine. This gave the two supported catalysts, **73-SBA** and **73-SiO₂**. Kim immobilized compounds **74a** and **74b** using SBA-15 that had been pre-reacted with 3-chloropropyltriethoxysilane yielding catalysts **74a-SBA** and **74b-SBA**. These immobilized catalysts were then tested in the asymmetric dihydroxylation of several olefins (Table 5).

As shown in Table 5, both **73-SBA** and **74b-SBA** are good mimics of the homogeneous catalyst. The silica gel supported alkaloid **73-SiO₂** is an effective ligand for the AD reaction, but consistently underperforms **73-SBA** and **74b-SBA**. Of the two ligands prepared by Kim, **74b**, with a longer spacer between the ligand and the support, performs consistently better than **74a-SBA**.

Table 5. Asymmetric dihydroxylation with supported cinchona alkaloids.

Substrate	62	73-SBA	73-SiO₂	74a-SBA	74b-SBA
Ph⟋⟍Ph	>99%	>99%	–	99%	>99%
Ph⟋⟍CO₂Et	95%	98%	94%	98%	98%
Ph⟋⟍Me	97%	98%	96%	95%	98%
Ph⟋⟍H	96%	92%	87%	87%	96%
Bu⟋⟍Bu	93%	–	–	75%	87%
Ph⟋⟍(Me)	93%	–	–	75%	90%

In both systems, the ligands are reusable but recharging with osmium is necessary after each use. Ligand **73-SBA** was reused 3 times without discernable loss of enantioselectivity. Ligands **74a/b-SBA** did display some decrease in enantioselectivity after multiple uses, with **74b-SBA** again being the better ligand. In the AD of stilbene, the ee decreased from 99% on the first run to 94% on the second. After six runs, the ee of the product was 92%. For **74a-SBA**, the ee of the product had dropped to 88% after 6 runs.

4.2.3   Immobilization of Osmium for Dihydroxylation Reactions

As noted above, and indeed in virtually all supported cinchona alkaloid systems for the AD reaction, the osmium is not retained on the support.[136] This is presumably due to the relatively low affinity of the tertiary nitrogen for osmium. Two elegant approaches to retain osmium have been reported recently. Jacobs and co-workers took advantage of two key facts concerning the catalytic dihydroxylation mechanism in order to prepare a catalyst that retains osmium. Firstly, tetrasubstituted olefins are not converted into diols *catalytically*, instead, they react with OsO_4 to give a stable osmium glycolate that is not easily hydrolyzed. Secondly, such osmium glycolates can function as dihydroxylation *catalysts*. It is this "second-cycle" that leads to lower selectivities in the Sharpless asymmetric system, and necessitates the use of $K_3Fe(CN)_6$ as a reoxidant.[129] The Jacobs group concluded that a tetrasubstituted olefin supported on silica gel would be capable of retaining osmium during a dihydroxylation reaction. The desired support was prepared as shown in Scheme 10.

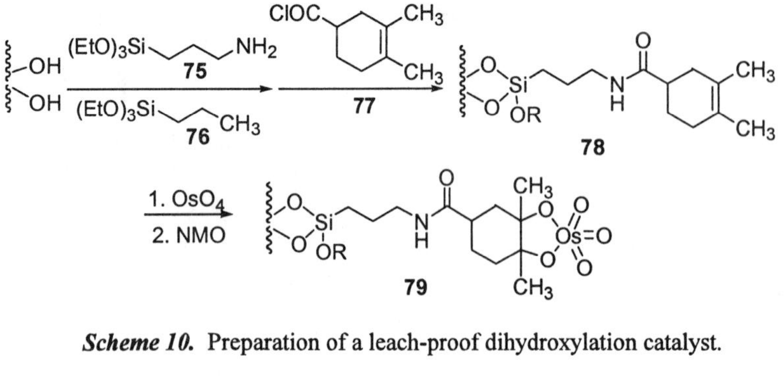

Scheme 10. Preparation of a leach-proof dihydroxylation catalyst.

Supported osmium catalyst **79** was employed in the dihydroxylation of a variety of olefins. The rate of the reaction was strongly dependent on the ratio of amino groups to propyl groups on the support (**75** : **76**). If only **75** was grafted, the resulting catalyst was inactive. At a 1:1 ratio, 5% conversion was obtained (24hr), while a 9:1 ratio in favour of **76** gives 60% conversion (24hr). Jacobs *et al.* attribute the difference in activity to site isolation promoted by the propyl groups, and to an overall lower loading of osmium in the 9:1 catalyst. In compound **78**, if the olefinic groups are too close together, reaction with osmium can occur to place two tetrasubstituted glycolates on the same osmium, rendering it inactive. The only drawback to this inventive and rational approach is that the dihydroxylation occurs without a chiral ligand, so the diols are racemic.

    Choudary has reported that layered double hydroxides treated with OsO_4^{2-} are able to affect the dihydroxylation of olefins without leaching.[137] The most intriguing aspect of this report is that Choudary also incorporated Pd into the support and used it to carry out a *sequential* Heck reaction/AD reaction (eq. 33). Although the reaction is enantioselective, the chiral ligand ($DHQD_2PHAL$) is not tethered to the catalyst, and thus is not recovered upon filtration.[138] Despite a considerable amount of innovative research in the area of supported dihydroxylation chemistry, a support that retains *both* osmium *and* chiral ligand is still needed.[139]

$$Ph\diagdown + ArI \xrightarrow[\substack{2.\ DHQD_2PHAL\ (1\text{-}2\%) \\ t\text{-}BuOH/H_2O,\ NMO,\ H_2O_2,\ RT,\ 12h}]{1.\ Os/Pd/LDH\ (1\%),\ NEt_3,\ 70°C,\ 8h} Ph\overset{OH}{\underset{OH}{\diagdown}}Ar \quad (33)$$

83-95% yield, 95-99% ee

5.   CONCLUSIONS AND FUTURE OUTLOOK

    The concept of immobilizing homogeneous transition metal complexes has advanced considerably since its inception 30 years ago. It has become clear that the microenvironment of the catalyst can have a significant impact on its activity, just as the individual amino acids in the active site of an enzyme are only partially responsible for the catalytic activity of the enzyme.

    Three general approaches to design of the active site have emerged. In the first, exemplified by molecular imprinting, the active site is specifically designed to have shaped cavities, and appropriately positioned functional groups/ligands. This is an intriguing approach, but the fact that the most active sites are also the least selective is a drawback that must be addressed.

    In the second approach, an attempt is made to reproduce the microenvironment of solution phase, which is the phase in which the catalyst was first optimized. This approach is typified by the use of soluble organic polymers. Although useful, this technique suffers from the fact that large quantities of solvent are often needed to precipitate the polymer, and only a limited number of solvents can be used for the reaction. Unique solutions to these problems have been provided by some research groups, for example, the thermomorphic polymers developed by Bergbreiter and co-workers.

    Finally, another approach is to neither actively design the active site, nor attempt to reproduce solution-phase conditions, but to support the catalyst on an ordered structure and thereby reduce the diversity of active sites. An amorphous support that has pores of different sizes and shapes presents several different microenvironments to the catalyst, while ordered supports such as zeolites and mesoporous molecular sieves presumably have less diversity of grafting sites, leading to a greater homogeneity of catalyst environments. The method of grafting in this case becomes important to ensure that each complex is in a similar site.

Significant advances have also been made recently in terms of methods for binding catalysts to recoverable supports. Supported aqueous phase chemistry (Davis), supported hydrogen-bonding (Bianchini), ionic immobilization (Broene, Tumas) and immobilization with the assistance of heteropolyacids (Augustine) are all extremely important additions to the literature. The last two are particularly impressive since modification of the ligand is not required. Although preliminary, the use of gold nanoclusters (MPC's) also holds significant promise.

Finally, the advent of reliable, reproducible methods for the preparation of mesoporous molecular sieves offers some exciting new opportunities in design of catalyst active sites. Johnson,[140] Bonneviot,[141] Stein[142] and Maschmeyer, Seddon and Bruce[143] have shown that it is possible to introduce catalysts or other compounds specifically on the inside of the pores. This is accomplished using either a selective protection of the external surface followed by functionalization,[142] or by using the template to deliver the desired group,[145] or finally, by displacing the template *with* the desired functional group.[143,144] As demonstrated by Johnson and Thomas,[5a] selective functionalization of the internal surface can lead to considerably different (improved) catalyst selectivity.

The use of bridged siloxanes such as $(EtO)_3SiCH_2CH_2Si(OEt)_3$ in place of $Si(OEt)_4$ provides structured organic/inorganic composites if the hydrolysis/condensation is carried out in the presence of a surfactant template, and the Si–C bond survives both the sol-gel conditions and the removal of the template. Inagaki,[144] Ozin,[145] Stein[146] and Sayari[147] have reported such materials. The preparation of chiral organic/inorganic composites with ordered pores and morphology would represent a considerable advance in the area of ordered materials.

ACKNOWLEDGMENTS

Merck Frosst and Dr. Robert Young are thanked for generous financial support of the research described herein, as are the Canada Foundation for Innovation, the Regional Economic Development Fund and the Natural Sciences and Engineering Research Council of Canada. My students and postdoctoral fellows are thanked for their tireless enthusiasm and dedication to chemistry.

REFERENCES AND NOTES

1. Kresge, C.T.; Leonowicz, M.E.; Roth, W.J.; Vartulli, J.C.; Beck, J.S. *Nature* **1992**, *359*, 710. Beck, J.S.; Vartulli, J.C.; Roth, W.J.; Leonowicz, M.E.; Kresge, C.T.; Schmitt, K.D.; Chu, C.T.-W.; Olson, D.H.; Sheppard, E.W.; McCullen, S.B.; Higgins, J.B.; Schlenker, J.L. *J. Am. Chem. Soc.* **1992**, *114*, 10834.

2. Zhao, D.; Feng, J.; Huo, Q.; Melosh, N.; Fredrickson, G.H.; Chmelka, B.F.; Stucky, G.D. *Science* **1998**, *279*, 548. Zhao, D.; Huo, Q.; Feng, J.; Chmelka, B.F.; Stucky, G.D. *J. Am. Chem. Soc.* **1998**, *120*, 6024.

3. Yanagisawa, T.; Shimizu, T.; Kuroda, K.; Kato, C. *Bull. Chem. Soc. Jpn.* **1990**, *63*, 988. Inagaki, S.; Koiwai, A.; Suzuki, N.; Fukushima, Y.; Kuroda, K. *Bull. Chem. Soc. Jpn.* **1996**, *69*, 1449. Inagaki, S.; Sakamoto, Y.; Fukushima, Y.; Terasaki, O. *Chem. Mater.* **1996**, *8*, 2089.

4. Reviews: Stein, A.; Melde, B.J.; Schroden, R.C. *Adv. Mater.* **2000**, *12*, 1403. Moller, K.; Bein, T. *Chem. Mater.* **1998**, *10*, 2950. Brunel, D.; Bellocq, N.; Sutra, P.; Cauvel, A.; Laspéras, M.; Moreau, P.; Di Renzo, F.; Galarneau, A.; Fajula, F. *Coord. Chem. Rev.* **1998**, *178-180*, 1085. Ying, J.Y.; Mehnert, C.P.; Wong, M.S. *Angew. Chem. Int. Ed.* **1999**, *38*, 56. Corma, A. *Chem. Rev.* **1997**, *97*, 2373. Thomas, J.M. *Angew. Chem. Int. Ed.* **1999**, *38*, 3588. He, X.; Antonelli, D. *Angew. Chem. Int. Ed.* **2002**, *41*, 214.

5. Asymmetric allylic amination: Johnson, B.F.G.; Raynor, S.A.; Shephard, D.S; Mashmeyer, T.; Thomas, J.M.; Sankar, G.; Bromley, S.; Oldroyd, R.; Gladden, L.; Mantle, M.D. *Chem. Commun.* **1999**, 1167. Asymmetric hydrogenation of enamines: Raynor, S.A.; Thomas, J.M.; Raja, R.; Johnson, B.F.G.; Bell, R.G.; Mantle, M.D. *Chem. Commun.* **2000**, 1925. Asymmetric addition of diethyl zinc to aldehydes: Laspéras, M.; Bellocq, N.; Brunel, D.; Moreau, P. *Tetrahedron: Asymm.* **1998**, *9*, 3053. Kim, S.-W.; Bae, S.J.; Hyeon, T.; Kim, B.M. *Micropor. Mesopor. Mater.* **2001**, *44-45*, 523. Bae, S.J.; Kim, S.-W.; Hyeon, T.; Kim, B.M. *Chem. Commun.* **2000**, 31. Asymmetric epoxidation: Zhou, X.-G.; Yu, X.-Q.; Huang, J.-S.; Li, S.G.; Li, L.-S.; Che, C.-M. *Chem. Commun.* **1999**, 1789. Kim, G.-J.; Kim, S.-H. *Catal. Lett.* **1999**, *57*, 139. Pauson Khand: Kim, S.W.; Son, S.U;. Lee, S.I.; Hyeon, T.; Chung, Y.K. *J. Am. Chem. Soc.* **2000**, *122*, 1550. Polymerization: Kageyama, K.; Tamazawa, J.-I.; Aida, T. *Science* **1999**, *285*, 2113. Hydroformylation: Nowotny, M.; Maschmeyer,T.; Johnson, B.F.G.; Lahuerta, P.; Thomas, J.M.; Davies, J.E. *Angew. Chem. Int. Ed.* **2001**, *40*, 955. Heck reaction: Mehnert, C.P.; Ying, J.Y. *Chem. Commun.* **1997**, 2215. Mehnert, C.P.; Weaver, D.W.; Ying, J.Y. *J. Am. Chem. Soc.* **1998**, *120*, 12289. Suzuki Reaction: Kosslick, H.; Mönnich, I.; Paetzold, E.; Fuhrmann, H.; Fricke, R.; Müller, D.; Oehme, G. *Micropor. Mesopor. Mater.* **2001**, *44-45*, 537. Epoxidation: Liu, C.-J.; Yu, W.-Y.; Li, S.-G.; Che, C.-M. *J. Org. Chem.* **1998**, *63*, 7364. Cheng, A.K.-W.; Lin, W.-Y.; Li, S.-G.; Che, C.-M.; Pang, W.-Q. *New J. Chem.* **1999**, *23*, 733. Butterworth, A.J.; Clark, J.H.; Walton, P.H.; Barlow, S.J. *J. Chem. Soc., Chem. Comm.* **1996**, 1859. Chisem, I.C.; Rafelt, J.; Shieh, M.T.; Chisem, J.; Clark, J.H.; Jachuck, R.; Macquarrie, D.; Ramshaw, C.; Scott, K. *J. Chem. Soc., Chem. Comm.* **1998**, 1949. Das, T.K.; Chaudhari, K.; Nandanan, E.; Chandwadkar, A.J.; Sudalai, A.; Ravindranathan, T.; Sivasanker, S. *Tetrahedron Lett.* **1997**, *38*, 3631. Maschmeyer, T.; Rey, F.; Sankar, G.; Thomas, J.M. *Nature* **1995**, *378*, 159. Ernst, S.; Selle, M. *Micropor. Mesopor. Mater.* **1999**, *27*, 355. Maschmeyer, T.; Oldroyd, R.D.;

Sankar, G.; Thomas, J.M.; Shannon, I.J.; Klepetko, J.A.; Masters, A.F.; Beattie, J.K.; Catlow, C.R.A. *Angew. Chem. Int. Ed.* **1997**, *36*, 1639. Arens, I.W.C.E.; Sheldon, R.A.; Wallau, M.; Schuchardt, U. *Angew. Chem. Int. Ed.* **1997**, *36*, 1145.

6.   Shyu, S.-G.; Cheng, S.-W.; Tzou, D.-L. *Chem. Commun.* **1999**, 2337.

7.   Crudden, C.M.; Allen, D.; Mikoluk, M.D.; Sun, J. *Chem. Commun.* **2001**, 1154.

8.   Liu, A.M.; Hidajat, K.; Kawi, S. *J. Mol. Catal. A: Chem.* **2001**, *168*, 303. See also ref. 5(a).

9.   Lasperas, M.; Lloret, T.; Chaves, L.; Rodriguez, I.; Cauvel, A.; Brunel, D. *Stud. Surf. Sci. Catal.* **1997**, *108*, 75. Subba Rao, Y.V.; De Vos, D.E.; Jacobs, P.A. *Angew. Chem. Int. Ed.* **1997**, *36*, 2661. Cauvel, A.; Renard, G.; Brunel, D. *J. Org. Chem.* **1997**, *62*, 749. Subba Rao, Y.V.; De Vos, D.E.; Jacobs, P.A. *Angew. Chem. Int. Ed.* **1997**, *36*, 2261. Jaenicke, S.; Chuah, G.K.; Lin, X.H.; Hu, X.C. *Micropor. Mesopor. Mater.* **2000**, *35–36* 143. Lin, X.; Chuah, G.K.; Jaenicke, S. *J. Mol. Catal. A: Chem.* **1999**, *150*, 287. Macquarrie, D.J.; Jackson, D.B.; Tailland, S.; Utting, K.A. *J. Mater. Chem.* **2001**, *11*, 1843. Demicheli, G.; Maggi, R.; Mazzacani, A.; Righi, P.; Sartori, G.; Bigi, F. *Tetrahedron Lett.* **2001**, *42*, 2401. Thoelen, C.; Van de Walle, K.; Vankelecom, I.F.J.; Jacobs, P.A. *Chem. Commun.* **1999**, 1841.

10.   Motorina, I.; Crudden, C.M. *Org. Lett.* **2001**, *3*, 2325.

11.   Lee, H.M.; Kim, S.-W.; Hyeon, T.; Kim, B.M. *Tetrahedron: Asymm.* **2001**, *12*, 1537.

12.   Noyori, R. *Asymmetric Catalysis in Organic Synthesis*, John Wiley & Sons: New York, 1994. Ojima, I. *Catalytic Asymmetric Synthesis*, 2nd Ed. VCH Publishers: New York, 2000. Jacobsen, E.N.; Pfaltz, A; Yamamoto, H. *Comprehensive Asymmetric Catalysis*, Springer-Verlag: Heidelberg, 1999.

13.   Another limitation of asymmetric catalysis is the lack of generality many catalytic reactions exhibit. Asymmetric dihydroxylation (Sharpless) is exceptionally general, as is the asymmetric reduction of ketones (Noyori) and alkenes with RuBINAP (Noyori).

14.   Orito, Y.; Imai, S.; Niwa, S. *J. Chem. Soc. Jpn.* **1980**, *4*, 670. Blaser, H.U.; Jalett, H.P.; Lottenbach, W.; Studer, M. *J. Am. Chem. Soc.* **2000**, *122*, 12675. von Arx, M.; Mallat, T.; Baiker, A. *Angew. Chem. Int. Ed.* **2001**, *40*, 2302. LeBlond, C.; Wang, J; Liu, J.; Andrews, A.T.; Sun, Y.-K. *J. Am. Chem. Soc.* **1999**, *121*, 4920. The hydrogenation of diketones can also be accomplished with high ee using tartaric acid modified Raney nickel: Tai, A.; Kikukawa, T.; Sugimura, T.; Inoue, Y.; Osawa, T.; Fujii, S. *J. Chem. Soc., Chem. Comm.* **1991**, 795.

15.   Dumont, W.; Poulin, J.-C.; Dang, T.-P.; Kagan, H.B. *J. Am. Chem. Soc.* **1973**, *95*, 8295.

16.   Bruner, H.; Bailar Jr., J.C. *Inorg. Chem.* **1973**, *12*, 1465.

17.   Grubbs, R.H.; Kroll, L.C. *J. Am. Chem. Soc.* **1971**, *93*, 3062.

18.   Collman, J.P.; Hegedus, L.S.; Cooke, M.P.; Norton, J.R.; Dolcetti, G.; Marquardt, D.N. *J. Am. Chem. Soc.* **1972**, *94*, 1789.

19.   Capka, M.; Svoboda, P.; Creny, M.; Hetflejs, J. *Tetrahedron Lett.* **1971**, 4787.

20.   Bailey, D.C.; Langer, S.H. *Chem. Rev.* **1981**, *81*, 109. Pittman, C.U., Jr. in *Comprehensive Organometallic Chemistry*; Wilkinson, G., Stone, F.G.A.; Abel, E.W. Eds.; Pergamon Press: Oxford, 1982, Vol. 8. Hartley, F.R.; Vezey, P.N. *Adv. Organomet. Chem.* **1977**, *15*, 189. Hartley, F.R. *Supported Metal Complexes*, Reidel, Dordrecht, 1985.

21.   Lindner, E.; Schneller, T.; Auer, F.; Mayer, H.A. *Angew. Chem. Int. Ed.* **1999**, *38*, 2154.

22.   For example: Holland, B.T.; Walkup, C.; Stein, A. *J. Phys. Chem. B* **1998**, *102*, 4301. Seen, A.J.; Townsend, A.T.; Bellis, J.C.; Cavell, K.J. *J. Mol. Catal. A: Chem.* **1999**, *149*, 233. Quiroga, M.E.; Cagnola, E.A.; Liprandi, D.A; L'Argentiere, P.C. *J. Mol. Catal. A: Chem.* **1999**, *149*, 147. Yang, H.; Gao, H.; Angelici, R.J. *Organometallics* **2000**, *19*, 622. Nagel, U.; Leipold, J. *Chem. Ber.* **1996**, *129*, 815.

23.   Pugin, B. *J. Mol. Catal. A: Chem.* **1996**, *107*, 273.

24.   Drago, R.; Pribich, D.C. *Inorg. Chem.* **1985**, *24*, 1983.

25.   Hodge, P. *Chem. Soc. Rev.* **1997**, *26*, 417.

26.   Merrifield, R.B. *J. Am. Chem. Soc.* **1963**, *85*, 2149. Merrifield, R.B. *Science* **1965**, *150*, 178.

27.   Phosphines are also known to break down by phosphorus-carbon bond cleavage, which can be a second contributor to leaching: Garrou, P.E. *Chem. Rev.* **1985**, *85*, 171. Carty, A.J. *Pure Appl. Chem.* **1982**, *54*, 113.

28.   Santini, R.; Griffith, M.C.; Qi, M. *Tetrahedron Lett.* **1998**, *39*, 8951. Shuttleworth, S.J.; Allin, S.M.; Sharma, P.K. *Synthesis* **1997**, 1217.

29.   Shemyakin, M.M.; Ovchinnikov, Yu. A.; Kiryushkin, A.A.; Kozhevnikova, I.V. *Tetrahedron Lett.* **1965**, *6*, 2323. Mutter, M.; Hagenmaier, H.; Bayer, E. *Angew. Chem. Int. Ed.* **1971**, *10*, 811. Andreatta, R.H.; Rink, H. *Helv. Chim. Acta* **1973**, *56*, 1205. Narita, M. *Bull. Chem. Soc. Jpn.* **1978**, *51*, 1477. Gravert, D.J.; Janda, K.D. *Chem. Rev.* **1997**, *97*, 489. Wentworth Jr., P.; Janda, K.D. J. Chem. Soc., Chem. Comm. **1999**, 1917.

30.   Bergbreiter, D.E.; Weatherford, D.A. *J. Org. Chem.* **1989**, *54*, 2726. Bergbreiter, D.E. *CHEMTECH* **1987**, 686. Bergbreiter, D.E.; Chandran, R. *J. Am. Chem. Soc.* **1987**, *109*, 174. Hartley, F.R. *Brit. Polym. J.* **1984**, *16*, 199.

31.   Bergbreiter, D.E.; Osburn, P.L.; Wilson, A.; Sink, E.M. *J. Am. Chem. Soc.* **2000**, *122*, 9058. Bergbreiter, D.E.; Osburn, P.L.; Frels, J.D. *J. Am. Chem. Soc.* **2001**, *123*, 11105.

32.   Nozaki, K.; Itoi, Y.; Shibahara, F.; Shirakawa, E.; Ohta, T.; Takaya, H.; Hiyama, T. *J. Am. Chem. Soc.* **1998**, *120*, 4051. Nozaki, K.; Itoi, Y.; Shibahara, F.; Shirakawa, E.; Ohta, T.; Takaya, H.; Hiyama, T. *Bull. Chem. Soc. Jpn.* **1999**, *72*, 1911.

33.   Taylor, R.A.; Santora, B.P.; Gagné, M.R. *Org. Lett.* **2000**, *2*, 1781. Santora, B.P.; Larsen, A.O.; Gagné, M.R. *Organometallics* **1999**, *18*, 3138. Santora, B.P.; White, P.S.; Gagné, M.R. *Organometallics* **1999**, *18*, 2557.

34.   Nestler, O.; Severin, K. *Org. Lett.* **2001**, *3*, 3907 and references cited therein.

35.   Fisher, L.; Muller, R.; Ekberg, B.; Mosbach, K. *J. Am. Chem. Soc.* **1991**, *113*, 9358.

36.   Beach, J.V.; Shea, K.J. *J. Am. Chem. Soc.* **1994**, *116*, 379. Wulff, G.; Gross, T.; Schönfeld, R. *Angew. Chem. Int. Ed.* **1997**, *36*, 1962. Liu, X.-C.; Mosbach, K. *Macromol. Rapid Commun.* **1997**, *18*, 609. Robinson, D. K.; Mosbach, K. *J. Chem. Soc., J. Chem. Soc., Chem. Comm.* **1989**, 969. Strikovsky, A.G.; Kasper, D.; Grün, M.; Green, B.S.; Hradil, J.; Wulff, G. *J. Am. Chem. Soc.* **2000**, *122*, 6295.

37.   Brunkan, N.M.; Gagné, M.R. *J. Am. Chem. Soc.* **2000**, *122*, 6217.

38.   Koh, J.H.; Larsen, A.O.; White, P.S.; Gagné, M.R. *Organometallics* **2002**, *21*, 7.

39.   Polborn, K.; Severin, K. *Chem. Eur. J.* **2000**, *6*, 4604. A seven-fold increase in rate was observed with Ru systems: Polborn, K.; Severin, K. *Eur. J. Inorg. Chem.* **2000**, 1687. Polborn, K.; Severin, K. *Chem. Commun.* **1999**, 2481.

40.   Matsui, J.; Nicholls, I.A.; Karube, I.; Mosbach, K. *J. Org. Chem.* **1996**, *61*, 5414. Fujii, Y.; Matsutani, K.; Kikuchi, K. *J. Chem. Soc., Chem. Comm.* **1985**, 415.

41.   For a discussion of Lemaire's work in this area, see section 2.2.2.

42.   AIBN = 2,2'-azobisisobutryonitrile

43.   Alper, H.; Hamel, N. *J. Chem. Soc., Chem. Comm.* **1990**, 135. See also a cellulose-based phosphite: Kawabata, Y.; Tanaka, M.; Ogata, I. *Chem. Lett.* **1976**, 1213.

44.   The use of peptides as ligands for asymmetric catalysis has the advantage that many different peptides can be prepared rapidly using combinatorial methods, unlike the preparation of chiral phosphines which can be laborious. For the use of peptides as ligands or as catalysts themselves see: Vasbinder, M.M.; Jarvo, E.R.; Miller, S.J. *Angew. Chem. Int. Ed.* **2001**, *40*, 2824. Mizutani, H.; Degrado, S.J.; Hoveyda, A.H. *J. Am. Chem. Soc.* **2002**, *124*, 779. Josephsohn, N.S.; Kuntz, K.W.; Snapper, M.L.; Hoveyda, A.H. *J. Am. Chem. Soc.* **2001**, *123*, 11594.

45.   Lère-Porte, J.P.; Moreau, J.J.E.; Serein-Spirau, F.; Wakim, S. *Tetrahedron Lett.* **2001**, *42*, 3073.

46.   Fan, Q.-H.; Ren, C.-Y.; Yeung, C.-H.; Hu, W.-H.; Chan, A.S.C. *J. Am. Chem. Soc.* **1999**, *121*, 7407.

47.   For an excellent review see Yu, H.-B.; Pu, L. *Chem. Rev.* **2001**, *101*, 757.

48.   Yu, H.-B.; Hu, Q.-S.; Pu, L. *Tetrahedron Lett.* **2000**, *41*, 1681.

49.   Yu, H.-B.; Hu, Q.-S.; Pu, L. *J. Am. Chem. Soc.* **2000**, *122*, 6500

50.   Saluzzo, C.; Halle, R.; Touchard, F.; Fache, F.; Schulz, E.; Lemaire, M. *J. Organomet. Chem.* **2000**, *603*, 30.

51.   Halle, R.; Colasson, B.; Schulz, E.; Spagnol, M.; Lemaire, M. *Tetrahedron Lett.* **2000**, *41*, 643.

52.   Locatelli, F.; Gamez, P.; Lemaire, M. *J. Mol. Catal. A: Chem.* **1998**, *135*, 89.

53.   Gamez, P.; Dunjic, B.; Pinel, C.; Lemaire, M. *Tetrahedron Lett.* **1995**, *36*, 8779.

54.   For a review on nanoporous and mesoporous organic materials see: Langley, P.J.; Hulliger, J. *Chem. Soc. Rev.* **1999**, *28*, 279. For the preparation of ordered, mesoporous carbon: Jun, S.; Joo, S.H.; Ryoo, R.; Kruk, M.; Jaroniec, M.; Liu, Z.; Ohsuna, T.; Terasaki, O. *J. Am. Chem. Soc.* **2000**, *122*, 10712.

55.   Wuest, J.D. in *Mesomolecules: From Molecules to Materials*, Mendenhall, G.D.; Greenberg, A.; Liebman, J.F., Eds. Chapman & Hall: New York, 1995.

56.   Miller, S.A.; Kim, E.; Gray, D.H.; Gin, D.L. *Angew. Chem. Int. Ed.* **1999**, *38*, 3022.

57.   Gu, W.; Zhou, W.-J.; Gin, D.L. *Chem Mater.* **2001**, *13*, 1949.

58.   Sing, K.S.W.; Everett, D.H.; Haul, R.A.W.; Moscou, L.; Pierotti, R.A.; Rouquérol, J.; Siemieniewska, T. *Pure Appl. Chem.* **1985**, *57*, 603.

59.   Clark, J.H.; Macquarrie, D.J. *Chem. Soc. Rev.* **1996**, 303. Clark, J.H.; Macquarrie, D.J. *Chem. Commun.* **1998**, 853. See also reference 20.

60.   Davis, M.E. *Micropor. Mesopor. Mater.* **1998**, *21*, 173 and references therein.

61.   Pinnavaia, T.J.; Raythatha, R.; Lee, J.G.-S.; Halloran, L.J.; Hoffman, J.F. *J. Am. Chem. Soc.* **1979**, *101*, 6891. Pinnavaia, T.J. *Science* **1983**, *220*, 365. Sento, T.; Shimazu, S.; Ichikuni, N.; Uematsu, T. *Chem. Lett.* **1998**, 1191.

62.   Kumar, K.R.; Choudary, B.M.; Jamil, Z.; Thyagarajan, G. *J. Chem. Soc., Chem. Comm.* **1986**, 130. Choudary, B.M.; Kumar, K.R.; Jamil, Z.; Thyagarajan, G. *J. Chem. Soc., Chem. Comm.* **1985**, 937. Kumar, K.R.; Choudary, B.M.; Jamil, Z.; Thyagarajan, G. *J. Chem. Soc., Chem. Comm.* **1985**, 937.

63.   Aldea, R.; Alper, H. *J. Org. Chem.* **1998**, *63*, 9425.

64. Pd clay prepared with BIPY ligands: Valli, V.L.K.; Alper, H. *J. Am. Chem. Soc.* **1993**, *115*, 3778.

65. Margalef-Catala, R.; Claver, C.; Salagre, P.; Fernandez, E. *Tetrahedron: Asymm.* **2000**, *11*, 1469.

66. Segarra, A.M.; Guerrero, R.; Claver, C.; Fernandez, E. *Chem. Commun.* **2001**, 1808.

67. Corma, A.; Martinez, A. *Adv. Mater.* **1995**, *7*, 137. Dartt, C.B.; Davis, M.E. *Catal. Today* **1994**, *19*, 151.

68. For the ship-in-a-bottle method of encapsulation see: Kowalak, S.; Weiss, R.C.; Balkus Jr., K.J. *J. Chem. Soc., Chem. Comm.* **1991**, 57. Zhang, Z.; Dai, S.; Hunt, R.D.; Wei, Y.; Qiu, S. *Adv. Mater.* **2001**, *13*, 493. Ernst, S.; Fuchs, E.; Yang, X. *Micropor. Mesopor. Mater.* **2000**, *35-36*, 137.

69. For the seminal contributions of the Davis group see: Davis, M.E.; Saldarriaga, C.; Montes, C.; Garces, J.; Crowder, C. *Nature* **1988**, *331*, 698. Freyhardt, C.C.; Tsapatsis, M.; Lobo, R.F.; Balkus, K.J.; Davis, M.E. *Nature* **1996**, *381*, 295. Wagner, P.; Yoshikawa, M.; Lovallo, M.; Tsuji, K.; Taspatsis, M.; Davis, M.E. *Chem. Commun.* **1997**, 2179.

70. Corma, A.; Iglesias, M.; del Pino, C.; Sanchez, F. *J. Chem. Soc., Chem Comm.* **1991**, 1253.

71. Katz, A.; Davis, M.E. *Nature* **2000**, *403*, 286. Davis, M.E. *Stud. Surf. Sci. Catal.* **2000**, *130*, 49.

72. For an excellent review see: Raimondi, M.E.; Seddon, J.M. *Liquid Cryst.* **1999**, *26*, 305.

73. For the "true-liquid-crystal-templating" procedure see: Attard, G.S.; Glyde, J.C.; Göltner, C.G.; *Nature* **1995**, *378*, 366. Göltner, C.G.; Antonietti, M. *Adv. Mater.* **1997**, *9*, 431.

74. Wingen, A.; Anastasievic, N.; Hollnagel, A.; Werner, D.; Schüth, F. *J. Catal.* **2000**, *193*, 248. See also references 4 and 5.

75. Jung, J.H.; Ono, Y.; Hanabusa, K.; Shinkai, S. *J. Am. Chem. Soc.* **2000**, *122*, 5008. Ono, Y.; Nakahima, K.; Sano, M.; Kanekiyo, Y.; Inoue, K.; Hojo, J.; Shinkai, S. *Chem. Commun.* **1998**, 1477. Jung, J.H.; Ono, Y.; Shinkai, S. *Angew. Chem. Int. Ed.* **2000**, *39*, 1862. Jung, J.H.; Kobayashi, H.; Masuda, M.; Shimizu, T.; Shinkai, S. *J. Am. Chem. Soc.* **2001**, *123*, 8785.

76. For Binol-based organic/inorganic composites see: Brethon, A.; Hesemann, P.; Réjaud, L.; Moreau, J.J.E.; Man, M.W.C. *J. Organomet. Chem.* **2001**, *627*, 239. Hesemann, P.; Moreau, J.J.E. *Tetrahedron: Asymm.* **2000**, *11*, 2183.

77. Moreau, J.J.E.; Vellutini, L.; Man, M.W.C.; Bied, C. *J. Am. Chem. Soc.* **2001**, *123*, 1509.

78. Adima, A.; Moreau, J.J.E.; Man, M.W.C. *J. Mater. Chem.* **1997**, *7*, 2331.

79. Junges, U.; Jacobs, W.; Voigt-Martin, I.; Krutzsch, B.; Schüth, F. *J. Chem. Soc., Chem. Comm.* **1995**, 2283. Corma, A.; Martinez, A.; Martinez-Soria, V. *J. Catal.* **1997**, *169*, 480. Luan, Z.; Kevan, L. *J. Phys. Chem. B* **1997**, *101*, 2020. Mulukutla, R.S.; Asakura, K.; Namba, S.; Iwasawa, Y. *Chem. Commun.* **1998**, 1425.

80. Basset, J.-M.; Lefebvre, F.; Santini, C.C. *Coord. Chem. Rev.* **1998**, *178-180*, 1703.

81. Amor Nait Ajjou, J.; Rice, G.L.; Scott, S.L. *J. Am. Chem. Soc.* **1998**, *120*, 13436. Amor Nait Ajjou, J.; Scott, S.L. *J. Am. Chem. Soc.* **2000**, *122*, 8968. Richmond, M.K.; Scott, S.L.; Alper, H. *J. Am. Chem. Soc.* **2001**, *123*, 10521.

82. Vidal, V.; Théolier, A.; Thivolle-Cazat, J.; Basset, J.-M. *Science* **1997**, *276*, 99. Maury, O.; Lefort, L.; Vidal, V.; Thivolle-Cazat, J.; Basset, J.-M. *Angew. Chem. Int. Ed.* **1999**,

38, 1952. Chabanas, M.; Quadrelli, E.A.; Fenet, B.; Coperet, C.; Thivolle-Cazat, J.; Basset, J.-M.; Lesage, A.W.; Emsley, L. *Angew. Chem. Int. Ed.* **2001**, *40*, 4493.

83.   Anwander, R.; Runte, O.; Eppinger, J.; Gerstberger, G.; Herdtweck, E.; Spiegler, M. *J. Chem. Soc., Dalton Trans.* **1998**, 847. Gerstberger, G.; Anwander, R. *Micropor. Mesopor. Mater.* **2001**, *44-45*, 303.

84.   Kobayashi, S.; Endo, M.; Nagayama, S. *J. Am. Chem. Soc.* **1999**, *121*, 11229. See also: Patchornik, A.; Ben-David, Y.; Milstein, D. *J. Chem. Soc., Chem. Comm.* **1990**, 1090, for the immobilization of a Rh complex in polystyrene by mixing the two in THF followed by treatment and extraction with MeOH.

85.   Rosenfeld, A.; Avnir, D.; Blum, J. *J. Chem. Soc., Chem. Comm.* **1993**, 583.

86.   Gelman, F.; Avnir, D.; Schumann, H.; Blum, J. *J. Mol. Catal. A: Chem.* **1999**, *146*, 123.

87.   Sertchook, H.; Avnir, D.; Blum, J.; Joo, F.; Katho, A.; Schumann, H.; Weimann, R.; Wernik, S. *J. Mol. Catal.: A Chem.* **1996**, *108*, 153. Rosenfeld, A.; Blum, J.; Avnir, D. *J. Catal.* **1996**, *164*, 363.

88.   Blum, J.; Rosenfeld, A.; Polak, N.; Israelson, O.; Schumann, H.; Avnir, D. *J. Mol. Catal.: A. Chem.* **1996**, *107*, 217.

89.   Vankelecom, I.; Wolfson, A.; Geresh, S.; Landau, M.; Gottlieb, M.; Hershkovitz, M. *Chem. Commun.* **1999**, 2407.

90.   Wolfson, A.; Janssens, S.; Vankelecom, I.; Geresh, S.; Gottlieb, M.; Herskowitz, M. *Chem. Commun.* **2002**, 388.

91.   Augustine, R.; Tanielyan, S.; Anderson, S.; Yang, H. *Chem. Commun.* **1999**, 1257. For an application of this method see: Burk, M.J.; Gerlach, A.; Semmeril, D. *J. Org. Chem.* **2000**, *65*, 8933.

92.   Dipamp = 1,2-bis(4-methoxylphenylphenylphosphino)ethane

93.   BPPM = 2-diphenylphosphinomethyl-4-diphenylphosphino-1-*t*-butoxycarbonyl-pyrrolidine

94.   Augustine suggests that reduction of the tungsten in the PTA may be taking place during the first run.

95.   de Rege, F.M.; Morita, D.K.; Ott, K.C.; Tumas, W.; Broene, R.D. *Chem. Commun.* **2000**, 1797.

96.   Note that reductions performed with **37** in methanol proceed with higher enantioselectivity (>99%).

97.   Bianchini, C.; Barbaro, P.; Dal Santo, V.; Gobetto, R.; Meli, A.; Oberhauser, W.; Psaro, R.; Vizza, F. *Adv. Syn. Catal.* **2001**, *343*, 41. Wagner, H.H.; Hausmann, H.; Hölderich, W.F. *J. Catal.* **2001**, *203*, 150.

98.   Tollner, K.; Popovitz-Biro, R.; Lahav, M.; Milstein, D. *Science* **1997**, *278*, 2100.

99.   Kuntz, E.G. Fr. Patent 2,550,202 to Rhone-Poulenc Recherches. Kalck, P.; Monteil, F. *Adv. Organomet. Chem.* **1992**, *34*, 219.

100.   Bianchini, C.; Dal Santo, V.; Meli, A.; Oberhauser, W.; Psaro, R.; Vizza, F. *Organometallics* **2000**, *19*, 2433. Bianchini, C.; Burnaby, D.G.; Evans, J.; Frediani, P.; Meli, A.; Oberhauser, W.; Psaro, R.; Sordelli, L.; Vizza, F. *J. Am. Chem. Soc.* **1999**, *121*, 5961.

101.   For other examples of supported hydroformylation catalysts that do not leach see: Sandee, A.J.; Reek, J.N.H.; Kamer, P.C.J.; van Leeuwen, P.W.N.M. *J. Am. Chem. Soc.* **2001**, *123*, 8468. See also reference 112.

102.   Arhancet, J.P.; Davis, M.E.; Merola, J.S.; Hanson, B.E. *Nature* **1989**, *339*, 454.

103.   Wan, K.T.; Davis, M.E. *Nature* **1994**, *370*, 449.

104.   Davis postulates that the chloro ligand is removed in water, leading to a less enantioselective catalyst. For the results in a water film see: Wan, K.T.; Davis, M.E. *J. Chem. Soc., Chem. Comm.* **1993**, 1262.

105.   Allum, K.G.; Hancock, R.D.; Howell, I.V.; McKenzie, S.; Pitkethly, R.C.; Robinson, P.J. *J. Organomet. Chem.* **1975**, *87*, 203. Czaková, M.; Capka, M. *J. Mol. Catal.* **1981**, *11*, 313. Zbirovsky, V.; Capka, M. *Collect. Czech. Chem. Comm.* **1986**, *51*, 836.

106.   It is estimated that each alkoxy silane is bound to the surface with approximately 2.5 Si–O bonds. Landmesser, H.; Kosslikc, H.; Storek, W.; Fricke, R. *Solid State Ionics* **1997**, *101-103*, 271. Merckle, C.; Blümel, J. *Chem. Mater.* **2001**, *13*, 3617.

107.   Feng, X.; Fryxell, G.E.; Wang, L.-Q.; Kim, A.Y.; Liu, J.; Kemner, K.M. *Science* **1997**, *276*, 923. Liu, J.; Feng, X.; Fryxell, G.E.; Wang, L.-Q.; Kim, A.Y.; Gong, M. *Adv. Mater.* **1998**, *10*, 161. van Rhijin, W.; De Vos, D.; Bossaert, W.; Bullen, J.; Wouters, B.; Grobet, P.; Jacobs, P. *Stud. Surf. Sci. Catal.* **1998**, *117*, 183.

108.   For imprint coating see: Dai, S.; Burleigh, M.C.; Shin, Y.; Morrow, C.C.; Barnes, C.E.; Xue, Z. *Angew. Chem. Int. Ed.* **1999**, *38*, 1235.

109.   Dubois, L.H.; Zegarski, B.R. *J. Am. Chem. Soc.* **1993**, *115*, 1190. Blumel, J. *J. Am. Chem. Soc.* **1995**, *117*, 2112.

110.   Collman, J.P.; Belmont, J.A.; Brauman, J.J. *J. Am. Chem. Soc.* **1983**, *105*, 7288.

111.   See reference 6 and section 4.2.3. Depending on the conditions, catalysts can use remaining hydroxyls on the surface to migrate across the surface. For an in-depth discussion of this see: Roveda, C.; Church, T.L.; Alper, H.; Scott, S.L. *Chem. Mater.* **2000**, *12*, 857.

112.   Bourque, S.C.; Alper, H.; Manzer, L.E.; Arya, P. *J. Am. Chem. Soc.* **2000**, *122*, 956.

113.   Arya, P.; Panda, G.; Venugopal Rao, N.; Alper, H.; Bourque, S.C.; Manzer, L.E. *J. Am. Chem. Soc.* **2001**, *123*, 2889.

114.   Petrucci, M.G.L.; Lebuis, A.M.; Kakkar, A.K. *Organometallics* **1998**, *17*, 4966. Petrucci, M.G.L.; Kakkar, A.K. *Chem. Mater.* **1999**, *11*, 269.

115.   Bemi, L.; Clark, H.C.; Davies, J.A.; Fyfe, C.A.; Wasylishen, R.E. *J. Am. Chem. Soc.* **1982**, *104*, 438. Blümel, J. *Inorg. Chem.* **1994**, *33*, 5050. Behringer, K.D.; Blümel, J. *Inorg. Chem.* **1996**, *35*, 1814.

116.   Li, H.; Luk, Y.-Y.; Mrksich, M. *Langmuir* **1999**, *15*, 4957.

117.   Shon, Y.S.; Mazzitelli, C.; Murray, R.W. *Langmuir* **2001**, *17*, 7735.

118.   Ingram, R.S.; Hostetler, M.J.; Murray, R.W. *J. Am. Chem. Soc.* **1997**, *119*, 9175.

119.   Personal communication, S.G. Shyu.

120.   Michalska, Z.M.; Strzelec, K. *J. Mol. Catal. A: Chem.* **2001**, *177*, 89. 121.

121.   The catalyst is named according to the complex immobilized (51), then the nature of the support (SBA), the temperature of grafting (25°C) and the presence or absence of EtSi(OEt)₃ during grafting (Et), hence this catalyst is **51-SBA(25)-Et**.

122.   Jacobsen, E.N.; Marko, I.; Mungall, S.; Schroeder, G.; Sharpless, K.B. *J. Am. Chem. Soc.* **1988**, *110*, 1968. Sharpless, K.B.; Amberg, W.; Beller, M.; Chen, H.; Hartung, J.; Kawanami, Y.; Lubben, D.; Manoury, E.; Ogino, Y.; Shibata, T.; Ukita, T. *J. Org. Chem.* **1991**, *56*, 4585. Kolb, H.C.; Van Nieuwenhze, M.S.; Sharpless, K.B. *Chem. Rev.* **1994**, *94*, 2483.

123.   Kim, B.M; Sharpless, K.B. *Tetrahedron Lett.*, **1990**, *31*, 3003.

124.   Pini, D.; Petri, A.; Nardi, A.; Rosini, C.; Salvadori, P. *Tetrahedron Lett.* **1991**, *32*, 5175.

125.   Pini, D.; Petri, A.; Salvadori, P. *Tetrahedron: Asymm.* **1993**, *4*, 2351.

126.   Lohray, B.B.; Thomas, A.; Chittari, P.; Ahuja, J.R.; Dhal, P.K. *Tetrahedron Lett.* **1992**, *33*, 5453.

127.   Salvadori, P.; Pini, D.; Petri, A. *J. Am. Chem. Soc.* **1997**, *119*, 6929.

128.   Pini, D.; Petri, A.; Salvadori, P. *Tetrahedron* **1994**, *50*, 11321. Petri, A.; Pini, D.; Salvadori, P. *Tetrahedron Lett.* **1995**, *36*, 1549. Song, C.E.; Roh, E.J.; Lee, S.-G.; Kim, I.O. *Tetrahedron: Asymm.* **1995**, *6*, 2687. Petri, A.; Pini, D.; Rapaccini, S.; Salvadori, P. *Chirality*, **1995**, *7*, 580.

129.   Different solvents are required depending on the co-oxidant. If NMO is used, a less polar solvent mixture can be employed (acetone/water, 10/1). However, higher enantioselectivities are obtained when KFe$_3$(CN)$_6$ is used as the oxidant, which requires a more polar solvent (*t*-BuOH/water, 1/1). Minato, M.; Yamamoto, K.; Tsuji, J. *J. Org. Chem.* **1990**, *55*, 766.

130.   Nandanan, E.; Sudalai, A.; Ravindranathan, T. *Tetrahedron Lett.* **1997**, *38*, 2577. Song, C.E.; Yang, J.W.; Ha, H.J.; Lee, S.-G. *Tetrahedron: Asymm.* **1996**, *7*, 645. Lohray, B.B.; Nandanan, E.; Bhushan, V. *Tetrahedron Lett.* **1994**, *35*, 6559.

131.   Canali, L.; Song, C.E.; Sherrington, D.C. *Tetrahedron: Asymm.* **1998**, *9*, 1029.

132.   Lohray, B. B.; Nandanan, E.; Bhushan, V. *Tetrahedron: Asymm.* **1996**, *7*, 2805.

133.   Bolm, C.; Maischak, A.; Gerlach, A. *Chem. Commun.* **1997**, 2353.

134.   Bolm, C.; Gerlach, A. *Angew. Chem. Int. Ed.* **1997**, *36*, 741. Bolm, C.; Gerlach, A. *Eur. J. Org. Chem.* **1998**, 21. Bolm, C.; Maischak, A. *Synlett* **2001**, 93.

135.   Han, H.; Janda, K.D. *J. Am. Chem. Soc.* **1996**, *118*, 7632. Han, H.; Janda, K. D. *Tetrahedron Lett.* **1997**, *38*, 1527. Han, H.; Janda, K.D. *Angew. Chem. Int. Ed.* **1997**, 36, 1731. Toy, P.H.; Janda, K.D. *Acc. Chem. Res.* **2000**, *33*, 546.

136.   For the use of polyvinylpyridine to bind OsO$_4$ see: Herrmann, W.A.; Kratzer, R.M.; Blümel, J.; Friedrich, H.B.; Fischer, R,W.; Apperley, D.C.; Mink, J.; Berkesi, O. *J. Mol. Catal. A: Chem.* **1997**, *120*, 197.

137.   Choudary, B.M.; Chowdari, N.S.; Madhi, S.; Kantam, M.L. *Angew. Chem. Int. Ed.* **2001**, *40*, 4620. Choudary, B.M.; Chowdari, N.S.; Kantam, M.L.; Raghavan, K.V. *J. Am. Chem. Soc.* **2001**, *123*, 9220.

138.   The ligand can be recovered by extraction or chromatography, but not by simple filtration as in the other cases.

139.   Wöltinger, J.; Henniges, H.; Krimmer, H.-P.; Bommarius, A.S.; Drauz, K. *Tetrahedron: Asymm.* **2001**, *12*, 2095.

140.   Shephard, D.S.; Zhou, W.; Maschmeyer, T.; Matters, J.M.; Roper, C.L.; Parsons, S.; Johnson, B.F.G.; Duer, M.J. *Angew. Chem. Int. Ed.* **1998**, *37*, 2719. See also reference 5(a).

141.   Badiei, A.-R.; Bonneviot, L. *Inorg. Chem.* **1998**, *37*, 4142.

142.   Aronson, B.J.; Blanford, C.F.; Stein, A. *Chem. Mater.* **1997**, *9*, 2842.

143.   Jervis, H.B.; Raimondi, M.E.; Raja, R.; Maschmeyer, T.; Seddon, J.M.; Bruce, D.W. *Chem. Commun.* **1999**, 2031.

144.   Inagaki, S.; Guan, S.; Fukushima, Y.; Ohsuna, T.; Terasaki, O. *J. Am. Chem. Soc.* **1999**, *121*, 9611.

145. Asefa, T.; MacLachlan, M.J.; Coombs, N.; Ozin, G. *Nature* **1999**, *402*, 867.
146. Melde, B.J.; Holland, B.T.; Blanford, C.F.; Stein, A. *Chem. Mater.* **1999**, *11*, 3302.
147. Sayari, A.; Hamoudi, S.; Yang, Y.; Moudrakovski, I.L.; Ripmeester, J.R. *Chem. Mater.* **2000,** *12*, 3857.

Chapter 6

DESIGN OF CHIRAL HYBRID ORGANIC-INORGANIC MESOPOROUS MATERIALS AS ENANTIOSELECTIVE EPOXIDATION AND ALKYLATION CATALYSTS

Daniel Brunel,* Monique Laspéras
Laboratoire des Matériaux Catalytiques et Catalyse en Chimie Organique-UMR-5618-CNRS -ENSCM, 8 rue de l'Ecole Normale, F-34296 -MONTPELLIER Cédex 05, France

Keywords: MCM-41, hybrid, organic-inorganic, enantioselective catalysis, asymmetric epoxidation, alkylation, tetraazamacrocycle, ephedrine, dialkylzinc

Abstract: The nineties have seen important advances in the synthesis of hybrid organic-inorganic mesoporous materials based on the functionalization of the surface of new micelle-templated mineral oxides such as MCM-41 type silicates.[1] The major advantages of these new supports are their large surface area, their regular system of monodisperse mesopores and homogeneity of chemical surface properties, which allow the preparation of well-defined mesoporous hybrid materials. Nevertheless, asymmetric catalysts grafted on supports are generally less enantioselective than their homogeneous counterparts. This chapter deals mainly with the design of more efficient supported catalysts using MCM-41-type silica for enantioselective epoxidation and C-C bond formation reactions. This design is based on control of the different modification steps in order to improve both catalytic dispersion and mineral surface coverage. Keeping these factors in mind, a new generation of catalysts can be prepared that display enantioselectivity equal to homogeneous catalysts.

Nanostructured Catalysts, edited by S. Scott *et al.*
Kluwer Academic/Plenum Publishers, 2003

1.   INTRODUCTION

In the past decade, considerable effort has been devoted to the preparation of supported chiral ligands in order to develop useful heterogeneous enantioselective catalysts. The major advantages expected from such a strategy are the ability to recover expensive catalysts for their reuse and the absence of toxic transition metals in the products, which are often intermediate compounds for pharmaceuticals.[2] Among the possible ways to reach such a goal, the addition of a chiral auxiliary to an already efficient heterogeneous catalyst has been extensively explored.[3] Another interesting strategy is to immobilize a chiral homogeneous catalyst onto a polymeric support through a covalent linkage in order to limit leaching.[4] It is noteworthy that the large majority of research in this area has focused on covalently anchoring chiral catalysts to organic polymers, rather than porous metal oxides which would provide higher mechanical stability and a larger accessible surface in addition to strong binding.

One of the first examples of the preparation of chiral catalysts covalently anchored onto a mineral surface was reported by Nagel *et al.* in 1986.[5] The rhodium complex of chiral 3,4-(*R,R*)-bis(diphenylphosphino)pyrrolidine was covalently grafted to the surface of 3-aminopropyl-modified silica via various linkers (R), Scheme 1. These supported Rh complexes were active for the hydrogenation of α-(acetylamino)cinnamyl derivatives, eq 1. The enantio-selectivities obtained with the heterogenized complexes increased as the link between the chiral catalytic site and the silica surface increased in length. The most effective catalyst was as enantioselective as its homogeneous counterpart (>90% e.e.), but the conversions were significantly lower and deactivation was observed during several reuses.

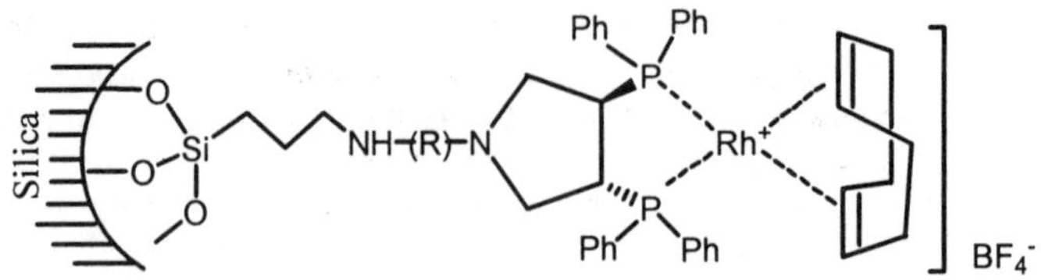

1a, R = -C(O)-C(O)-
1b, R = -C(O)-(CH$_2$)$_3$-C(O)-
1c, R = -C(O)-C$_6$H$_4$-C(O)-

Scheme 1. Chiral rhodium complex anchored to silica surface with various linkers **1a, 1b, 1c**, according to ref. 5.

$$>90\% \ ee \tag{1}$$

The authors postulated that interaction of the metal complex and the surface limited the access of the reactant to the catalytic site, leading to a less active catalyst. This type of interaction can also be detrimental to enantioselectivity if small amounts of leached (achiral) catalyst in the solvent phase become more active than the surface-deactivated supported catalyst. More recently, Pugin has investigated the effect of site isolation on the activity of neutral Ir complexes of similar structure.[6] Although cationic diphosphine rhodium complexes exhibit no tendency to interact with each other, closely spaced neutral complexes react to give inactive dimers. Hence the activity of the neutral complex increased as the catalyst loading decreased. In contrast, Collman *et al.* have demonstrated that a step in the catalytic cycle of hydroformylation of styrene involves two rhodium centers.[7] They assigned this step to the dinuclear elimination of aldehyde which is favored by a high loading of silica-supported rhodium diphosphine complexes. These results highlight the possible role of interactions between catalytic sites, which depend strongly on where the functionalized chains are located on the surface.

Few examples have been reported on the beneficial role of the support in the catalytic reaction. Battioni *et al.* have shown that silica-supported Mn-porphyrins prepared by strong adsorption of tetracationic Mn[*meso*-tetra(4-*N*-methylpyridiniumyl)porphyrin](Cl)$^{4+}$4Cl$^-$ are more efficient than soluble Mn-porphyrins or the catalyst supported on alumina for the PhIO–promoted epoxidation of cyclooctene and the hydroxylation of alkanes.[8] The epoxidation took place in excellent yield (80 to 100%), as did the hydroxylation. More importantly, a markedly higher alcohol:ketone ratio was obtained using the supported catalyst compared with the homogeneous counterpart. Other beneficial effects of the silica support were good stability of the catalyst, which completed 30,000 turnovers, and easy recovery.

More recently the same research group covalently bound manganese porphyrins to silica and K10 montmorillonite, Scheme 2. Interestingly, catalyst **2** was able to transfer an oxygen atom from H_2O_2 to a substrate with between 65 and 95% yield (based on H_2O_2), and **3** was a remarkable catalyst for the hydroxylation of linear alkanes. These results show that it is possible to prepare metalloporphyrin-based supported catalysts for alkene oxidation with strong suppression of non-productive H_2O_2 decomposition.[9]

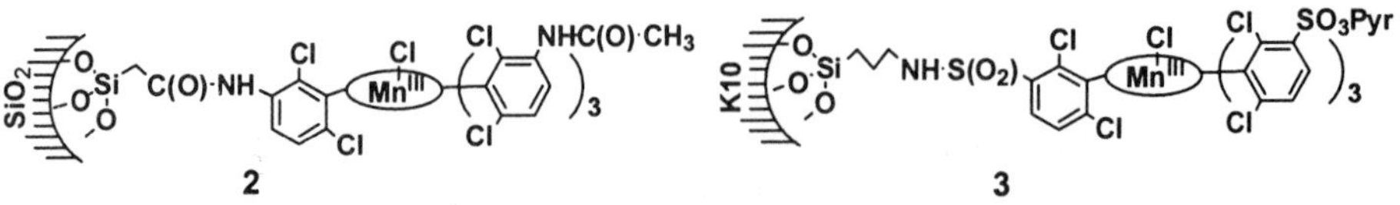

⬭ = tetrapyrrole ring. Only the meso-aryl substituents are indicated here.

Scheme 2. Manganese porphyrin complex linked to silica surface, **2,** and to K10 montmorillonite surface, **3.**

The improved catalytic performance of immobilized catalysts in the case of porphyrin complexes might be explained by the prevention of self-aggregation of metal complexes in solution through π-π interactions. For example, Fe[M(4-*N*-MePy)TFPP]Cl$_2$ gives slighly better product yields when it is immobilized compared with homogeneous conditions, even though immobilization was only through electrostatic interactions.[10] Oxidation catalysts often show greater activity when supported because dimerization through oxo-bridges leads to lower activity. However, it is difficult to generalize this behavior to all systems, because the mechanisms of oxidation processes are very diverse. For example, Sorokin and Tuel have shown that a dimeric μ-oxo iron terasulfophthalocyanine grafted onto amino-modified silica or MCM-41 is a selective catalyst for the oxidation of 2-methylnaphthalene to 2-methylnaphthaquinone (vitamin K3) and of 2,3,6-trimethylphenol to trimethylbenzoquinone.[11]

Another example of beneficial effects of the mineral framework upon catalytic performance has been reported by Corma *et al.*[12] These authors covalently anchored Rh^I, Ru^{II}, Co^{II} and Ni^{II}-complexes onto silica and onto the walls of the supermicropores contained in ultrastable zeolite Y, Scheme 3, and used these supported catalysts **4** for the hydrogenation of α-(acetylamino)cinnamyl derivatives. They proposed that the higher activity of the complexes anchored to the zeolite surface relative to silica-supported catalysts or homogeneous complexes was due to a concentration effect or a specific interaction of the substrate with the electrostatic field in the zeolite. On the other hand, the improved enantioselectivity was attributed to confinement of the chiral catalyst inside the supermicropores. This confinement effect has been also stressed by various researchers studying "ship-in-a-bottle" type catalysts for

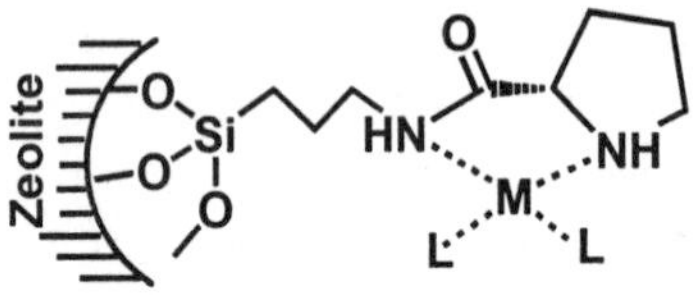

M = Rh(I),Ru(II),Co(II),Ni(II)

L = COD, Acac, CO

Scheme 3. Chiral transition metal complex-containing zeolite.

alkane oxidation, achiral alkene epoxidation and Diels-Alder reactions.[13] As with epoxidation, this strategy has been also been developed for the entrapment of chiral Jacobsen-type complexes.

The field of supported catalysis was greatly rejuvenated by the development of reliable methods for the synthesis of hybrid organic-inorganic mesoporous micelle-templated mineral oxides such as MCM-41.[14] The use of supported catalysts based on hybrid organic-inorganic mesoporous materials has been recently reviewed by several authors.[15] The major advantages of these new supports, initially disclosed by Mobil researches,[16] are their large surface areas and their regular system of monodisperse mesopores, as described in the chapter by Andreas Stein and Rick Schroden in this book. Moreover the uniqueness of their chemical surface properties, which has been carefully investigated,[17] provides new opportunities for their application as model supports for molecular or supramolecular confinement.[1,14b,15c,18] In particular, these supports allow the design of mesoporous hybrid materials in which the organic component bears well-defined catalytic sites having steric and chemical environments that are easily controlled.[1,14b,15,19] The confinement effect has been illustrated by the work of Thomas *et al.,* who reported that significant increases in regioselectivity and enantiomeric excess were observed for the allylic amination of cinnamyl acetate when a Pd[II]/1,1'-bis(diphenylphosphino)ferrocene complex was selectively anchored on the inner walls of MCM-41 versus its homogeneous counterpart or its surface-bound analogue attached to a non-porous silica.[20] We will describe more precisely the effect of the chemical environment and the accessibility of the catalytic sites in the part of this chapter devoted to enantioselective alkylation reactions.

2.   ENANTIOSELECTIVE EPOXIDATION

The initial breakthrough in the area of homogeneous asymmetric epoxidation came, in 1980, from the work of Katsuki and Sharpless, who showed that allylic alcohols could be epoxidized in high yield and with high e.e. using a chiral tartrate/titanium alkoxide combination.[21] The use of 3Å or 4Å molecular sieves (zeolites) substantially increases the scope of the Ti(IV)-catalyzed asymmetric epoxidation, giving higher conversions and /or higher selectivities. The role of the zeolite was ascribed to protection of the catalyst from adventitious water in the reaction medium.[22]

The attempted heterogenization of the catalyst on either polymers[23] or mineral oxides[24] led to less enantioselective catalysts in the first case (e.e. of 66%) or the leaching of at least one of the constituents in the second case (e.e. 90-98%). Tartrate esters have also been immobilized on soluble polymers such as polyethylene glycol monoethyl ether.[25] Recently, Wang *et al.* reported that in the presence of Ti(OPr^i)$_4$ (Ti:tartrate 1:2.4), the soluble polymer-supported tartrate ester **5** gave high chemical yields and good enantiomeric excesses (up to 93% e.e.) for the epoxidation of *trans*-hex-2-en-1-ol using *t*-BuOOH as the oxygen donor, eq. 2.[26] In order to obtain these results, the molar ratios of allylic alcohol to Ti were either 20:1 or 5:1. Since the enantiomeric excesses strongly decreased as a function of these different ratios, further efforts are needed to optimize the catalyst and to evaluate the recyclable characteristics of this soluble polymer-supported Sharpless epoxidation catalyst.

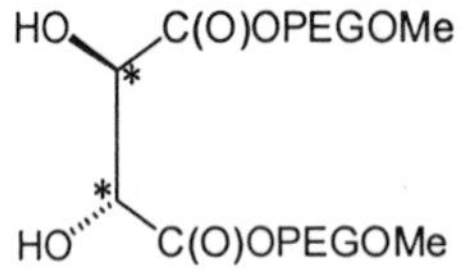

Silica-supported tantalum ethoxides **6** and **7** were prepared in two steps by Basset *et al.,* Scheme 4.[27] In the presence of chiral diisopropyltartrate, **6** and **7** catalyze the asymmetric epoxidation of alkenes with *t*-BuOOH. In the presence of added molecular sieves, supported catalyst **8** displayed activity (TON = 10) and enantioselectivity (90% ee) in the same range as the homogeneous system for the identical substrate (TON = 16; 96% ee).

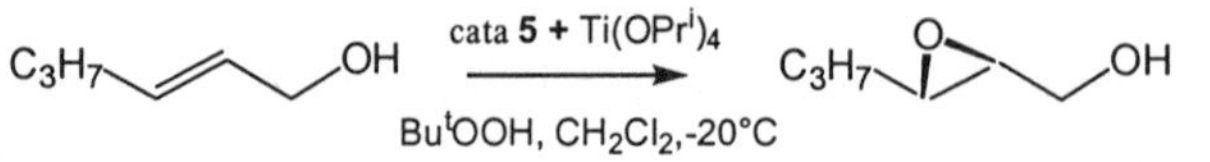

The next major breakthrough in the area of catalytic asymmetric epoxidation came from Jacobsen's group in 1990. They reported a new Mn salen-type catalyst **9** prepared from chiral substituted diamines with substituted salicylaldehyde derivatives, which displayed the highest turnovers ever observed and very high enantioselectivity for the epoxidation of unfunctionalized (*cis*) alkenes, eq. 3.[28]

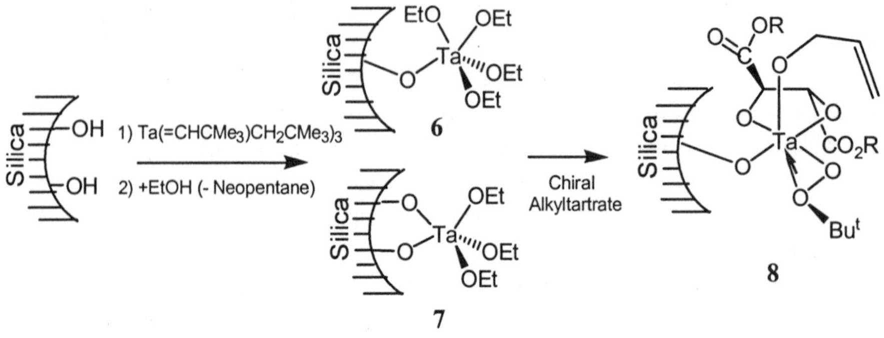

Scheme 4. Preparation of supported chiral tantalum-tartrate adduct.

$$R_1 \underset{R_2}{\overset{}{\diagdown}} R_3 \quad \xrightarrow[\text{oxidant-cooxidant}]{\text{cata } \mathbf{9} : 0.5\text{-}5\%} \quad R_1 \underset{R_2}{\overset{O}{\diagup\!\!\triangle\!\!\diagdown}} R_3 \quad (3)$$

9

Since the initial report of this remarkable reaction, considerable effort has been directed towards immobilizing the Jacobsen catalysts. Entrapment in zeolites through a "ship-in-the-bottle" procedure was one of the first useful methods for the non-covalent immobilization of the chiral Mn complex described independently and simultaneously by Corma *et al.*[29] and Bein *et al.*[30] In the first approach, a Mn(III)/Jacobsen complex **10** was prepared by condensation of chiral diaminocyclohexane around a Mn(II) ion of Mn-exchanged faujasite zeolite, the zeolite oxygen framework being the counterion for the cationic Mn(III), Scheme 5. In the second method, metalation with Mn(II) was performed after encapsulation of the chiral salen-type ligand in EMT topology, which is more accessible (hypercage: 1.3 x 1.3 x 1.4 nm with three elliptical windows : 0.69 x 0.74 nm and two 0.74 nm circular apertures) than the FAU zeolite (almost spherical cavities of: 1.3 nm interconnected through smaller apertures of 0.74 nm). Subsequent oxidation in the presence of LiCl afforded the Mn(III) chloro[N,N'-bis(salicylidene) cyclohexanediamine. The latter catalyst was highly enantioselective for the epoxidation of *cis*-methylstyrene (88% ee) with NaOCl as the stoichiometric oxidant, possibly because the zeolite framework ensures isolation of the catalytic sites. Despite this fact, the catalyst degraded upon reuse, likely because of radical oxidation. Furthermore, the entrapped catalyst was less active than the homogeneous counterpart, possibly because of slower diffusion of the substrate.

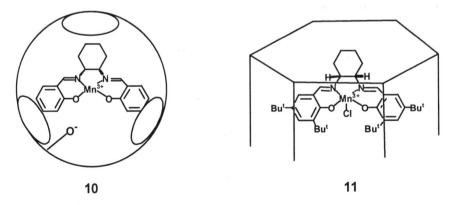

10   **11**

Scheme 5. Mn(III)/Jacobson complex entrapped in zeolite micropore **10** and embedded in MCM-41 mesopore **11**.

If diffusion and accessibility are truly responsible for decreased activity in the previous examples, the large accessible pores of MCM-41 would make it a considerably better support. In fact, Kosslick showed that simply embedding a chiral salen Mn(III) complex into MCM-41 silica solely by host/guest interactions gave a catalyst (11) with similar activity and selectivity to the homogeneous system.[31] This suggests that the reactant molecules have free access to the embedded Jacobsen catalyst, as they do in the homogeneous case. The high stereoselectivity observed with *cis*-olefins would be consistent with the proposed side-on attack of the Mn-oxo moiety. Thus it is important that the configuration of the adsorbed complexes, which are probably aligned with the walls of the mesopores, does not block approach of the substrate along this trajectory.

Hutchings *et al.* have prepared MCM-41 containing an immobilized Jacobsen catalyst by adsorption of the chiral ligand into the mesopores of Mn-exchanged Al-MCM-41.[32] Although the yield and enantioselectivity for the epoxidation of *cis*-stilbene into *trans*-stilbene oxide were reasonably consistent with the homogeneous catalyst in the first run (~70% yield and ~70% e.e. (compare: 71% yield and 78% e.e. under homogeneous conditions), the catalytic performance decreased by more than a factor of two during the second run.

Kim *et al.* have also encapsulated new unsymmetrical chiral Mn(III), Co(III) and Ti(IV) salen complexes 12 into the mesopores of Al-MCM-41 by ion exchange from the Na form, Scheme 6.[33] The Mn complex prepared by this method gave high conversion but moderate enantiomeric excess in the epoxidation of styrene and α-methylstyrene. However, the immobilized Ti complex grafted on purely siliceous MCM-41 was particularly efficient for the enantioselective trimethylsilylcyanation of benzaldehyde, while the Co complex on Al-MCM-41 was very effective for the enantioselective hydrolysis of racemic styrene epoxide. High enantioselectivities (~ 97%ee) and high product yields (~ 75%) of the corresponding diols were obtained.

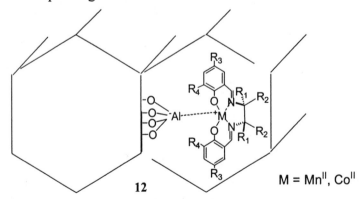

$M = Mn^{II}, Co^{II}$

Scheme 6. Chiral transition metal complexes immobilized in Al-MCM-41.

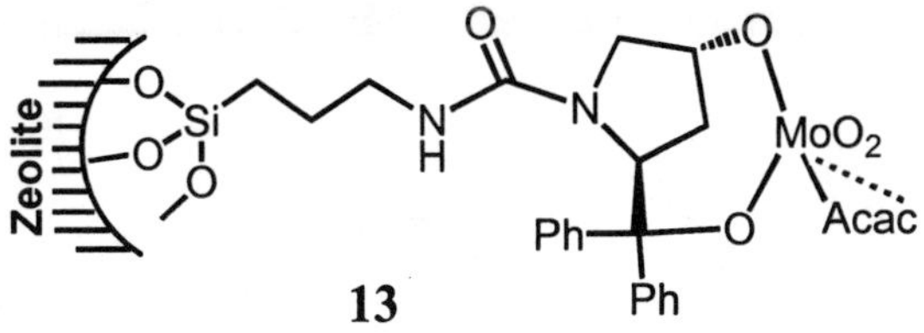

Scheme 7. Mo/proline complex anchored on the surface of dealuminated Y zeolite.

In order to avoid leaching of the supported catalyst immobilized through ionic interactions, oxidation catalysts have also been anchored by covalent linkages. First, Corma *et al.* functionalized zeolite surfaces with an anchored chiral complex based on a Mo/proline derivative system (**13**), Scheme 7, for enantioselective epoxidation, using the methodology already mentioned for the hydrogenation of olefins.[34]

Following this work, Salvadori *et al.* prepared the first example of a silica-grafted Jacobsen complex, **14**,[35] using the elegant method to anchor the catalyst previously described by the same group for binding the chiral Mn(III)(salen) complex to an organic polymer, *i.e.*, radical *anti*-Markovnikov addition of a thiol to a pendant vinylic group on the Jacobsen ligand, Scheme 8.[36] Epoxidation of prochiral aromatic olefins was carried out using *m*-chloroperbenzoic acid/*N*-methylmorpholine-*N*-oxide as the oxidant system.

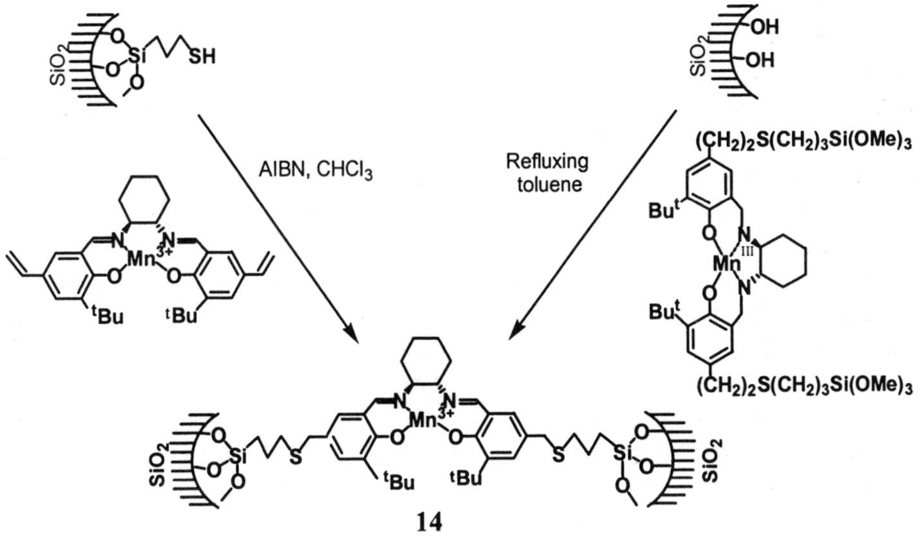

Scheme 8. Preparation of Jacobson-type complex grafted on silica surface **14**.

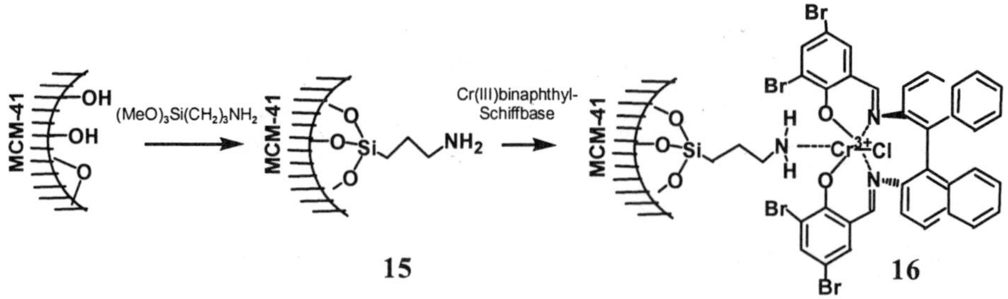

Scheme 9. Immobilization of chiral Cr/Jacobson-type complex on modified MCM-41 surface.

High enantioselectivity was observed in the epoxidation of phenylcyclohexene (53-58%). Although these values are lower than those observed under homogeneous conditions, they are significantly better than the results obtained with the same catalyst bound to an organic polymer. It is further worth noting that, regardless of the anchoring technique used (post anchorage of Mn-salen complex or grafting with organic silane bearing Mn-salen complex), the activity and stereoselectivity were not significantly different.

A Jacobsen-type complex based on the chromium binaphthyl Schiff base complex **16** was recently immobilized on 3-aminopropyl-modified MCM-41 **15**, Scheme 9. Immobilization is proposed to occur by formation of a dative bond between the amine and an apical site on Cr.[37] This strategy was previously employed by the same group for the preparation of an achiral Ru porphyrin catalyst immobilized on modified MCM-41 for alkene epoxidation.[38]

A truly covalent anchoring of a Jacobsen-type complex was accomplished by an elegant multigrafting method described by Kim *et al*.[39] These new chiral salen complexes bearing different substituents were synthesized and supported on mesoporous MCM-41 **18** through the monocondensation of **15** with 2,6-diformyl-4-*tert*-butylphenol, then condensation with chiral diamine followed by another condensation with a 3,5-disubstituted salicylaldehyde to give the ligand **17**, Scheme 10. The immobilized optically active Co(II) salen complexes gave very high enantioselectivity in the asymmetric borohydride reduction of aromatic ketones. The immobilized chiral Co(II) salen complexes were stable during the reaction and exhibited a relatively high enantioselectivity for the reduction of ketones, compared to homogeneous salen-type catalysts.

Interestingly, these new chiral immobilized salen-type ligands **17** can also be metalated with Mn(II), then oxidized to (Cl)Mn(III) and used for the enantioselective epoxidation of styrene and α-methylstyrene.[40] Epoxidation reactions with catalyst **18** in the presence of *m*-chloroperbenzoic acid and *N*-methylmorpholine-*N*-oxide led to styrene oxide and α-methylstyrene oxide with

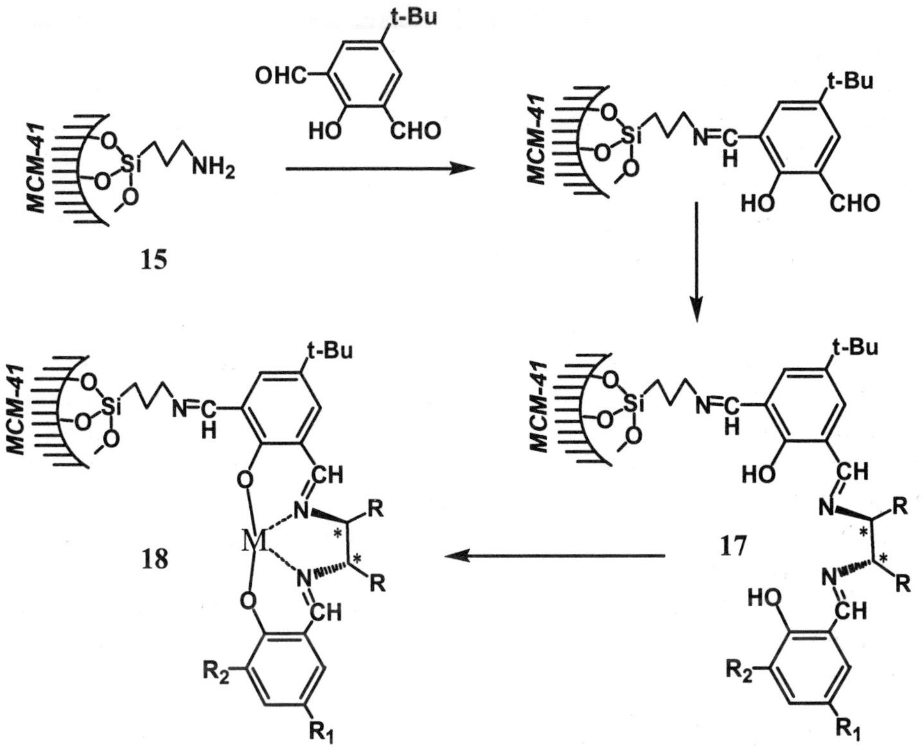

Scheme 10. Grafting of chiral salen-type complexes by multi-step methodology.

e.e.'s of 89% and 72%, observed at 92% and 70% conversion, respectively. More recently, Kim's group has successfully applied these catalysts to the asymmetric hydrolysis of epoxides to diols.[41]

The same strategy was adopted by Corma *et al.* for the preparation of Cr complexes anchored on MCM-41 and ITQ-2 as catalysts for the enantioselective ring opening of epoxides with trimethylsilylazide.[42] No leaching was observed during epoxide ring opening, in contrast to the behavior of catalysts anchored by apical coordination on aminopropyl-functionalized materials. The latter showed high enantioselectivities in the epoxidation reaction (up to 70% e.e.). However, the enantioselectivities for epoxide opening obtained with the Corma catalysts were modest (less than 20% e.e.), and significantly lower than attainable by homogeneous catalysis under comparable conditions (above 50% e.e.).

Our group has previously anchored Mn(III) Salpr complexes via the grafting of free ligand 3-(*N*,*N*'-bis-3-salicyldenaminopropylamine) on the MCM-41 surface. After complexation with Mn(III), the supported complex catalyzed the epoxidation of styrene oxide with PhIO in 58% selectivity.[43] The use of PhIO as the stoichiometric oxidant is necessary in this system, since more economical oxygen donors such as *t*-BuOOH or H_2O_2 gave poor yields due to peroxide decomposition and/or complex alteration. De Vos *et al.* have anchored dimethyl-1,4,7-triazacyclononane on the MCM-41 surface and have shown that its Mn(II) complex is capable of heterolytically activating peroxides with oxidation from Mn(II) to Mn(IV) in the presence of coligands such as oxalate. In this case, not only was the yield improved relative to the amount of peroxide used, but the diol was also formed stereoselectively, providing a new route for the direct *cis* dihydroxylation of olefins.[44] The formation of diols was surprising as mentioned by De Vos *et al.*[45]

Using a different strategy, Bohm *et al.* reported the first results demonstrating the general feasibility of asymmetric oxidation with a manganese complex based on 1,4,7-triazacyclononane bearing *N*-substituents which possess optically active centers in the β-position.[46] In particular Mn/1,4,7-triazacyclononane, in which two nitrogen atoms are substituted with chiral 2-propanol, was effective in the homogeneous enantioselective epoxidation of *cis*-β-methyl-styrene with hydrogen peroxide, leading to chiral *trans*-1-phenylpropylene oxide with 40% e.e.

Making use of the synthesis of a new chiral tetraazacyclododecane, eq. 4, containing a diaminocyclohexane unit **19** adapted from Riley *et al.*,[47] we have anchored this ligand on MCM-41 through substitution of the chlorine atom of a 3-chloropropylsilane chain previously grafted on to the mesopore surface. The anchoring reaction was performed on chloropropylated-MCM-41 **20** that was previously passivated with hexamethyldisilazane in order to eliminate any residual silanols, Scheme 11. Such sites could subsequently act as ligands for manganese salts during the metalation step and also as hydrophilic patches which can degrade the selectivity towards epoxide. In fact, the performance of Ti-SiO$_2$ and Ti-MCM-41 in the epoxidation of alkene with H_2O_2 was enhanced when they were hydrophobized in this manner. Sato *et al.* have also shown that an Fe porphyrin immobilized on a hydrophobic silica surface by means of imidazole as an apical ligand demonstrated high activity in the epoxidation of alkenes with H_2O_2. It is likely that the hydrophobic environment prevents or inhibits decomposition of the FeTPP complex by preventing extended contact with concentrated H_2O_2. The high activity and selectivity observed can also be attributed to prevention of undesirable radical chemistry by the use of imidazole

(4)

19

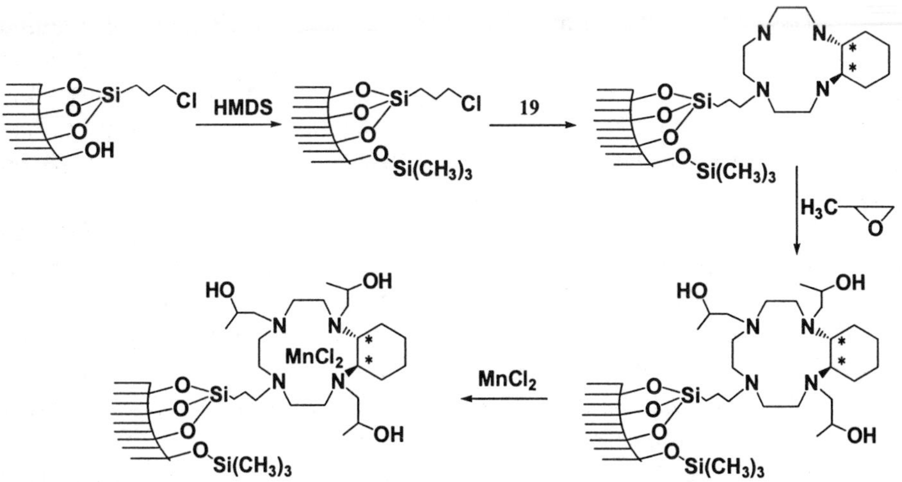

Scheme 11. Anchoring of chiral Mn/tetraazacyclodecane complex on MCM-41 surface.

ligation which promotes heterolysis of H_2O_2.[48] Chiral complex **21** catalyzed the epoxidation of *trans*-β-methylstyrene using PhIO as an oxygen donor with modest activity, but high enantioselectivity (76% e.e.).

3. ENANTIOSELECTIVE ALKYLATION

In the field of stereoselective C-C bond forming reactions, the enantioselective addition of dialkylzinc species to carbonyl compounds, eq. 5, is one of the most promising methods for the preparation of chiral alcohols. Since the pioneering work by Ogumi and Omi in 1984,[49] numerous chiral auxiliaries, particularly β-aminoalcohols, have been employed as catalysts in this reaction.[50] The preparation of supported chiral auxiliaries for this reaction would provide the

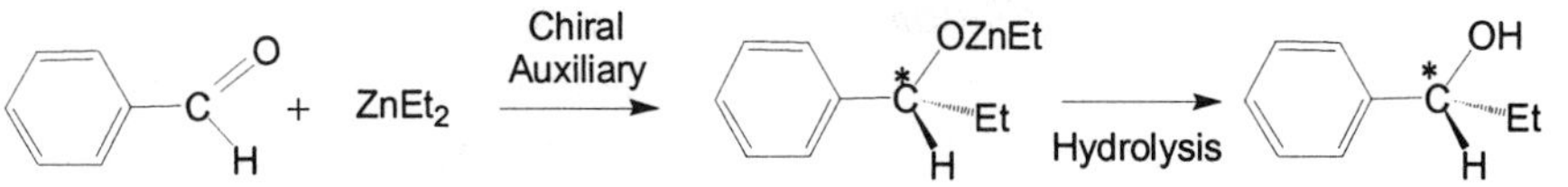

advantages of recovery and reuse of the catalyst as previously discussed. In this case, immobilization of the chiral auxiliary on a polymeric support provides enantioselectivities comparable to those obtained under homogeneous conditions.[51] However, rates are lower and limitations relating to the mechanical stability of the support and stirring difficulties have been described.[52] Inorganic oxides may be preferred as supports because of their rigid structure and their improved physical stability.[53,54]

The first results reported by Soai *et al.* using silica gel and alumina supported (–)-ephedrine showed that lower enantioselectivities were obtained relative to the homogeneous case.[53,54] The chiral aminoalcohol (–)-ephedrine, grafted on *M*esoporous *T*emplated *S*ilica (MTS), also reacted with decreased enantioselectivity.   End-capping of the surface and dilution of the catalytic sites[55,56] failed to improve the enantioselectivity significantly.

Kim *et al.* have described the preparation and application of prolinol derivatives on mesoporous silicas.[57]   In this case, higher enantioselectivities (75% e.e.) were obtained mainly due to the higher intrinsic efficiency of the chiral auxiliary employed (homogeneous e.e. 93%) and to the addition of butyllithium prior to ZnEt$_2$. The enantiomeric excesses obtained on supports are actually far lower than those obtained with the identical β-aminoalcohol under homogeneous conditions or anchored to an organic polymer support.

In order to understand the reason for the observed decrease in selectivity, it is important to consider the mechanism of the reaction. According to Soai,[58] Noyori[59] and Corey,[60] the initial formation of a 1:1 chelate between the β-aminoalcohol and alkylzinc is of prime importance. This monoalkyl chelate **22** is the actual catalytic site which coordinates to benzaldehyde prior to addition of a second molecule of dialkylzinc. Stereoselection occurs during alkyl transfer from the complexed ZnR$_2$ molecule to the coordinated aldehyde, controlled by the geometry of the complex **23**. A key feature of this proposal is that complexation of the aldehyde to the chiral zinc alkoxide **22** activates it towards attack by a second molecule of ZnEt$_2$, Scheme 12. Thus it is not surprising that reactions

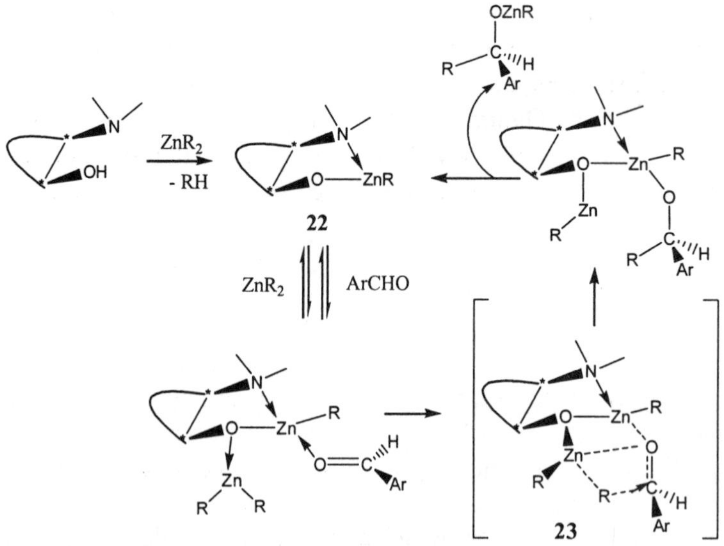

Scheme 12. Pathway of enantioselective alkylation of benzaldehyde with dialkylzinc using chiral β-aminoalcohol.

between the surfaces of mineral oxides and dialkylzinc reagents would produce species that catalyze the alkyation of benzaldehyde with a reaction rate comparable to that observed with supported ephedrine. This reaction pathway produces racemic 1-phenylpropan-1-ol.

Reaction by this path is even more likely when one considers that, after grafting by silylation, eq. 6, hydroxyl groups remain on the surface in hydrophilic islands mainly consisting of adjacent silanol groups.

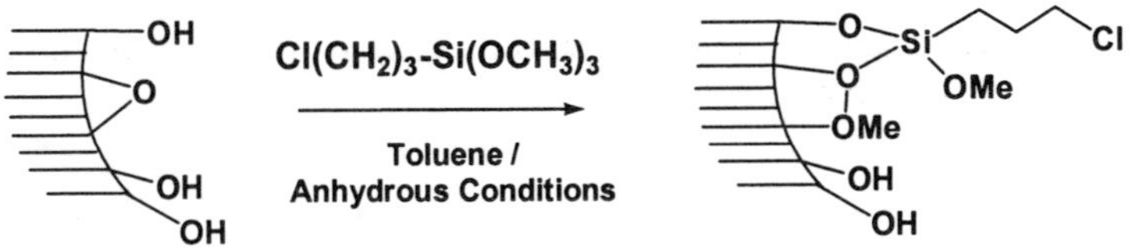

$$(6)$$

Thus the low e.e.'s obtained with ephedrine immobilized on silica supports result from the activation of dialkylzinc by the naked surface, leading to achiral catalytic sites **24**, Scheme 13.[61]

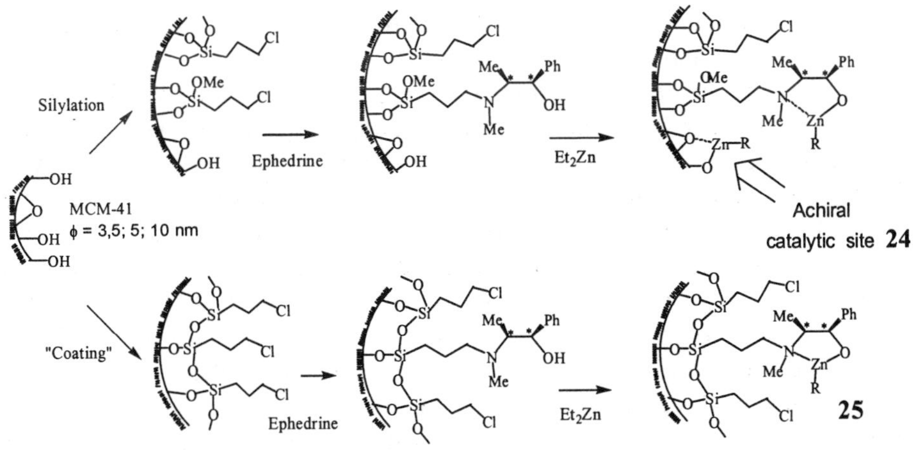

Scheme 13. Preparation of alkylating catalyst through either silylation or coating methodology.

Recently, we have shown that MCM-41 can be functionalized by a new sol-gel method involving prehydrolysis of the alkoxy groups of the silylating agent and control of the condensation reaction between monolayer-packed silylating agent.[62] This method affords high surface coverage with monolayer anchored alkylsilane chains by sol-gel polymerization at the surface **26** without anarchical polymerization of the alkoxysilane **27**, Scheme 14. When this coverage procedure was performed before the ephedrine anchoring step, the undesirable background activity of the uncovered surface was significantly

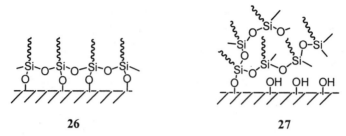

                                                 26   27

Scheme 14. Horizontal surface sol-gel polymerization of alkyltrialkoxysilane **26** and anarchical sol-gel polymerization of alkyltrialkoxysilane, **27**.

decreased.[63] This new generation of catalysts **25** demonstrated remarkable enantioselectivity at the level of the homogeneous catalysts.

Another interesting example of the effect of surface polarity somes from the group of Seebach.[64] Ti-TADDOL'ate species anchored on a silica surface that was totally hydrophobized catalyzed the addition of Et_2Zn to benzaldehyde with >99% conversion and 98% enantioselectivity.

Changes in the composition of the support and the pore size[65] have been investigated in order to understand why there is apparently no correlation between the rates of heterogeneously-catalyzed reactions and their observed enantioselectivities. Only a few reports have been published describing the kinetics of enantioselective catalytic reactions, although this aspect of asymmetric catalysis is one of the most important factors in a successful reaction, as enunciated by Noyori in 2000: "Asymmetric catalysis is four-dimensional chemistry. High efficiency can only be achieved through a combination of both an ideal three-dimensional structure (x,y,z) and suitable kinetics."[66]

The use of well-ordered nanostructured silicates possessing different pore sizes as supports was studied in order to discriminate between the role of diffusion and the special effect of the mineral surface. Thus two kinds of mesoporous MCM-41 were used as supports: one possessing a well-ordered structure (▲,△,▼) and another with irregular structure (●,○,■). These materials were then functionalized with 3-chloropropyltriethoxysilane alone (▲,●) or diluted with nonfunctional butylsilane chains (△,○). Finally, the use of chlorobenzylsilane as a rigid tether to anchor ephedrine provided a linked auxiliary which could not interact with the surface. These rigid linkers were diluted between phenylsilane chains(▼,■). The activities of the supported catalysts were compared with those of homogeneous catalysts (♦), Figure 1.[67]

When ephedrine was supported on MCM-41 possessing irregular mesopores of diameter 3.5 nm, rates were limited by the diffusion of the reactants

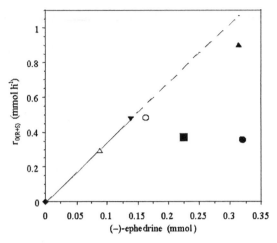

Figure 1. Rate of alkylation reaction as a function of ephedrine loading using either well-structured MCM-41 (▲,△,▼) or irregular MCM-41 (●,○,■). Reprinted from *Tetrahedron: Asymmetry* **2002**, *13*, 357. Copyright 2002, with permission from Elsevier Science.

at high loadings. In contrast, catalysts supported on MCM-41 possessing a regular structure and larger mesopores (5 and 8.3 nm) reacted with a turnover frequency of *ca.* 3.4 hr^{-1} regardless of the loading, the dispersion and the environment of the catalytic sites. Hence, although these supported catalysts demonstrated lower activity than the homogeneous species (TOF = 7 hr^{-1}), it is noteworthy that they demonstrated intrinsic activity and enantioselectivity in the absence of diffusion limitations.

4. CONCLUSIONS AND OUTLOOK

The discovery of highly structured and ordered silicates such as M-41-S materials and others has opened up new possibilities for their use as nanostructured supports for heterogenized catalysts. In particular, these materials offer larger accessible surfaces than zeolites, and therefore present considerable advantages for their applications as catalysts in fine organic chemistry, particularly in enantioselective catalysis. The true effect of the nanostructured nature of these supports has not been conclusively demonstrated at this time. Despite this fact, several new reports in the recent literature lend credence to the idea that well-ordered materials with large monodisperse pores may possess the key properties needed to solve various problems related to diffusion limitations or contact time of the reactants that can adversely affect many heterogeneous catalytic processes.

One drawback of these materials, which must be overcome before industrial development is possible, is their chemical instability towards hydrolysis or solvolysis. At this time, some materials obtained by various post-synthesis treatments exhibit considerable improvement in their chemical stability and their hydrolytic stability due to optimum coverage of their surface by organic moieties. Materials treated in this manner have the chemical nature of the surface so deeply modified that it becomes totally hydrophobic.[68] The major consequence of such functionalization is that it protects the mineral surface from attack by water or protic solvent. It is remarkable to consider that supports based on mineral oxides benefit from being organically modified by surface sol-gel polymerization in order to enhance their chemical stability while conversely, organic polymer supports are sometimes doped with mineral oxides to improve their mechanical resistance.

ACKNOWLEDGEMENTS

Our students, Anne Cauvel, Pierre Sutra, Nathalie Bellocq, Sébastien Abramson and Alexandre Blanc are thanked for their generous and strong contributions to the research cited here. Dr. Cathleen Crudden is thanked for editing and suggestions. We are very indebted to her for encouragement and invaluable help before and during the preparation of this manuscript.

REFERENCES

1.   a) Corma, A. *Chem. Rev.* **1997**, *97*, 249 b) Moller, K.; Bein, T. *Chem. Mater.* **1998**, *10*, 2950. c) Brunel, D. *Micropor. Macropor. Mater.* **1999**, *27*, 329.
2.   *Chiral Catalyst Immobilization and Recycling*, De Vos, D. E.; Vankelecom, I. F. J.; Jacobs, P.A. Eds, Wiley-VCH: New York, **2000.**
3.   Baiker, A. *J. Mol.Catal.* **1997**, *115*, 473.
4.   a) Itsuno, S.; Frechet, J.M.J. *J. Org. Chem,* **1987**, *52*, 4140. b) Zengpu, Z.; Hodge, P.; Stratford, P.W. *React. Polymers* **1991**, *15*, 71. c) Soai, K.; Niwa, S.; Watanabe, M. J. *J. Org. Chem.* **1988**, *53*, 927. d) De, B.B.; Lohray, B.B.; Sivaram, S.; Dhal, P. K. *Tetrahedron: Asymmetry* **1995**, *6*, *2105.* e) Minutolo, F.; Pini, D.; Petri, A.; Salvatori, P. *Tet. Lett.* **1996**, *37*, 3375. f) Canali, L.; Cowan, E.; Deleuze, H.; Gibson, C.L.; Sherrington, D. C. *Chem. Commun.* **1998**, 2561. g) Breysse, E.; Pinel, C.; Lemaire, M. *Tetrahedron: Asymmetry* **1998**, *9*, 897.
5.   Nagel, U.; Kinzel, E. *J. Chem. Soc., Chem. Comm.,* **1986**, 1098.
6.   Pugin, B. *J. Mol. Catal. A.* **1996**, *107*, 273.
7.   Collman, J. P.; Belmont, J. A.; Brauman, J. I. *J. Am. Chem. Soc.* **1983**, *105*, 7288.
8.   Battioni, P.; Lallier, J.-P.; Barloy, L.; Mansuy, D. *J. Chem. Soc., Chem. Comm.* **1989**, 1149.
9.   Martinez-Lorente, M. A.; Battioni, P.; Kleemiss, W.; Bartoli, J. F.; Mansuy, D. *J. Mol. Catal. A* **1996**, *113*, 343.

10. Prado-Manso, C. M. C.; Vidoto, E. A.; Vinhado, F. S.; Sacco, H. C.; Ciuffi, K. J.; Martins, P.R.; Ferreira, A. G.; Lindsay-Smith, J. R.; Nascimento, O.R.; Iamamoto, Y. *J. Mol. Catal. A* **1999**, *150*, 251.
11. Sorokin, A. B.; Tuel, A. *New J. Chem.* **1999**, *23*, 473.
12. a) Corma, A.; Iglesias, M.; del Pino, C.; Sanchez, F. *J. Chem. Soc., Chem.Commun.* **1991**, *1253*. b) Sanchez, F. Iglesias, M; Corma, A.; del Pino, C. *J. Mol. Catal.* **1991**, *70*, 369. c) Corma, A.; Iglesias, M.; del Pino, C.; Sanchez, F. *J. Organomet. Chem.* **1992**, *431*, 233. d) Corma, A.; Iglesias, M; Mohino, F.; Sanchez, F. *J. Organomet. Chem.* **1997**, *544*, 147.
13. a) Romanovsky, B.V. *Proc. 8th Congr. Catal.* Verlag Chemie: Weinhein, **1984**, *4*, 657. b) Herron, N.; Stucky, G. D.; Tolman, C. A. *J. Chem. Soc., Chem. Comm.* **1986**, 1521. c) Kimura, T.; Fukuoka, A.; Ichikawa, M. *Catal. Lett.* **1990**, *4*, 279. d) Bowers, C.; Dutta, P. K. *J. Catal.* **1990**, *122*, 271. e) Parton, R. F.; Uytterhoeven, L.; Jacobs, P. A. *Stud. Surf. Sci. Catal.* **1991**, *59*, 395. f) Knops-Gerrits, P. P.; De Vos, D., Thibault-Starzyk, F.; Jacobs, P. A. *Nature*, **1994**, *369*, 543.
14. a) Corma, A. *Top. Catal.* **1997**, *4*, 249. b) Ying, J. Y.; Mehnert, C. P.; Wong, M. S., *Angew. Chem. Chem. Int. Ed.*, **1999**, *38*, 56.
15. a) Tron On, D.; Desplantier-Giscard, D.; Danumah, C.; Kaliaguine, S. *Appl. Catal. A* **2001**, *222*, 299. b) Sayari, A.; Hamoudi, S. *Chem. Mater.* **2001**, *13*, 3151. c) Brunel, D.; Blanc, A. C.; Galarneau, A.; Fajula, F. *Catal. Today* **2002**, *73*, 139.
16. a) Kresge, C. T.; Leonowicz, M.E.; Roth, W. J.; Vartuli, J. C. *Nature*, **1992**, *359*, 710. b) Beck, S.; Chu, C. T.-W.; Chu, I. D.; Kresge, C. T.; Leonowicz, M. E.; Roth, W. J.; Vartuli, J. C. *WO. Pat.* **1991**, *91*/11390.
17. a) Cauvel, A.; Brunel, D.; Di Renzo, F.; Fubini, B.; Garrone, E. *Langmuir*, **1997**, *13*, 2773. b) Sutra, P.; Fajula, F.; Brunel, D.; Lentz, P.; Daelen, G.; Nagy, J. B. *Colloid Surf.* **1999**, *158*, 21. c) Zhao, X. S.; Lu, G. Q. *J. Phys. Chem. B* **1998**, *102*, 1556.
18. Thomas, J. M. *J. Mol. Catal.* **1999**, *146*, 77.
19. Johnson, B. F. G.; Raynor, S. A.; Shephard, D. S.; Mashmeyer, T.; Thomas, J. M.; Sankar, G.; Bromley, S.; Oldroyd, R.; Gladden, L.; Mantle, M. D. *Chem. Commun.* **1999**, 1167.
20. Thomas, J. M.; Mashmeyer, T.; Johnson, B. F. G.; Shephard, D. S. *J. Mol. Catal.* **1999**, *141*, 139.
21. a) Katsuki, T.; Sharpless, K. B. *J. Am. Chem. Soc.* **1980**, *102*, 5976. b) Johnson, R. A.; Sharpless, K. B., in *"Catalytic Asymmetric Synthesis"* Ojima, I. Ed., VCH:New York, 1993.
22. Gao, Y.; Hanson, R. M.; Klunder, J. M.; Ko, S.Y.; Masumune, H.; Sharpless, K. B., *J. Am. Chem. Soc.,* **1987**, *109*, *5765*.
23. a) Canali, L.; Karjalainen, D. C.; Sherrington, D. C.; Horni, O. E. O., *Chem. Commun.* **1997**, 123. b) Karjalainen, D.C.; Horni, O.E.O.; Sherrington, D.C. *Tetrahedron: Asymm.* **1998**, *9*, 2019.
24. a) Farral, M. J.; Alexis, M.; Tracarten, M. *New J. Chem.* **1983**, *7*, 449. b) Choudary, B. M.; Valli, V. L. K.; Prasad, A. D. D. *J. Chem. Soc., Chem. Comm.* **1990**, 1186.
25. Ikeda, N.; Arai, I.; Yamamoto, H. *J. Am. Chem. Soc.* **1986**, *108*, 483.
26. Guo, H.; Shi, X.; Quiao, Z.; Hou, S.; Wang, M. *Chem. Commun.* **2002**, 118.
27. Meunier, D; Piechaczyk. A.; de Mallmann, A.; Basset, J.-M. *Angew. Chem. Int. Ed.*, **1999**, *38*, 3540.
28. Zhang, W.; Loebach, J. L.; Wilson, S. R.; Jacobsen, E. N. *J. Am. Chem. Soc.* **1990**, *112*, 2801.
29. Sabater, M. J.; Corma, A.; Domenech, A.; Fornés, V.; Garcia, H. *Chem. Commun.* **1997**, 1285.
30. Ogunwumi, S.B.; Bein, T. *Chem. Commun.* **1997**, 901.

31.   Frunza, L.; Kosslick, H.; Landmesser, H.; Höft, E.; Fricke, R. *J. Mol. Catal.* **1997**, *123*, 179.

32.   a) Piaggio, P.; McMorn, P.; Langham, C.; Bethel, D.; Buman-Page, P. C.; Hancock, F. E.; Hutchings, G. J. *New J. Chem.* **1998**, 1167. b) Piaggio, P; Langham, C.; McMorn, P.; Bethel, D.; Bulman-Page, P. C.; Hancock, F. E.; Sly, C.; Hutchings, G. J. *J. Chem. Soc., Perkin Trans. 2* **2000**, 143. c) Piaggio, P; McMorn, P.; Murphy, D.; Bethel, D.; Bulman-Page, P. C.; Hancock, F. E.; Sly, C.; Kerton, O. J.; Hutchings, G. J. *J. Chem. Soc., Perkin Trans. 2* **2000**, 2008.

33.   Kim, G.-J.; Shin, J.-H. *Catal. Lett.* **1999**, *63*, 83.

34.   Corma, A.; Iglesias, M.; Obispo, J.R.; Sanchez, F. in *Chiral Reactions in Heterogeneous Catalysis*, Jannes, G.; Dubois, V. Eds., Plenum:New York, 1995.

35.   Pini, D.; Mandoli, A.; Orlandi, S.; Salvadori, P. *Tetrahedron: Asymm.* **1999**, *10*, 3883.

36.   Minutolo, F.; Pini, D.; Petri, A.; Salvadori, P. *Tetrahedron: Asymm.* **1996**, *7*, 2293.

37.   Zhou, X.-G.; Yu, X.-Q.; Huang, J.-S.; Li, S.-G.; Li, L. S.; Che, C.-M. *Chem. Commun.* **1999**, 1789.

38.   Liu, C.-J.; Li, S.-G.; Pang, W.-Q.; Che, C.-M. *Chem. Commun.* **1997**, 67.

39.   Kim, G.-J.; Shin, J.-H. *Catal. Lett.* **1999**, *63*, 205.

40.   Kim, G.-J.; Shin, J.-H. *Tet. Lett.* **1999**, *40*, 6827.

41.   Kim, G.-J.; Park, D.-W. *Catal. Today* **2000**, *63*, 537.

42.   Baleizao, C.; Gigante, B.; Sabater, M.J.; Garcia, H.; Corma, A. *App. Catal. A* **2002**, *228*, 279.

43.   Sutra, P.; Brunel, D. *Chem. Commun.* **1996**, 2485.

44.   a) Rao, Y. V. S.; De Vos, D. E., Bein, T.; Jacobs, P. A. *Chem. Commun.* **1997**, 355. b) De Vos, D. E.; De Wildeman, S.; Sels, B. F.; Grobet, P. J.; Jacobs, P. A. *Angew. Chem., Int. Ed. Engl.* **1999**, *38*, 1033.

45.   De Vos, D. E.; Sels, B. F.; Jacobs, P. A. *Adv. Catal.* **2001**, *46*, 1.

46.   Bohm, C.; Kadereit, D.; Valacchi, M. *Synlett,* **1997**, 687.

47.   Riley, D. P.; .Henke, S. L.; Lennon, P. J.; Weiss, R. H.; Neumann, W .L.; Rivers, Jr., W. J.; Asto, K. W.; Sample, K. R.; Rahman, H.; Ling, C-S.; Shieh, J-J.; Busch, D. H.; Szulbinski, W. *Inorg. Chem.* **1996**, *35*, 5213.

48.   Miki, K.; Sato, Y. *Bull. Chem. Soc. Jpn.* **1993**, *66*, 2385.

49.   Ogumi, N.; Omi, T. *Tetrahedron Lett.* **1984**, *25*, 2823.

50.   Soai, K.; Niwa, S. *Chem. Rev.* **1992**, *92*, 833.

51.   a) Itsuno, S.; Frechet, J. M. J. *J. Org. Chem.* **1987**, *52*, 4140. b) Soai, K.; Niwa, S.; Watanabe, M. *J. Org. Chem.* **1988**, *53*, 927. c) Hodges, P;. Sung, D. W. L.; Stratford, P. W. *J. Chem. Soc., Perkin Trans. 1*, **1999**, 2335.

52.   Itsuno, S.; Sakurai, Y.; Ito, K.; Marumaya, T.; Nakahama, S.; Frechet, J. M. J. *J. Org. Chem.* **1990**, *55*, 304.

53.   Soai, K.; Watanabe, M.; Yamamoto, A. *J. Org. Chem.* **1990**, *55*, 4832.

54.   Laspéras, M.; Bellocq, N.; Brunel, D.; Moreau, P. *Tetrahedron: Asymm.* **1998**, *9*, 3053.

55.   Bellocq, N.; Abramson, S.; Laspéras, M.; Brunel, D.; Moreau, P. *Tetrahedron: Asymm.* **1999**, *10*, 3229.

56.   a) Abramson, S.; Bellocq, N.; Laspéras, M. *Top. Catal.* **2000**, *13*, 339. b) Bellocq, N.; Abramson, S.; Laspéras, M.; Brunel, D.; Moreau, P. *Tetrahedron: Asymm.* **1999**, *10*, 3229.

57.   Bae, S. J.; Kim, S-W.; Hyeon T.; Kim, B. M. *Chem. Commun.* **2000**, 31.

58.   Soai, K.; Niwa, S.; Watanabe, M. *J. Chem. Soc., Perkin Trans 1*, **1989**, 109.

59.   Yamakawa, M.; Noyori, R. *J. Am. Chem. Soc.* **1995**, *117*, 6327.

60.   a) Corey, E. J.; Hannon, F. J. *Tetrahedron Lett.* **1987**, *28*, 5233. b) Corey, E. J.; Hannon, F. J. *Tetrahedron Lett.* **1987**, *28*, 5237.

61.   Brunel, D.; Cauvel, A.; Di Renzo, F.; Fajula, F.; Fubini, B.; Onida, B.; Garrone, E. *New J. Chem.* **2000**, *24*, 807.
62.   a) Martin, T.; Galarneau, G.; Brunel, D.; Izard, V.; Hulea, V.; Blanc, A. C.; Abramson, S.; Di Renzo, F.; Fajula, F. *Stud. Surf. Sci. Catal.* **2001**, *135*, 2902 . b) Martin, T.; Brunel, D.; Galarneau, A.; Hulea, V.; Blanc, A.C; Abramson, S.; Di Renzo, F.; Fajula, F., *Proc. Ind. Int. Conf. Silica Sci. Techn. Silica* **2001**, *108.* c) Lefevre, B.; Gobin, P. F.; Martin, T.; Galarneau, A.; Brunel, D. *Materials Research Society* 2002, April 1-5.
63.   Abramson, S.; Laspéras, M.; Galarneau, A.; Desplantier-Giscard, D.; Brunel, D. *Chem. Commun.* **2000**, 1773.
64.   Heckel, A.; Seebach, D. *Angew. Chem., Int. Engl. Ed.* **2000**, *39*, 163.
65.   Di Renzo, F.; Chiche, B.; Fajula, F.; Viale, S.; Garrone, E. *Stud. Surf. Sci. Catal.* **1996**, *101*, 851.
66.   Noyori, R. in *Architectural and Functional Molecular Engineering for Asymmetric Catalysis*, presented during the conference of Société Française de Chimie, F- Rennes, **2000**, Sept 18-22.
67.   Abramson, S.; Laspéras, M.; Brunel, D. *Tetrahedron: Asymm.* **2002**, *13*, 357.
68.   Martin, T.; Lefevre, B.; Brunel, D.; Galarneau, A.; Di Renzo, F.; Fajula, F.; Gobin, P. F.; Quinson, J. F.; Vigier, G. *Chem. Commun.* **2002**, 24.

Chapter 7

CHIRAL NANOSTRUCTURES AT METAL SURFACES: A NEW VIEWPOINT ON ENANTIOSELECTIVE CATALYSIS

R. Raval
Leverhulme Centre for Innovative Catalysis and Surface Science Research Centre, Department of Chemistry, University of Liverpool, Liverpool L69 7ZD, UK.

Keywords: chiral surfaces, enantioselective control, heterogeneous asymmetric catalysis, tartaric acid, supramolecular assemblies

Abstract: Asymmetric catalysis producing single enantiomer products remains an important industrial and academic goal. A significant future strategy is the development of heterogeneous catalytic systems that allow sub-stoichiometric productions of chiral products. Currently, extensive attention is being focussed on the potential use of metals that have been modified by preadsorption of chiral molecules as heterogeneous enantioselective catalysts. In this chapter, recent surface science results on model systems are reviewed which reveal a new viewpoint on these modified metal surfaces; in particular, the dynamic nature of the modified interface and the creation of well-organized chiral nanostructures that may be central to enantioselective control will be reviewed.

1.   INTRODUCTION

Since the thalidomide tragedy, which starkly demonstrated the different physiological responses a biological system has to opposite enantiomers, there has been a major momentum within the chemicals and pharmaceuticals industries to establish enantioselective catalytic methods[1] whereby pure enantiomeric forms of products can be produced. The application of such methods goes beyond pharmaceuticals, embracing agrochemicals, flavors, fragrances and new materials. In terms of chemical transformations, the most important reactions which create asymmetric centres within a molecule involve the oxidation of C=C groups, the hydrogenation of C=C, C=O and C=N groups, and the creation of C-C groups. At present, most commercial-scale operations are based on homogeneous catalytic methods[1] which provide high enantioselectivity and yield for a wide number of reactions. In contrast, heterogeneous enantioselective catalysis remains a lesser-tested and less-developed alternative but promises to deliver the significant advantages of non-stoichiometric production of chiral molecules, combined with ease of handling and separation. A number of different approaches have been adopted for creating heterogeneous chiral catalysts,[2-5] some of which are identified below:

- immobilizing chiral transition metal complexes at surfaces;
- attaching chiral auxiliaries to the reactant to induce an asymmetry in the surface reaction;
- creating intrinsically chiral solids and surfaces in which asymmetric configurations of the reactive surface are displayed;
- adsorbing chiral molecules at achiral solid surfaces to introduce asymmetry.

This review will concentrate on the last approach, namely utilizing metals modified by chiral molecules to catalyze enantioselective reactions. At present, such systems have had their best success reported for the hydrogenation of β-ketoesters over supported Ni catalysts.[6-10] The central feature of these, and related systems, is that enantioselectivity is only manifested if the metal surfaces are first modified by the adsorption of chiral organic "modifier" molecules. For hydrogenation reactions at Ni, Cu and Co surfaces,[3-7,9] modifiers based on the chiral dicarboxylic acid tartaric acid are well-known for stereodirecting the hydrogenation of methylacetoacetate (MAA) to give R-methyl-3-hydroxybutrate or S-methyl-3-hydroxybutrate, as shown in Figure 1. Since MAA is a planar molecule, the optical activity of the product is defined simply by the molecular face at which the hydrogen attack occurs. For the unmodified metal, the probability of hydrogen attack on the prochiral substrate is identical for each face, yielding a racemic product mixture. However, the final product created on modified metal surfaces is governed by the presence of the modifier, for example, modification of a nickel surface by R,R-tartaric acid yields the R-product in >90% enantiomeric excess (*ee*), while S,S-tartaric acid favors the S-product.

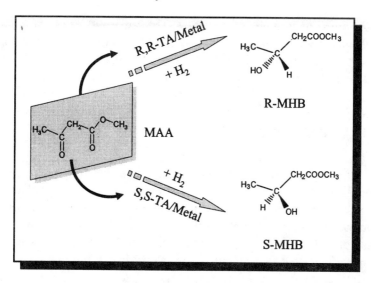

Figure 1. The stereodirected hydrogenation of methylacetonacetate (MAA) to give methyl-3-hydroxybutrate (MHB) on tartaric acid (TA) chirally-modified Ni catalysts. Reprinted with permission from *J. Catal.* **2002**, *205*, 123. Copyright 2002 Elsevier Science.

Undoubtedly, the nature of the modified metal surface lies at the heart of of this enantioselective process. However, the complexity of a working catalyst and the difficulty of probing such an interface with suitable spectroscopies combine to obscure the molecular nature of this stereocontrol. To overcome these problems, and to provide a starting point for a detailed understanding, the behavior of the modifier *R,R*-tartaric acid on the two-dimensionally anisotropic Cu(110) surface was investigated. Model modified surfaces were created by adsorption of optically pure modifiers on the defined metal single crystal surface in ultra-high vacuum conditions and then scrutinised using the formidable array of surface analytical tools that are now available. Three main *in situ* techniques were employed to probe the modifier-metal interface: Reflection Absorption Infrared Spectroscopy (RAIRS), Low Energy Electron Diffraction (LEED) and Scanning Tunnelling Microscopy (STM). Each technique provides complementary information on the modified surface, enabling a detailed picture of the interface to be constructed. The high sensitivity, high resolution vibrational data obtained by Fourier-transform RAIRS provides direct information on the chemical nature of the adsorbed modifier molecules and their perturbation by the surface, while application of the strict dipole selection rule that operates for the technique (*i.e.*, only vibrations that produce a dynamic dipole normal to the metal surface are observed) allows the orientation of the species to be determined. LEED monitors the long-range two-dimensional periodicity of the adlayer while STM provides information on the local arrangements of the modifier molecules within the domains formed at the surface. The sensitivity of all these techniques allows the detection of submonolayers of modifier at the metal surface which, on a typical single crystal sample with a surface area of 1 cm^2, represents the ability to sense a nanomole or less of the modifier.

2.   EXPERIMENTAL

Experiments were carried out in two separate surface analysis instruments, each with a base pressure of better than 2 x 10^{-10} mbar. The first chamber was interfaced, via auxiliary optics, with a Mattson 6020 FTIR spectrometer to allow RAIRS experiments to be conducted by the single reflection of IR light, at near-grazing incidence, from the Cu(110) surface. RAIR spectra were recorded throughout a continuous dosing regime as sample single beam infrared spectra and subsequently ratioed against a reference background single beam spectrum from the clean Cu(110) surface in order to obtain $\Delta R/R^o$ spectra. A liquid N$_2$-cooled HgCdTe detector allowed IR data to be collected over the 4000-800 cm^{-1} region and a polariser placed in front of the detector ensured that only p-polarised light was detected. All spectra were recorded at 4 cm^{-1} resolution with the coaddition of 256 scans. The chamber was also equipped with LEED, quadrupole mass spectrometry, Auger Electron Spectroscopy (AES) and sample cleaning facilities. LEED patterns displayed on the phosphor screen in the chamber were captured and digitized by a CCD video camera interfaced to a computer.

STM experiments were carried out in a separate Omicron Vakuumphysik chamber which was also equipped with STM, LEED, AES and sample cleaning facilities. The STM experiments were carried out by creating the required adlayer by specific exposures of the modifier molecules at the required temperature and then cooling to room temperature to record the data. All STM images were acquired in constant current mode. Depending on the tolerance of the adlayer to electron beam damage, LEED experiments were conducted before or after the STM experiments, in order to provide a direct correlation between the two sets of data.

Prior to all experiments, the Cu(110) crystal was cleaned by cycles of Ar$^+$ sputtering and annealing at 600 K. The sample cleanliness and surface order were monitored by AES and LEED, respectively.

Pure enantiomers of tartaric acid (99%) were obtained from Fluka Chemical Company and were used without further purification. The required modifier was contained in a small electrically heated glass tube separated from the main vacuum chamber by a gate valve and differentially pumped by a turbomolecular pump. Before sublimation, the modifier sample was pumped for a few hours and outgassed at a temperature of ~350 K. The modifier was then heated to a temperature of ~370 K and exposed to the copper crystal. During sublimation the main chamber pressure was ~ 2 x 10^{-9} mbar, ensuring ultra-clean deposition conditions.

Modifier coverage at the surface is given in terms of fractional monolayers (ML), quoted with respect to the number density of surface metal atoms. The adlayer unit mesh is given in standard matrix notation as follows and quoted in the text as (G_{11} G_{12} , G_{21} G_{22}):

$$\begin{pmatrix} a' \\ b' \end{pmatrix} = \begin{pmatrix} G_{11} & G_{12} \\ G_{21} & G_{22} \end{pmatrix} \begin{pmatrix} a \\ b \end{pmatrix}$$

where **a**', **b**' are the overlayer net vectors and the underlying metal surface mesh is defined by **a**, the unit vector along the <1$\bar{1}$0> direction and **b**, the unit vector along the <001> direction.

3. RESULTS AND DISCUSSION

The adsorption characteristics of pure enantiomers of tartaric acid on Cu(110)[11-15] led to a number of new observations which significantly alter previous perceptions of the heterogeneous active site.[6-10, 16-18] The centre-point of this new perspective is that the modifier/metal system occupies a complex and varied phase space. Two main factors emerge as contributors towards this behavior:

- Although the modifier possesses an intrinsically simple molecular form, it shows great versatility in the local chemical, bonding and orientational structure that is adopted at a surface, leading to an overall complex and dynamic behaviour;
- A very new, and hitherto unsuspected, aspect of modifier behavior is that they display a remarkable ability for self-organization at a surface, leading to the creation of a series of nanostructures, each possessing a different ordered, crystalline architecture. Some of these chiral nanostructures are described in detail in the following sections.

Overall, at least six different types of monolayer phases are fashioned in the course of adsorbing *R,R*-tartaric acid on Cu(110),[12] depending on adsorption temperature, coverage and holding time. Of these various phases, only the two low coverage phases are involved in the enantioselective hydrogenation of β-ketoesters, since these are the only ones that can accommodate methylacetoacetate, the simplest reactant for this reaction, at the surface. The spectroscopic data for these phases have been described in detail elsewhere,[12] so their structures are only briefly outlined here. Each phase possesses a different two-dimensional organisation at the surface, described as the (4 0, 2 3) and the (9 0, 1 2) structures in matrix notation.

3.1   *R,R*-Tartaric Acid on Cu(110): The (4 0, 2 3) Structure

This structure is formed directly when adsorption is carried out on a clean Cu(110) surface at room temperature (300 K). STM data, Figure 2, show that the nucleation of *R,R*-tartaric acid islands occurs in the early stages of adsorption and is often located at step edges. Increasing coverage leads to the steady growth of these islands until the entire surface is covered in this phase.

Detailed information on the nature of this island phase has been gathered using a combination of STM, LEED and RAIRS data. First, LEED photographs of this phase, Figure 3a, show sharp diffraction spots indicating that the islands possess very ordered arrangements of the modifier molecules in a c(4x6), or in matrix notation, a (4 0, 2 3) structure, on the Cu surface. This is actually a rather large unit cell, but STM data, Figure 3b, reveal that there are two other molecules within this unit cell, giving a much more packed structure with a local coverage of 0.25 ML.

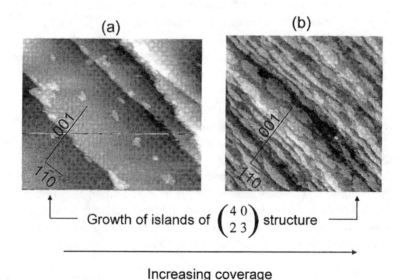

Increasing coverage

Figure 2. STM images showing the main stages of growth of (4 0, 2 3) islands of *R,R*-tartaric acid on Cu(110) at 300 K: **a)** initial nucleation stage (500 × 500 Å) [V_{tip}: -2.05 volts; I_t: 0.46 nA]; and **b)** growth of islands (1000 ×1000 Å) [V_{tip}: -1.90 volts; I_t: 0.46 nA]. Adapted from *J. Phys. Chem B* **1999**, *103*, 10661. Copyright 1999 American Chemical Society.

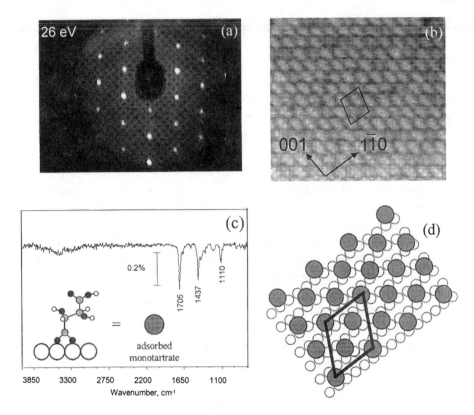

Figure 3. The (4 0, 2 3) phase of *R,R*-tartaric acid on Cu(110): **(a)** LEED pattern obtained at 26 eV; **(b)** STM image (80 × 75 Å) (V_{tip}: -1.52 volts, I_t: 1.25 nA) showing the position of individual adsorbates; **(c)** RAIRS data monitoring the nature of the adsorbed species; and **(d)** a structural model of the phase with the unit cell outlined. Adapted from *J. Catal.* **2002**, *205*, 123. Copyright 2002 Elsevier Science.

It should be noted that STM data only image the adsorbed entities as an unresolved single feature, yielding no information on their local chemicalstructure. This detail is, instead, provided by RAIR spectra obtained for this phase, Figure 3c, with detailed analysis of the vibrational features[12] revealing that the *R,R*-tartaric acid is adsorbed as a monotartrate species which is bound to the surface via the deprotonated carboxylate group. In addition, the free and intact COOH acid group is held away from the surface and perturbations in the frequency of the ν(C=O) vibration of this group suggest that it is involved in considerable intermolecular H-bonding interactions with the alcohol groups of neighbouring monotartrate species, leading to a strong tendency for island growth in this phase. Combining all these different pieces of information, a fairly complete description of this adsorbed phase is constructed, Figure 3d.

3.2 *R,R*-Tartaric Acid on Cu(110): The (9 0, 1 2) Structure

If one starts with an adlayer consisting of islands of the (4 0, 2 3) structure, a further transformation of this structure to a new one can be effected by an increase in temperature. At 300 K, this transformation is very slow and undetectable during the normal course of an experiment. However, at 350 K the relaxation can be followed easily by IR spectroscopy, Figure 4a, which shows the gradual attenuation of the ν(C=O) vibration at 1700 cm^{-1}, accompanied by the appearance of new bands in the ν_s(COO$^-$) region at 1430 and 1410 cm^{-1}. This transformation can be interpreted clearly as a deprotonation of the monotartrate form into a bitartrate form,[12,14] in which the second acid group is converted into a carboxylate functionality. Importantly, this transformation extends beyond the local level and STM data, Figure 4b, provide a striking record of this phase evolution. It can be seen that, over time, molecules in the high local density (4 0, 2 3) structure migrate out onto clean metal areas and form a new, lower density phase with a new structure. At 300 K, the mobility of the modifier molecules is so low that the (4 0, 2 3) islands remain kinetically frozen in. However, as the temperature is increased to 350 K, significant mobility is initiated and over 15 to 20 minutes, the transformation is seen to occur in the RAIRS and the STM experiments. When the adsorption is carried out at 430 K, the mobility of the

(a)   (b)

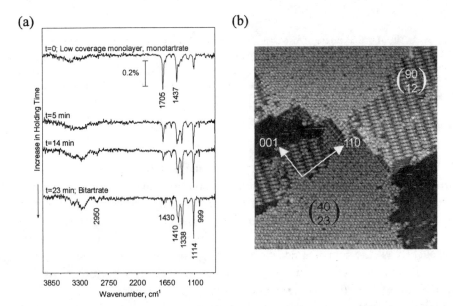

Figure 4. **a)** RAIRS; and **b)** STM data, showing the changes occuring at the local level and in the long-range order during the transformation of the (4 0, 2 3) monotartrate structure into the bitartrate (9 0, 1 2) phase. Adapted with permission from *J. Catal.* **2002**, *205*, 123. Copyright 2002 Elsevier Science.

impinging molecules allows the new structure to form immediately. Clearly, this process, in which chemical transformation, molecular mass transport and expansion in the adsorption area occurs, involves a significant activation barrier. We have recently measured this barrier to be 73 ± 2 kJ mol^{-1}.[14]

Again, the detailed nature of this new structure may be constructed from a comprehensive analysis of the LEED, STM and RAIRS data. LEED data, Figure 5a, confirm that the long-range ordering of modifiers at the surface has now altered from the original (4 0, 2 3) matrix to yield a new (9 0, 1 2) structure which is consistent with a very large unit cell possessing the dimensions 23.04 Å x 7.68 Å, $\alpha = 19.47°$. High resolution STM images, displayed in Figure 5b, show that there are three bitartrate molecules per unit cell, resulting in a fractional coverage of $^1/_6$. The STM images also reveal that rows of three bitartrate molecules assemble at the surface to form long chains, which are aligned along the <1$\bar{1}$4> surface direction. We have already discussed how the RAIRS data from the (9 0, 1 2) phase, Figure 4a, show that it consists of the bitartrate species. Detailed application of the dipole selection rule that operates in the RAIRS technique suggests that both carboxylate functionalities of the bitartrate molecule are

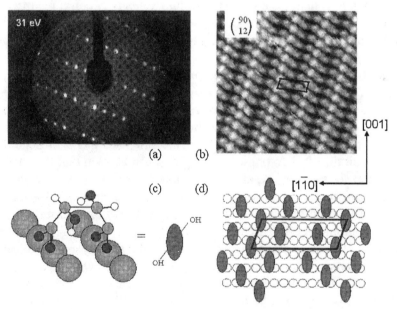

(a)   (b)

(c)   (d)

Figure 5. Details of the (9 0, 1 2) chiral phase created at low coverages and high temperatures showing **(a)** LEED pattern obtained at 31 eV; **(b)** 1500 × 200 Å STM image of the chiral surface showing 'trimers' of bitartrate molecules aligned in columns directed along the chiral <1$\bar{1}$4>direction; **c)** local bonding description of the bitartrate unit; and d) schematic model of the overlayer constructed from the STM, LEED and RAIRS data.

bonded at the Cu surface so that their two oxygens atoms are held equidistant from the metal, forcing the C_2-C_3 bond almost parallel to the surface, Figure 5c. In this way the modifier has a two-point bonding interaction with the metal, yielding a fairly rigid adsorption geometry.

Perhaps the most important characteristic of the bitartrate (9 0, 1 2) phase is that the organization and growth directions of the adsorbed species are such that all symmetry planes of the underlying Cu(110) surface are annihilated, thus bestowing *extended* two-dimensional chirality onto the surface.[13] The extended surface chirality arises as a direct result of the self-organization of the modifier molecules, generating a unique growth direction which lies along a non-symmetry vector (in this case the <1$\bar{1}$4> direction). It is important to note that the same growth direction is maintained over the entire surface and as a result *a truly extended 'handed' surface is created which is non-superimposable on its mirror image!* At present, we attribute this directional growth behaviour to non-covalent intermolecular chiral lateral interactions between adsorbed modifier molecules. For the bitartrate (9 0, 1 2) phase created by *R,R*-tartaric acid, we believe the growth direction is dictated by the spatial positioning of the α-hydroxy groups on the adsorbed bitartrate species, Figure 5c, with the optimum interaction leading to a <1$\bar{1}$4> chain alignment. On this basis, a better description of this extended chiral structure is that it is a *supramolecular assembly* of chiral modifiers.

3.2.1 Creation of Chiral Docking Spaces within Supramolecular Assemblies

So far, the discussion has centred largely on where the molecules are located on the Cu(110) surface. However, for the purposes of heterogeneous catalysis, a more pertinent factor is where the modifier molecules are *not*, since this is where bare metal sites, known to be essential for the catalysis, are available to effect the hydrogenation reaction. From the structural model presented of the extended chiral (9 0, 1 2) phase, Figure 5d, it can be seen that the structure is open enough to reveal empty, nanosized chiral channels and spaces within which the underlying metal is exposed. The genesis of these empty chiral spaces is still under investigation but early indications suggest they arise from a mismatch between supramolecular assembly dimensions and those of the underlying surface, which leads to strain-breaks in the modifier overlayer arrangement and, possibly, reconstruction of the underlying metal surface. Our most recent work suggests that these nanosized chiral spaces constitute the actual active enantioselective sites in which the reactant molecule docks in a preferential orientation, forcing the hydrogenation to occur at one reactant face only. On this basis, an important conclusion is that the heterogeneous enantioselective site is manufactured by groups or ensembles of modifiers acting cooperatively to create chiral nanostructures that confer asymmetry to the reactive metal sites.

3.2.2 Creating Mirror Chiral Nanostructures

An important aspect of the asymmetric hydrogenation of β-ketoesters on tartaric acid-modified surfaces is that switching the chirality of the modifier switches enantioselectivity from the *R*- to the *S*-reaction product,[7] Figure 1. Therefore, one would expect the opposite enantiomer of the modifier to create a very different surface structure. This can be tested readily by adsorbing *S,S*-tartaric acid on Cu(110) and creating the parallel extended chiral structure. A sharp LEED pattern is obtained from such an adlayer, confirming long-range order. However, the positions of the diffraction spots are switched, consistent with a mirror (9 0, -1 2) phase. The molecular detail of this chiral switching is better illustrated by the STM images, Figure 6, where the (9 0, -1 2) phase of *S,S*-tartaric acid on Cu(110) is revealed to be a true mirror image of the (9 0, 1 2) phase obtained for *R,R*-tartaric acid! The *S,S*-tartaric acid adlayer now shows chain growth in the <$\bar{1}$14> direction and the alignment of molecules in each row of three has also been switched to the mirror orientation. Importantly, the nano-channels and nanospaces within this chiral surface are now also reversed in handedness and should, therefore, naturally elicit an opposite enantioselective response from the reaction.

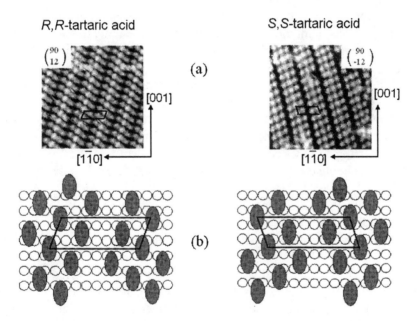

Figure 6. Switching extended surface chirality: *R,R*-tartaric acid versus *S,S*-tartaric acid. This chiral switching is illustrated by **(a)** 108 Å × 108 Å STM images; and **(b)** schematic models of mirror adlayers created when *R,R*-tartaric acid and *S,S*-tartaric acid are adsorbed on Cu(110). Reprinted with permission from *Nature* **2000**, *404*, 376. Copyright 2000 MacMillan Publishing.

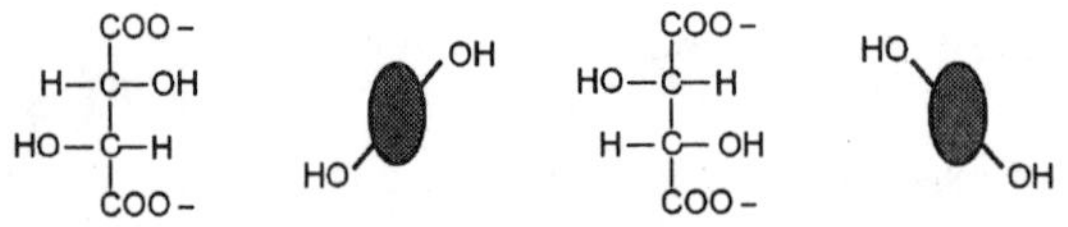

R,R-tartaric acid   S,S-tartaric acid

Figure 7. Schematic views of the adsorbed bitartrate unit showing the spatial orientation of the OH groups when viewed from above.

This surface chiral switching can be explained in terms of how opposite enantiomers guide the supramolecular assembly directions. In the chiral phases created by tartaric acid, Figure 6, the growth direction of the adlayer is dictated by the conformation of the α-hydroxy groups on the adsorbed bitartrate species.

Both the *R,R*- and *S,S*-bitartrate species possess defined and rigid adsorptiongeometries arising from their two-point bonding with the metal surface. As a result, they exhibit one major difference, namely the spatial orientation of their OH groups, Figure 7. Looking down at the surface, it can be seen that the positions of the OH groups on *R,R*-tartaric acid are oriented in a mirror configuration to those of *S,S*-tartaric acid. Thus, the enantiomers adopt mirror growth directions, with intermolecular interactions leading to a molecular chain alignment along the <1$\bar{1}$4> direction for *R,R*-tartaric acid, but the mirror <$\bar{1}$14> direction for *S,S*-tartaric acid.

Finally, the (9 0, 1 2) and the (9 0, -1 2) structures are some of the first examples demonstrating that a truly chiral surface, in which a single handedness is sustained across the entire surface, can be generated.[13] Generally, extended chirality has been very difficult to attain in surface chemistry, with most examples simply referring to local chirality.[19-20] We have identified the following three conditions as being central for sustaining single-handedness in these systems:

- chirality of the modifier molecule;
- a rigid and defined adsorption geometry;
- directional and anisotropic lateral interactions.

It can be seen that the chirally active bitartrate phases discussed in this paper fulfill all these requirements. The inherent chirality of these molecules and their two-point bonding at the surface uniquely defines the 'footprint' the molecule casts at a surface and dictates the position of *all* its functional groups in space. Once this is achieved, the intermolecular interactions between the modifiers control the placement of neighboring molecules and thus confine supramolecular assembly along particular growth directions. All these factors ensure that only one chiral unit mesh is created and that its mirror twin is never energetically favored. The same factors ensure that when the opposite enantiomer has been adsorbed, the functional groups are reflected in space and the opposite growth direction and unit mesh is nucleated, thus reversing the handedness of the surface and the system's enantioselectivity.

4. CONCLUSIONS AND NEW VIEWPOINTS

The fundamental understanding that has been obtained from a model metal surface modified by a chiral molecule provides a good framework for rationalizing a number of catalytic observations. First, the revelation from our work of the complex and dynamic phase space inhabited by the modifier/metal interface explains why achieving a modified surface which is capable of high enantioselectivity is so incredibly difficult, with the temperature of modification, temperature of reaction and concentration of modifier all having very significant effects.[3-9,21,22] It can be clearly appreciated that as experimental parameters are varied, a different surface phase is created, each providing a different interaction with the incoming prochiral substrate, with maybe only one or two phases extracting an enantioselective response.

Catalytic observations[6,7] also show that the size of metal particles can affect the *ee*, with higher *ee*'s being recorded for larger particles, reaching a maximum at particles of 100-200 Å and then levelling off. Such large particles would be perfectly capable of sustaining two-dimensional molecular assemblies of the type imaged on our single crystal surfaces - this is certainly a very new viewpoint on the nature of the heterogeneous enantioselective site, *i.e.*, that it is defined by a number of modifiers acting in concert. Presumably, this many-to-one nature of modifier-substrate interaction serves to provide a suitably restricted docking site for the incoming substrate. The supramolecular assembly that underpins the creation of these chiral nanostructures and chiral spaces can be switched in handedness by adsorption of the mirror modifier enantiomer, leading to the opposite configuration of the chiral cavity and, therefore, docking in the reversed sense. Finally, such fine matching between substrate dimensions and the enantioselective active site dimensions explains why such modifier systems are so substrate-specific,[3,4] making it difficult to translate the success of one reaction to another related one.

5.   FORWARD LOOK

In terms of heterogeneous enantioselective catalysis, one of the main drawbacks in creating new successful catalysts with modified metals is that there does not, at present, exist a critical volume of information at the molecular level on how such organic entities behave at a metal surface. Although the work presented here is confined to a defined Cu(110) surface studied under idealized conditions, some general and fundamental aspects of modifier behaviour have been gleaned which must manifest themselves in the real catalytic system, albeit in a modified form when solvents and promoters, *etc.*, are present. Therefore, to move to rational designs for a new generation of heterogeneous enantioselective catalysts, the catalytic scientist will need to incorporate the inherent complexities of the modifier surface phase diagram, the cooperativity of modifier surface chemistry with its attendant 2-dimensional architectures, and the requirements that must be placed on the modifier structure to generate extended chiral surfaces and chiral cavities that exhibit a single-handedness across an entire surface. We are still at an early stage in tracking all the fundamental factors that control enantioselective control in modified metal systems. The studies reported here are some of the first mappings of such systems and already a number of important parameters and phenomena have been revealed. No doubt other studies will reveal other important parameters, especially the balances that exist between modifier-modifier, modifier-metal, substrate-modifier and substrate-metal interactions. Once a database of the type that is available to the organic chemist or the homogeneous chemist is established for the heterogeneous catalytic chemist, the generic principles on which to base future design strategies may become clearer.

ACKNOWLEDGEMENTS

I am grateful to the EPSRC and The University of Liverpool for research support. Most importantly, I would like to thank my co-workers and research students, Maria Ortega-Lorenzo, Sam Haq, Chris Baddeley, Chris Muryn and Paul Murray, for their substantial contributions to this research effort. Also, thank you to Vincent Humblot for his help and patience in preparing a number of the illustrations.

REFERENCES

1.   Sheldon, R. A. *Chirotechnology* Dekker: New York, 1993.
2.   *Chiral Reactions in Heterogeneous Catalysis* Dubois, V.; Jannes, G. Eds. Plenum: New York, 1995, 33.
3.   Baiker, A.; Blaser, H. U. in *Handbook of Heterogeneous Catalysis*, Vol 5, Ertl, G. H.; Knoezinger, H.; Weinheim, J., Eds.VCH: New York, 1997, p. 2422.
4.   Blaser, H. U. *Tetrahedron: Asymm.* **1991**, *2*, 843.
5.   Baiker, A. *Curr. Opin. Solid State Mater. Sci.* **1998**, *3*, 86.
6.   Izumi, Y. *Adv. Catal.* **1983**, *32*, 215.
7.   Tai, A.; Harada, T. in *Tailored Metal Catalysts* Iwasawa, Y., Ed. Reidel: Tokyo, 1986, p. 265.
8.   Orito, Y.; Imai, S.; Niwa, S. *J. Chem. Soc. Jpn.* **1980**, 670.
9.   Webb, G.; Wells, P. B. *Catal. Today* **1992**, *12*, 319.
10. Blaser; H. U.; Jalett, H, P.; Muller, M.; Studer, M. *Catal. Today* **1997**, *37*, 441.
11. Raval, R; Baddeley, C. J.; Haq, S.; Louafi, S.; Murray, P.; Muryn, C.; Ortega Lorenzo, M.; Williams, J. *Stud. Surf. Sci. Catal.* **1999**, *122*, 11.
12. Ortega Lorenzo, M.; Haq, S.; Murray, P.; Raval, R.; Baddeley, C. J. *J. Phys. Chem. B* **1999**, *103*, 10661.
13. Orteza Lorenzo, M.; Baddeley, C. J.; Muryn, C.; Raval, R. *Nature* **2000**, *404*, 376.
14. Ortega Lorenzo, M.; Humblot, V.; Murray, P.; Baddelley, C. J.; Haq, S.; Raval, R. *J. Catal.* **2002**, *205*, 123.
15. Raval, R. CATTECH **2000**, *4*, 1.
16. Groenewegen, J. A; Sachtler; W. M. H. *J.Catal.* **1975**, *38*, 501.
17. Yasumori, I.; Yokozeki. M.; Inoue, Y. *Farad. Disc. Chem Soc.* **1981**, *72*, 385.
18. Klabunovskii, E. I.; Vedenyapin, A. A.; Krapeiskaya, E. I.; Parrlon, V. A. *Proc. 7th Int. Congr. Catal. Tokyo (1980)*, Elsevier: Tokyo, 1981, 390.
19. Lopinski, G. P.; Moffat, D. J.; Wayner; D. D. M.; Wolkov, R. A. *Nature* **1998**, *392*, 909.
20. Viswanathn, R.; Zasadzinski, J. A.; Schwartz, D. K. *Nature* **1994**, *368*, 440.
21. Keane, M. A.; Webb, G. *J. Catal.* **1992**, *136*, 1.
22. Keane, M. A. *Langmuir* **1997**, *13*, 41.

Chapter 8

ON THE STRUCTURE OF COBALT-SUBSTITUTED ALUMINOPHOSPHATE CATALYSTS AND THEIR CATALYTIC PERFORMANCE

Gopinathan Sankar,* Robert Raja
Davy Faraday Research Laboratory, The Royal Institution of Great Britain, 21 Albemarle Street, London , W1S 4BS, UK

Keywords: CoAlPO, aluminophosphate catalysts, EXAFS, XRD, selective oxidation,

Abstract: Microporous, transition metal ion-substituted aluminophosphate molecular sieves have attracted considerable attention over the last decade owing to their shape-selective catalytic properties for both acid-catalyzed conversion of methanol to lower olefins and aerobic oxyfunctionalization of saturated hydrocarbons. By controlling the types of transition metal ions that are substituted in place of Al(III) ions in the framework of the aluminophosphate structure and the pore structure of the microporous materials, it is now possible to perform both reactant- and product-selective catalytic reactions. Detailed structures of the molecular sieves, local structures of the active (transition metal ion) sites and their relationship to catalytic properties are discussed.

1. INTRODUCTION

Since the discovery of the open framework structures of aluminophosphates (which are similar to microporous zeolites) and their metal ion-substituted variants,[1,2] intense research activity has been undertaken in the areas of synthesis, characterization and applications, in particular in the field of shape-selective catalysis. Many novel microporous aluminophosphates have been synthesized over the last twenty years, some of which are structural analogues of

Nanostructured Catalysts, edited by S. Scott *et al.*
Kluwer Academic/Plenum Publishers, 2003

either minerals or aluminosilicates (zeolites), whilst others are new open framework structures. Some of the structures are shown in Figure 1.

Aluminophosphates are neutral solids consisting of corner-shared AlO_4 and PO_4 tetrahedra which can be converted to solid acid catalysts, similar to aluminosilicates, by substituting some of the framework Al(III) or P(V) ions by lower valent ions. The solids are usually prepared using structure-directing agents (SDA) such as amines or quaternary ammonium ions. These materials, if stable upon removal of the organic SDA, may function as shape-selective catalysts for both acid-catalyzed dehydration and isomerization reactions, and more importantly for the oxyfunctionalization of inert alkanes.[3-5] For example, when divalent ions such as Zn(II) or Mg(II) are incorporated in place of some of the Al(III) ions, an acid catalyst is produced upon removal of the organic template, with H^+ ions being introduced to maintain the neutrality of the framework.[6]

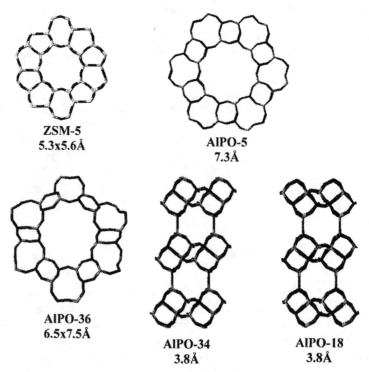

ZSM-5
5.3x5.6Å

AlPO-5
7.3Å

AlPO-36
6.5x7.5Å

AlPO-34
3.8Å

AlPO-18
3.8Å

Figure 1. Structural characteristics of an aluminosilicate (ZSM-5) and some microporous aluminophosphate catalysts. Typical pore sizes are also given.

If the divalent ions are Co(II) or Mn(II), they are usually oxidized to Co(II) or Mn(III) at the temperatures conventionally used for removing the organic template by calcination in air. Subsequent reduction of these calcined materials, for example, in hydrogen, results in the formation of Co(II) or Mn(II), yielding an acid catalyst which can readily be identified by the presence of an OH stretching frequency typical for a bridging hydroxyl in the infrared spectrum (see Figure 2).[6-8] This type of redox chemistry of cobalt-, manganese- and iron-containing materials has attracted considerable attention in the literature for use as selective oxidation catalysts for the functionalization of a variety of alkanes and alkenes.[5,9-11] The structural chemistry of cobalt-, manganese- and iron-containing aluminophosphate catalysts and some of their catalytic applications are reported here. In addition, the effect of the preparative method, in particular the types of organic SDA that influence the stability of the microporous structure and consequences for the redox chemistry and catalytic applications, are also discussed.

Early work in the 80's and early 90's was first directed towards the characterization of the redox chemistry of the substituted metal ions. Concurrently, attempts were made to use these systems for oxidizing alkanes. Both endeavors were restricted mainly to a few types of structures prepared by a variety of research groups. Several characterization techniques, in particular UV-Vis,[12] ESR,[13,14] NMR[15] and X-ray techniques, were used to understand the structures of these materials. Although X-ray diffraction (XRD) studies provided

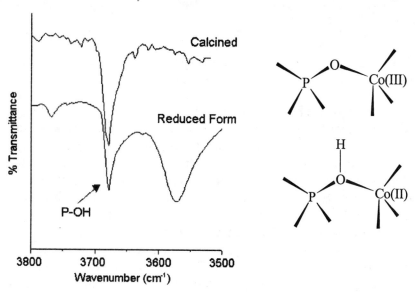

Figure 2. Typical IR spectra obtained for the calcined and reduced CoAlPO-18 catalyst. Corresponding local structure around the cobalt is shown schematically on the right.

vital structural information in atomic detail, this technique could not be used to determine the structures of the substituted heteroatoms which are responsible for catalytic function, since these ions are present only in small amounts and are randomly substituted. For these reasons, X-ray absorption fine structure (XAFS) is employed to determine the local structure of the catalytic centres, since this technique does not depend on long-range order and, more importantly, it is atom-specific.[16,17] Thus, the combined information derived from the two techniques, preferably performed in a single experiment, is extremely advantageous to fully characterize these materials. With the use of synchrotron radiation, it is now possible to carry out combined XRD/XAS measurements and, more importantly, these can be done *in situ* to determine accurately the state of the catalyst under operating conditions (see Figure 3).[16,17]

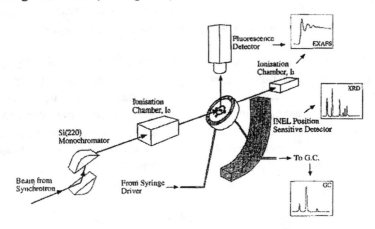

Figure 3. Schematic diagram of the typical combined XRD/QuEXAFS set-up employed at the SRS Daresbury laboratory.

Our group at the Royal Institution has explored several of these materials, systematically investigating their structures and the local environment of the heteroatom, in particular for cobalt-substituted aluminophosphates (CoAlPO).[18-20] First, we established that the microporous structures of some CoAlPO catalysts are stable (an important property for shape-selective reactions) upon removal of the organic template molecule.[8] Initial studies clearly showed that catalysts containing cobalt concentrations below *ca.* 10% retained their microporous structure. For example, the thermal stability of CoAlPO-44 is poor (see Figure 4, top). The characteristic d-spacings of this structure disappear rather abruptly at temperatures slightly higher than 200°C, giving rise (at *ca.* 350°C and above) to the characteristic X-ray pattern of the tridymitic (AlPO) analogue of the well-known silica polymorph. By contrast, the thermal stability of the same open-structure CoAPSO-44, containing some framework Si(IV) ions in place of P(V) in addition to the presence of cobalt, is considerably enhanced (see Figure 4, bottom).[21]

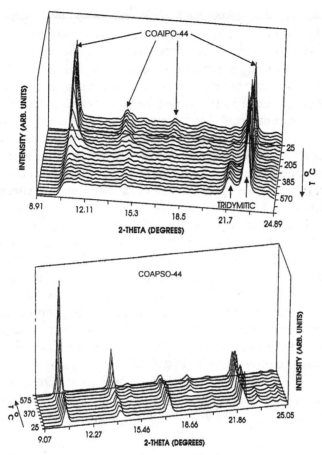

Figure 4. Stacked XRD of CoAlPO-44 (top) and CoAPSO-44 (bottom) recorded during calcination in oxygen using *in situ* combined XRD/XAS technique. Note that the CoAlPO-44 catalyst loses its microporous structure upon removal of the template and transforms to a dense phase related to the tridymite structure, whereas the CoAPSO-44 sample retains its crystallinity upon removal of the organic template, cyclohexylamine.[21] Adapted from *J. Phys. Chem. Solids* **1995**, *56*, 1395. Copyright 1995 Elsevier Science.

2. LOCAL ENVIRONMENT OF COBALT IN CoAlPOs AND ITS RELATION TO CATALYTIC PERFORMANCE

Several of the stable CoAlPO catalysts were investigated in detail by *in situ* Co K-edge XAFS, since one of the contentious issues in the literature pertains to whether the substituted Co(II) ions undergo a change in oxidation state to Co(III) upon calcination in oxygen at temperatures of *ca.* 550 °C.[14,22,23] Whilst

Iton and co-workers suggested the loss of the ESR signal intensity for calcined CoAlPO-34 was due to the formation of Co(III) species,[22] Kevan and co-workers[23,24] argued that a change in coordination geometry around cobalt, upon calcination, can explain the decrease in the ESR signal.

The as-prepared materials are all blue in colour, and the EXAFS-derived Co-O distances are close to 1.94 Å (see Table 1), which is typical for Co(II) in a tetrahedral environment.[18-20,25,26] Upon calcination, one would expect all the Co(II) ions to be converted to Co(III) ions, irrespective of the structure. However, the average Co-O distances determined by EXAFS differed for various structures, even though identical conditions were used for calcination and EXAFS measurements (see Table 1). It must be noted that materials containing Co(III) ions in a tetrahedral geometry are rare,[27] the only reported systems being a polyoxometalate, $K_6Co(III)W_{11}O_{39} \cdot 20H_2O$, and a garnet. The crystal structure of the polyoxometalate revealed that the Co(III)-O bond distance is *ca.* 1.87 Å,[29] which is much longer than that determined for the calcined CoAlPO-18 catalyst (see Table 1). In addition, the Co(III)-O distance reported[29] for the polyoxometalate ion is unlikely to be correct, since it is much higher than that estimated from the bond valence sum rule. Furthermore, some of the W-O and O-O separations reported in that work are much shorter than the expected typical distances. The differences in the Co-O bond distances for the calcined materials can be rationalized by considering that all of the Co(II) ions in certain structures are completely converted to Co(III), while in other structures only some are oxidized. It was proposed[20] that the unoxidized Co(II) may be associated with an oxygen ion vacancy in order to maintain the neutrality of the framework (see Scheme 1).

Table 1. Average Co-O distances of cobalt-substituted aluminophosphates in the as-prepared and calcined forms, derived from Co K-edge EXAFS.

System	Average Co-O distance (Å)		Fraction of Co(III), X	
	as prepared	calcined	X^a	X^b
CoAlPO-5	1.94	1.92	0.2	0.15
CoAlPO-11	1.94	1.91	0.3	0.25
CoAlPO-36	1.93	1.87	0.5	0.4
CoAlPO-18	1.93	1.82	1	0.8
CoAlPO-34	1.93	1.82	1	0.8
CoAlPO-44	1.94	1.85	0.8	0.65

[a] Estimated by taking the Co-O distance of CoAlPO-18 (1.82 Å) to be representative of $R_{Co(III)-O}$.[20]
[b] Fraction calculated by taking the average Co-O distance (1.80 Å) obtained for
  $K_5Co(III)W_{12}O_{40} \cdot 20H_2O$.[28]

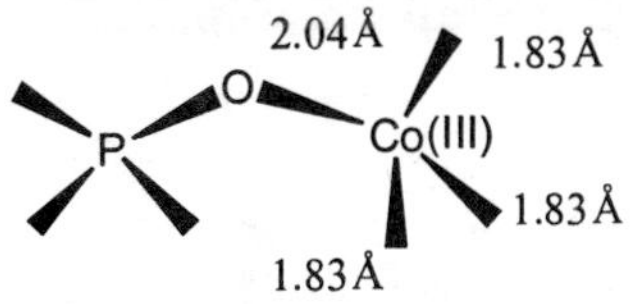

Oxidized Part

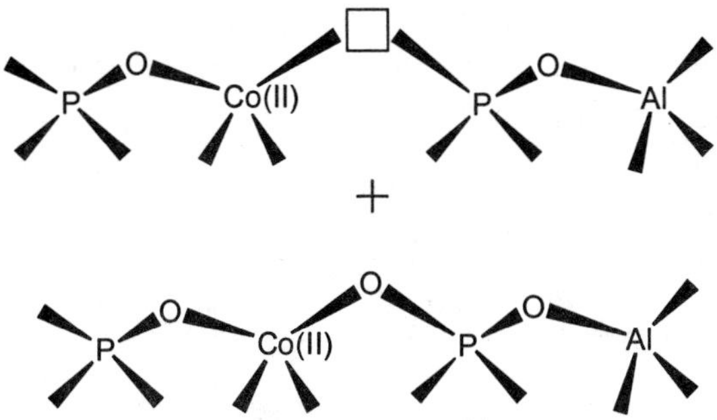

+

Unoxidized Part

Scheme 1. Proposed structures of Co sites in calcined CoAlPO's. Adapted from *J. Phys Chem.* **1996**, *100*, 8977. Copyright 1996 American Chemical Society.

The fraction of cobalt ions which undergo such redox reactions can be estimated by employing the linear relationship given below:

$$R_{EXAFS} = X \, R_{Co(III)-O} + (1-X)R_{Co(II)-O} \qquad (1)$$

where X is the fraction of Co(III), R_{EXAFS} is the average Co-O distance obtained from the EXAFS study of the calcined material and $R_{Co(III)-O}$ is the distance obtained from the crystallographic data for $CoAl_2O_4$.[30] In the absence of a proper model compound that represents Co(III) in tetrahedral geometry, the X values reported in our earlier work[20] were based on the $R_{Co(III)-O}$ distance in calcined CoAlPO-18 material as representative of the Co(III)-O distances. Recently, we synthesized and reported the structure of Co(III) in tetrahedral coordination in $K_5Co(III)W_{12}O_{40} \cdot 20H_2O$.[28] The Co(III)-O distances derived from this single crystal study were found to be *ca.* 1.80 Å. The fraction of Co(III) in the various CoAlPO structures has now been recalculated based on this single crystal study and are also listed in Table 1. Although they are slightly different from the values reported earlier, the trend in the fraction of Co(III) formed remains unaltered.

Thus, it appears from the EXAFS study that CoAlPO-5, the most widely studied material for the oxidation of alkanes, is unlikely to be the best catalyst since CoAlPO-36 has a higher fraction of Co(III) ions. Although CoAlPO-18

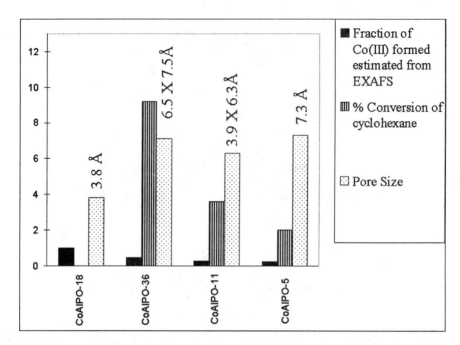

Figure 5. Comparison of the pore dimensions (derived from XRD) of various microporous solids and the fraction of cobalt raised to a higher oxidation state (estimated from EXAFS using Eqn. (1)) with catalytic conversion of cyclohexane in dry air (1.5 MPa) at 130 °C.[31]

contains the highest fraction of Co(III) ions, this catalyst may not oxidize cyclohexane simply because of its small pore dimensions (*ca.* 3.8 Å) which restrict access of the cyclohexane molecule to the active site.[20,31] A systematic study of catalytic cyclohexane oxidation, with air as the source of oxygen, for several cobalt-substituted AlPO's revealed that the overall conversion and turnover frequencies are consistent with the EXAFS results, with CoAlPO-36 showing the highest conversion (see Figure 5).[31]

Upon re-examination of the EXAFS results, it is clear that CoAlPO-18 catalysts have the highest amount of Co(III) ions[8,20] and, considering the pore dimensions derived from XRD studies, it was possible to design a new catalytic reaction with linear alkanes as the substrate.[32] Comparison of the overall catalytic performance of CoAlPO-18 with CoAlPO-36 showed that the former is superior. In addition, due to the restriction imposed by the small pore CoAlPO-18 catalyst, we observed a regioselective conversion of *n*-hexane to hexanoic acid by oxidation predominantly at the terminal carbon atom. CoAlPO-36, being a large pore channel system, did not show any regioselectivity; the energetically favorable secondary and tertiary carbon atoms were oxidized (see Figure 6). Thus this study clearly demonstrates the potential application of small pore microporous aluminophosphates for regioselective conversion of linear hydrocarbons.

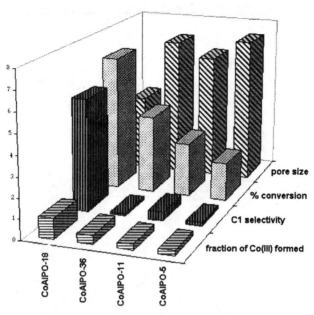

Figure 6. Comparison of the pore dimensions of various CoAlPO's and the fraction of cobalt raised to a higher oxidation state (estimated from EXAFS using Eqn. 1) with selectivity and conversion of *n*-hexane in dry air.[32]

3. SMALL PORE CoAlPO-18 AND CoAlPO-34

AlPO-18 is a small pore microporous material that can be synthesized using either tetraethylammonium hydroxide (TEAOH) or *N,N*-diisopropyl ethylamine as the structure-directing template. When TEAOH is used as the template molecule, above a certain concentration of the heteroatom (transition metal ions or silicon) only the AlPO-34 structure is formed.[33] (AlPO-34 has a structure identical to that of the mineral Chabazite). This was clearly seen in a detailed study using the *in situ* Energy Dispersive X-ray diffraction (EDXRD) technique.[34] The entire diffraction pattern can be collected on a timescale of a few seconds to a few minutes depending upon the system being studied. A schematic of the setup is shown in Figure 7.

The main advantage of this technique is that a real autoclave (slightly modified to improve the signal-to-noise ratio) can be used to study the crystallization of microporous or mesoporous materials under hydrothermal conditions.[35,36] We employed the EDXRD technique to follow the formation of the small pore microporous aluminophosphate catalyst using TEAOH as the

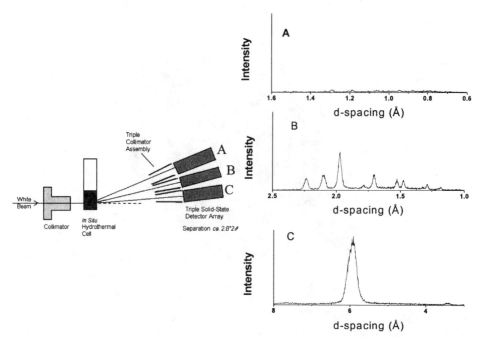

Figure 7. Schematic illustration of the three-element detector setup used for measuring Energy Dispersive X-ray Diffraction (EDXRD) patterns. Each detector is separated by a fixed angle, enabling us to cover a wider energy range (or d-spacing). This is illustrated on the right hand side of the Figure, where we show the EDXRD patterns of CoAlPO-5.[16,34]

template molecule, as a function of cobalt concentration.[34] At low cobalt concentrations (below Co/P=0.06), the AlPO-18 structure is formed in competition with AlPO-5. Note that by altering the temperature and pH, pure AlPO-18 can be produced (see Figure 8). However, above 6% cobalt (a Co/P ratio of 0.06), only the CoAlPO-34 structure is formed (see Figure 8). When *N,N*-diisopropylethylamine is used as the SDA, only the CoAlPO-18 structure is produced for cobalt concentrations up to ca 10%.

The main difference between the AlPO-18 and AlPO-34 structures is the way in which the double six ring units are oriented (see Figure 8). Although the structures are different crystallographically, they both possess a similar cage-type structure with pore openings consisting of 8-membered rings resulting in a pore entry size of *ca.* 3.8 Å. A detailed combined Co K-edge XAS/XRD study of both 10% cobalt-containing CoAlPO-18 and CoAlPO-34 showed that, similar to the CoAlPO-18 catalyst with a cobalt concentration of 4%, their microporous structures were stable upon removal of the SDA and, more importantly, all the Co(II) ions were converted to Co(III) irrespective of the structure and composition (see Figure 9).[37]

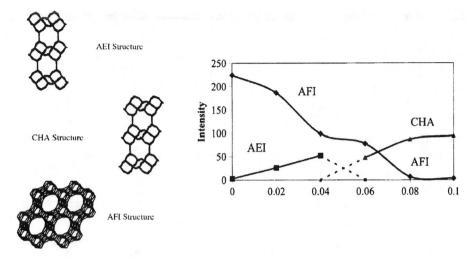

Figure 8. On the right, plot of the intensity of the main reflections (from the data recorded 60 minutes after start of the reaction, with bottom detector) for the structures (shown on the left) AFI, AEI and CHA vs Co/P ratio.[34] Reproduced with permission from *Phys. Chem. Chem. Phys.* **2000**, *2*, 3523. Copyright 2000 Royal Society of Chemistry.

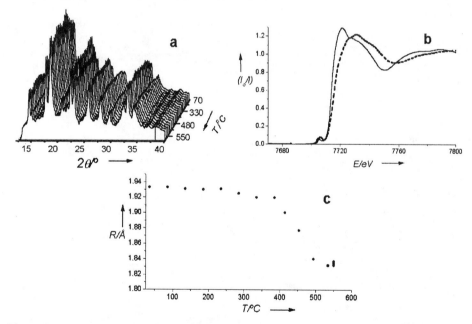

Figure 9. *In situ* combined XRD/XAS data recorded during the calcination of CoAlPO-18 (10% cobalt-containing catalyst): (a) stacked XRD data, (b) Co K-edge X-ray absorption near edge (XANES) data for the as-prepared (solid line) and calcined (dashed curve) catalysts. In (c), variation of the Co-O distance with temperature is shown. Both the shift in the Co K-edge position (see (b)) and decrease in Co-O distance from 1.93 to *ca.* 1.83 Å clearly show the formation of Co(III) upon calcination.[37] Reproduced with permission from *Angew. Chem. Int. Ed.* **1997**, *36*, 2675. Copyright 1996 Wiley Publishers.

A closer look at the Co-O bond distance derived from detailed analysis of the EXAFS data recorded during the calcination process revealed that a decrease from 1.93 to 1.90 Å took place around 400°C, with a further reduction to 1.83 Å seen above 500°C. By combining the information obtained from the shift in the X-ray absorption edge (which is known to be sensitive to the oxidation state of the metal ion), and *in situ* FTIR measurements, it was possible to rationalize that, upon breakdown of the template at *ca.* 400°C, Brønsted acid sites are formed, since cobalt remains in the same oxidation state as that of the starting material (Co(II)). Further heating in air to above 500°C resulted in the formation of Co(III) (see Figure 10).[16]

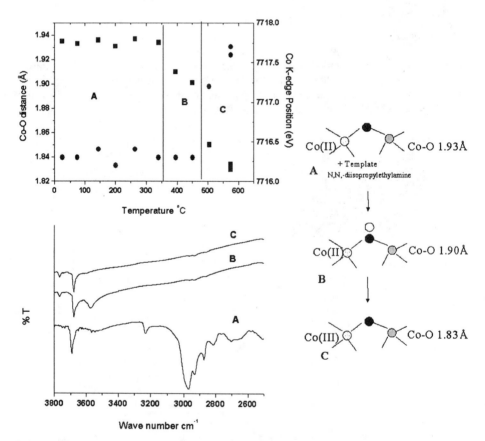

Figure 10. On the top left, the variation of Co-O distance and Co K-edge position derived from Co K-edge XANES data measured during calcination of CoAlPO-18 (4% cobalt) catalyst. On the bottom left, IR spectra of CoAlPO-18 measured at room temperature after heating this material in oxygen at three different temperatures: A-100°C; B-400°C and C-550°C. On the right, schematic local structural arrangements (with average Co-O distances derived from EXAFS) around cobalt in CoAlPO-18 representing the three states labelled A, B and C.[16]

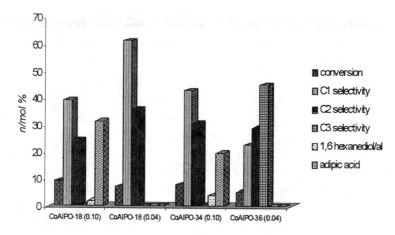

Figure 11. Comparison of the selectivity and activity for the aerial oxidation of *n*-hexane using various CoAlPO catalysts.[37]

Although the structural integrity and redox chemistry were found to be the same for the CoAlPO-18 catalysts containing 4% or 10% cobalt (or CoAlPO-34), product selectivities in the oxidation of *n*-hexane were found to be different. The catalyst containing 4% cobalt produced mainly hexanoic acid, whereas the catalyst containing 10% cobalt yielded adipic acid in addition to hexanoic acid (see Figure 11).[37] We argued that there may be two cobalt centres present in the cage positioned in such a way that both terminal carbons of *n*-hexane are oxidized simultaneously, yielding adipic acid (see Figure 12).

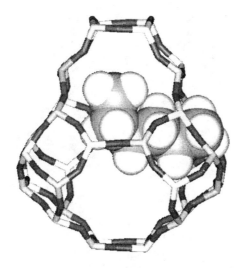

Figure 12. Computer graphic representation of the energy-minimized configuration of an *n*-hexane molecule inside an AlPO-18 structure.[37]

4.   EFFECT OF COBALT CONCENTRATION AND TYPE OF SDA ON THE STABILITY OF SMALL PORE CoAlPO-34

CoAlPO-34 catalysts can be prepared using several organic template molecules. Some of the commonly used SDA's are triethylamine, TEAOH, morpholine and cyclohexylamine (used to produce CoAlPO-44, which has the same chabazitic structure). Recently, we found two other template molecules, namely 4,4-piperidinopiperidine (the microporous material produced was designated as DAF-5) which was designed based on a detailed computational study (using ZEBEDDE),[38,39] and 1,3,3-trimethyl-6-azabicyclo[3.2.1]octane (from a rational synthesis approach), that are efficient in producing CoAlPO's with the chabazitic structure. The structures derived from a micro-crystal diffraction study of the solids prepared by these two template molecules are shown in Figures 13 and 15.

Combined XRD/XAS measurements (Figures 14 and 15) conducted during the calcination process, similar to that described above, showed that catalysts prepared using cyclohexylamine, morpholine and 4,4-piperidinopiperidine are unstable upon removal of the organic template. The microporous structure is lost above 400°C (see Figure 14) and, more importantly, the Co-O distance increases from 1.93 to *ca.* 1.98 Å, suggesting that there is no oxidation from Co(II) to Co(III). Close examination of the cobalt content of all

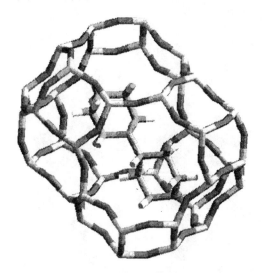

Figure 13. Structure of DAF-5 (a cobalt-substituted aluminophosphate prepared using 4,4-piperidinopiperidine as the SDA) obtained from a micro-crystal diffraction study. The location of the organic template within the chabazitic cage is shown. Reproduced with permission from *Angew. Chem. Int. Ed.* **1996**, *36*, 2675. Copyright 1996 Wiley Publishing.

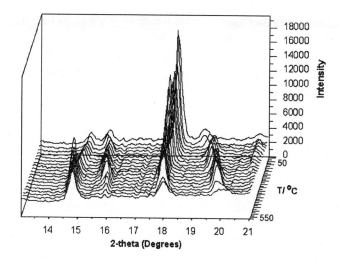

Figure 14. Stacked XRD pattern, recorded using the combined XRD/XAS technique, during calcination of DAF-5. The gradual loss in intensity and changes in the peak positions indicate a loss of the microporous structure that takes place during removal of the SDA.

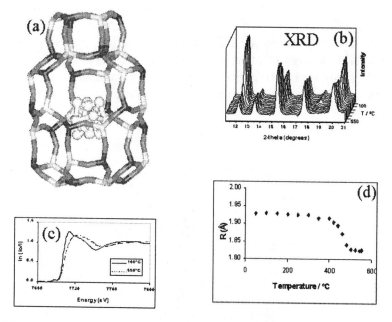

Figure 15. (a) Structure of CoAlPO-34 (prepared using 1,3,3-trimethyl-6-azabicyclo[3.2.1]octane as SDA) derived from a micro-crystal diffraction study. (b) XRD, (c) Co K-edge XANES of as-prepared (solid line) and calcined (dashed curve) materials, and (d) variation in Co-O distance with temperature, recorded during the calcination of the catalyst, employing the combined XRD/XAS technique. Note that this catalyst, similar to CoAlPO-18, also shows complete oxidation to Co(III) upon calcination.

the samples revealed that the catalysts which showed good thermal stability and redox chemistry have cobalt concentrations of *ca.* 10% or less. Materials containing a higher concentration of cobalt (*ca.* 20%) were found to be unstable. Thus, by this procedure we were able to screen for the material best suited for catalytic applications.

5.   AlPO-5 STRUCTURE WITH OTHER METAL IONS

Although small pore microporous materials show good regioselectivity and catalytic activity, large pore materials are necessary to convert bulkier substrates such as cyclohexane. The structure based on AlPO-36 is ideal, but synthesis of an AlPO-5 type structure is comparatively much easier than AlPO-36. Thus, the substitution of other metal ions such as manganese and iron in the AlPO-5 structure was investigated. The average metal-oxygen distances of the calcined MnAlPO-5 and FeAlPO-5 catalysts, derived from *in situ* EXAFS studies, clearly show that Mn(III) and Fe(III) ions are present in the framework of the calcined catalyst. Once again, we found that only some of the Mn(II) ions are converted to Mn(III), but all the iron ions are present as Fe(III). Comparison of the fraction of M(III) (M = Mn , Co or Fe) ions present in the calcined catalyst (estimated from the distances derived from EXAFS data and using Eqn. (1)) of various MAlPO's suggest that manganese- and iron-containing AlPO-5 materials should show higher catalytic performance for the oxidation of cyclohexane, consistent with the experimental results shown in Figure 16.[40]

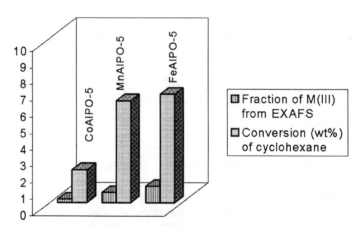

Figure 16. Comparison of the fraction of metal ions that are raised to a higher oxidation state (estimated from EXAFS using Eqn.1) with catalytic conversion of cyclohexane in dry air (1.5 MPa) at 130 °C.[40]

6. SUMMARY

In summary, characterization of metal ion-substituted microporous materials has enabled the detailed understanding of the structure of active sites, the associated redox properties, and the design of new catalytic reactions using these catalysts.

ACKNOWLEDGEMENTS

EPSRC and Leverhulme trust are thanked for financial support, as well as CCLRC for the facilities at the Daresbury laboratory. I also thank Professor Sir John Meurig Thomas, Professor C. Richard A. Catlow, Dr David Gleeson, Dr Philip A. Barrett and Mr Andrew M. Beale for their comments and useful discussions.

REFERENCES

1. Wilson, S. T.; Lok, B. M.; Messina, C. A.; Cannan, T. R.; Flanigen, E. M. *J. Am. Chem. Soc.* **1982**, *104*, 1146,.
2. Wilson, S. T.; Flanigen, E. M. in *ACS Symp. Ser. Zeolite Synthesis,* Occelli, M. L. Robson, H. E., Eds. ACS: Washington D.C. **1989**, *398*, 329.
3. Thomas, J. M. *Phil. Trans. Royal Soc. Lond. Ser. A* **1990**, *333*, 173.
4. Thomas, J. M. *Angew. Chem., Int. Ed. Engl.* **1994**, *33*, 913.
5. Thomas, J. M.; Raja, R.; Sankar, G.; Bell, R. G. *Acc. Chem. Res.* **2001**, *34*, 191.
6. Chen, J. S.; Thomas, J. M. *J. Chem. Soc., Chem. Comm.* **1994**, 603.
7. Marchese, L.; Chen, J. S.; Thomas, J. M.; Coluccia, S.; Zecchina, A. *J. Phys. Chem.* **1994**, *98*, 13350.
8. Thomas, J. M.; Greaves, G. N.; Sankar, G.; Wright, P. A.; Chen, J. S.; Dent, A. J. Marchese, L. *Angew. Chem., Int. Ed. Engl.* **1994**, *33*, 1871.
9. Sheldon, R. A.; Dakka, J. *Catal. Today* **1994**, *19*, 215.
10. Sheldon, R. A. *Proc. 3rd World Congress on Oxidation Catalysis* **1997**, *110*, 151.
11. Vanoppen, D. L., Devos, D. E.; Genet, M. J.; Rouxhet, P. G.; Jacobs, P. A. *Angew. Chem., Int. Ed. Engl.* **1995**, *34*, 560.
12. Weckhuysen, B. M., Rao, R. R.; Martens, J. A.; Schoonheydt, R. A.; *Eur. J. Inorg. Chem.* **1999**, 565.
13. Hartmann, M.; Kevan, L. *Chem. Rev.* **1999**, *99*, 635.
14. Weckhuysen, B. M.; Verberckmoes, A. A.; Uytterhoeven, M. G.; Mabbs, F. E.; Collison, D.; de Boer, E.; Schoonheydt, R. A. *J. Phys. Chem. B* **2000**, *104*, 37.
15. Canesson, L.; Boudeville, Y.; Tuel, A. *J. Am. Chem. Soc.* **1997**, *119*, 10754.
16. Sankar, G.; Thomas, J. M. *Top. Catal.* **1999**, *8*, 1.
17. Sankar, G., Thomas, J. M.; Catlow, C. R. A. *Top. Catal.* **2000**, *10*, 255.
18. Chen, J. S.; Sankar, G., Thomas, J. M.; Xu, R. R.; Greaves, G. N.; Waller, D. *Chem. Mater.* **1992**, *4*, 1373.
19. Sankar, G.; Thomas, J. M.; Chen, J.; Wright, P. A.; Barrett, P. A.; Greaves, G. N.; Catlow, C. R. A. *Nucl. Instrum. Meth. B* **1995**, *97*, 37.

20.   Barrett, P. A., Sankar, G.; Catlow, C. R. A.; Thomas, J. M. *J. Phys. Chem.* **1996**, *100*, 8977.
21.   Barrett, P. A., Sankar, G.; Catlow, C. R. A.; Thomas, J. M. *J. Phys. Chem. Solids* **1995**, *56*, 1395.
22.   Iton, L. E.; Choi, I.; Desjardins, J. A.; Maroni, V. A. *Zeolites* **1989**, *9*, 535.
23.   Kurshev, V.; Kevan, L.; Parillo, D. J.; Pereira, C.; Kokotailo, G. T.; Gorte, R. J. *J. Phys. Chem.* **1994**, *98*, 10160.
24.   Prakash, A. M.; Hartmann, M.; Kevan, L. *J. Phys. Chem. B* **1997**, *101*, 6819.
25.   Moen, A.; Nicholson, D. G.; Ronning, M.; Lamble, G. M.; Lee, J. F.; Emerich, H. *J. Chem. Soc., Farad. Trans.* **1997**, *93*, 4071.
26.   Thomson, S.; Luca, V.; Howe, R. *Phys. Chem. Chem. Phys.* **1999**, *1*, 615.
27.   Montes, C.; Davis, M. E.; Murray, B.; Narayana, M. *J. Phys. Chem.* **1990**, *94*, 6425.
28.   Muncaster, G.; Sankar, G.; Catlow, C. R. A.; Thomas, J. M.; Coles, S. J.; Hursthouse, M. *Chem. Mater.* **2000**, *12*, 16.
29.   Yannoni, N. F. PhD thesis, Univ. Boston, 1961; also see ICSD data base reference number 17580.
30.   Toriumi, K.; Ozima, M.; Akaogi, M; Saito, Y. *Acta Crystal. B* **1978**, *34*, 1093.
31.   Sankar, G.; Raja, R.; Thomas, J. M. *Catal. Lett.* **1998**, *55*, 15.
32.   Thomas, J. M.; Raja, R.; Sankar, G.; Bell, R. G. *Nature* **1999**, *398*, 227.
33.   Chen, J. S.; Wright, P. A.; Thomas, J. M.; Natarajan, L.; Marchese, L.; Bradley, S. M.; Sankar, G.; Catlow, C. R. A.; Gaiboyes, P. L.; Townsend, R. P.; Lok, C. M. *J. Phys. Chem.* **1994**, *98*, 10216.
34.   Muncaster, G.; Davies, A. T; Sankar, G.; Catlow, C. R. A.; Thomas, J. M.; Colston, S. L.; Barnes, P. I.; Walton, R. I.; O'Hare, D. *Phys. Chem. Chem. Phys.* **2000**, *2*, 3523.
35.   Walton, R. I.; O'Hare, D. *J. Chem. Soc., Chem. Comm.* **2000**, 2283.
36.   Francis, R. J.; O'Hare, D. *J. Chem. Soc., Dalton Trans.* **1998**, 3133.
37.   Raja, R.; Sankar, G.; Thomas, J. M. *Angew. Chem. Int. Ed.* **2000**, *39*, 2313.
38.   Sankar, G.; Wyles, J. K.; Jones, R. H.; Thomas, J. M.; Catlow, C. R. A.; Lewis, D. W.; Clegg, W.; Coles, S. J.; Teat, S. J. *J. Chem. Soc., Chem. Comm.* **1998**, 117.
39.   Lewis, D.W.; Sankar, G.; Wyles, J. K.; Thomas, J. M.; Catlow, C. R. A.; Willock, D. J. *Angew. Chem. Int. Ed.* **1997**, *36*, 2675.
40.   Raja, R.; Sankar, G.; Thomas, J. M. *J. Am. Chem. Soc.* **1999**, *121*, 11926.

Chapter 9

CATALYTIC ACTIVITY OF Pt AND TUNGSTO-PHOSPHORIC ACID SUPPORTED ON MCM-41 FOR THE REDUCTION OF NO

A. Jentys,[a] H. Vinek[b]

[a] Technische Universität München, Institute for Chemical Technology, Lichtenbergstr. 4, 85747 Garching, Germany
[b] Technische Universität Wien, Institut für Physikalische und Theoretische Chemie, Veterinärplatz 1, A-1210 Wien, Austria

Keywords: NO_x reduction, MCM-41, polyoxometalate, environmental catalysis

Abstract: The catalytic reduction of NO_x is one of the most rapidly growing applications in the field of environmental catalysis, an area which accounts in total for more than one-third of the world catalyst market today. For NO_x emissions from stationary sources, V_2O_5/TiO_2 catalysts and NH_3 as reducing agent are regularly used, however, extensive research is necessary to develop catalysts in which more environmentally more acceptable molecules such as hydrocarbons can be used as reducing agents. Initially, research focused on transition metal-loaded zeolites, but the drastic loss in activity when H_2O and/or SO_2 are present during the reaction has led to the development of alternative catalysts such as Pt group metals supported on metal oxides (SiO_2 and Al_2O_3). In order to combine the advantages of zeolites with oxide-based systems, the catalytic properties of noble and transition metals supported on mesoporous molecular sieves with MCM-41 type structure were investigated. To generate strong Brønsted acid sites, the catalysts were co-impregnated with tungstophosphoric acid. These catalysts showed promising properties with respect to activity and stability in presence of water vapor. The mesoporous molecular sieve was found to act as a high surface area support for the metal and acid clusters, which were, due to the special characteristics of MCM-41 type materials, highly stabilized against thermal sintering and pore blocking. The structure and the activity/selectivity

Nanostructured Catalysts, edited by S. Scott et al.
Kluwer Academic/Plenum Publishers, 2003

213

of these catalysts with respect to the influence of the water vapor are discussed in order to explain the unique properties of these materials in the reduction of NO_x with hydrocarbons in the presence of water vapor.

1.   INTRODUCTION

Nitric oxide, formed during combustion processes in power plants, waste incinerators and combustion engines, is among the major air pollutants leading to the formation of photochemical smog and acid rain. NO is rapidly oxidized in the atmosphere by ozone and free radicals (HO• and HO_2•) to other oxides such as NO_2, HNO_3 and HO_2NO_2.[1] Thus, reduction of NO emissions represents a worldwide effort to transform combustion processes into an environmentally friendly technology.[2] At present, catalytic processes are most frequently applied to exhaust gas treatment. For standard gasoline combustion engines, a noble metal catalyst is typically used for the simultaneous reduction of NO and oxidation of CO and hydrocarbons. This becomes possible when the O_2 concentration in the exhaust gas is adjusted to the stoichiometric level ($\lambda=1$). Due to the different operating conditions of diesel and lean burn engines, significantly higher O_2 concentrations are present in their exhaust gases and, therefore, a reducing agent has to be added to achieve the required NO conversion. This technology is already well-established for NO reduction in flue-gas streams of stationary sources, where catalysts based on titanium- and vanadium-oxides together with ammonia-based reducing agents are typically used.[3] For mobile sources, however, these reducing agents face technical as well as economic limitations. Extensive research is currently being carried out worldwide to find alternative catalytic systems in order to replace NH_3 with more environmentally acceptable reducing agents such as hydrocarbons.[4,5] Catalysts based on transition metal-containing zeolites, for example Cu/ZSM-5[6] and Co/ZSM-5,[7,8] were investigated initially, but their sharp decrease in activity when H_2O and/or SO_2 are present[9,10] led to the development of catalysts such as Fe/ZSM-5[12,13] and Pt/ZSM-5.[14] Alternatives are Pt group metals supported on SiO_2 and Al_2O_3,[15,16] which show, compared to the zeolite-based catalysts, higher activity at lower temperatures and better stability in the presence of H_2O vapor. The major drawback of these catalysts, however, is their significant formation of N_2O, which contributes strongly to the greenhouse effect and thus limits the potential for application of these catalysts.

In order to combine the advantages of zeolites and oxide-based systems, we studied the potential of noble metals supported on mesoporous molecular sieves with MCM-41 type structures as catalysts for the reduction of NO with hydrocarbons. The concentration of metallic particles and acid sites was varied over a wide range, including the controlled deposition of tungstophosphoric acid clusters, a procedure known to generate strong Brønsted acidic sites on MCM-41 type materials.[17]

2. EXPERIMENTAL

2.1 Materials

The synthesis of siliceous MCM-41 was carried out according to ref. 18 using fumed silica (SiO_2) as the Si-source, hexadecyltrimethylammonium bromide (CTABr) as template and tetramethylammonium hydroxide pentahydrate (TMAOH) as complexing agent. The gel composition was SiO_2:TMAOH: CTABr:H_2O = 1:0.19:0.27:40. The synthesis gel was prepared by dissolving TMAOH and CTABr in distilled water with stirring. After a clear solution was obtained, SiO_2 was added. After an initial aging at room temperature for 24 hr, condensation of the solid was carried out at 423 K for 48 hr in a Teflon autoclave. The product was filtered, washed with distilled water and dried in air at 323 K. The template was removed by calcination in synthetic air at 813 K for 10 hr using a heating rate of 1 $K \cdot min^{-1}$.

The mesoporous support was impregnated with an aqueous solution of $PtCl_4$,[19] of $H_3PW_{12}O_{40}$,[17] and of $PtCl_4$ and $H_3PW_{12}O_{40}$.[20] The Pt loading of the catalysts was 1.6 wt%, while the $H_3PW_{12}O_{40}$ loading varied between 0 and 60 wt%. After impregnation, the catalysts were dried at 373 K overnight in air and calcined in synthetic air at 773 K for 3 hr. In this chapter, the catalysts are denoted as Pt/MCM-41, HPW(X)/MCM-41 and Pt/HPW(X)/MCM-41 (X being the $H_3PW_{12}O_{40}$ loading, in wt%).

2.2 Characterization of Structural Properties

The structures of the mesoporous MCM-41 type support and of the catalysts before and after loading with the metal and $H_3PW_{12}O_{40}$ were characterized by X-ray powder diffraction and N_2 sorption. The BET surface area of the MCM-41 support, determined from N_2 adsorption isotherms at 77 K, was found to be 1006 m^2/g.

The nature of the acid sites was determined by benzene and pyridine adsorption after activation of the samples in vacuum ($< 10^{-6}$ mbar) at 673 K.[21] The spectra were recorded on a BRUKER IFS28 spectrometer using a resolution of 4 cm^{-1}, and normalized to the intensity of the structural vibrations of MCM-41 between 2100 cm^{-1} and 1770 cm^{-1}. The number of accessible metal atoms was determined by H_2 chemisorption carried out after reduction at 773 K for 2 hr in

Table 1. Composition and structural properties of the catalysts.

Sample	HPW (wt%)	Pt (wt%)	H/Pt (H$_2$ sorption)	CO/Pt (CO sorption)
Pt/MCM-41	-	1.61	0.63	
Pt/HPW(15)/MCM-41	15	1.61	-	0.16
Pt/HPW(30)/MCM-41	30	1.60	-	0.31
Pt/HPW(45)/MCM-41	60	1.64	-	0.10

flowing H$_2$ in a volumetric system. Due to partial reduction of H$_3$PW$_{12}$O$_{40}$ under the conditions of the H$_2$ chemisorption experiment, the number of accessible Pt atoms for HPW-containing catalysts was characterized by adsorption of CO followed by IR spectroscopy (after reduction in H$_2$ at 773 K for 1 hr). The number of accessible Pt atoms was calculated from the integral intensity of the CO band at 2074 cm^{-1} using Pt/MCM-41 (where the number of surface atoms was determined by H$_2$ chemisorption) as the standard.

The chemical nature of the metallic sites under reaction conditions was studied by X-ray absorption spectroscopy *in situ* during NO reduction. In these experiments, the NO conversion and changes in the XAS were measured simultaneously. Qualitative changes in the oxidation state of Pt were determined from the changes in the intensity of the peaks above the X-ray absorption edge, assuming that the XANES observed results from the superposition of the contributions of the metal oxide and the reduced metal.

The structural data for all catalysts investigated here are summarized in Table 1.

2.3 Catalytic Activity

The catalytic activity was studied in a quartz reactor using 100 mg of catalyst. Before the reaction, the catalyst was pressed into pellets, crushed in a mortar and sieved. Grain sizes smaller than 180 μm were used for the kinetic experiments. A total flow of 100 cm^3/min (NPT) used for all experiments resulted in a space velocity of 11,000 hr^{-1} (W/F = 6×10^{-2} g·s·cm^{-3}) over the catalyst bed. The reaction gas mixture consisted of 1010 ppm NO (about 91 ppm NO were directly oxidized in the reaction system to NO$_2$), 1012 ppm propene and 4.9 vol% O$_2$; the balance gas used was He. Up to 8 vol% water vapor could be added to the gas stream using a syringe pump.

Reactants and products were analyzed with a chemiluminescence NO/NO$_2$ analyzer and a gas chromatograph using TCD and FID detectors. Before each reaction, the catalysts were activated in He at 773 K for 1 hr. The conversion was measured in a temperature range between 453 and 773 K.

3. RESULTS

3.1 Structural Properties of the Catalysts

3.1.1 Surface Hydroxyl Groups on MCM-41

The isomorphous substitution of Si^{4+} by Al^{3+} creates, for zeolites and other microporous molecular sieves, bridging hydroxyl groups (Si-OH-Al groups) that act as strong Brønsted acids.[22,23] In comparison, on mesoporous molecular sieves the concentration of Brønsted acid sites was found to be very low even at high Al concentrations such as Si/Al $\approx$ 2.[23,24] The IR spectra of the hydroxyl groups of activated MCM-41 samples are shown in Figure 1. For comparison, three Al-containing samples with Si/Al ratios of 13, 50 and 100 are shown.[25]

For all MCM-41 samples, three hydroxyl bands at 3745, 3715 and 3530 cm^{-1} were observed. With increasing Al concentration, the bands at 3715 and at 3530 cm^{-1} increased in intensity, while the intensity of the band at 3745 cm^{-1} did not change. Changes in the position of the bands or the appearance of additional bands as a function of the Si/Al ratio were not observed.

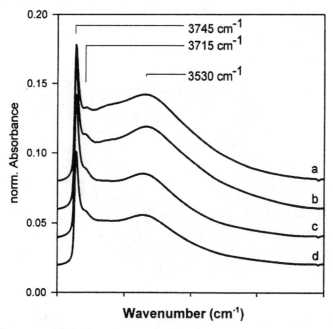

Figure 1. IR spectra of the hydroxyl groups in (**a**) MCM-41 (Si/Al=13); (**b**) MCM-41 (Si/Al=50); (**c**) MCM-41 (Si/Al=100); and (**d**) MCM-41 (siliceous).

The sharp band at 3745 cm^{-1} is assigned to isolated SiOH groups[26] terminating the particles. The position of the SiOH band at 3715 cm^{-1}, which is shifted to lower wavenumbers compared to isolated hydroxyl groups, indicates the presence of a lateral interaction with electron (pair) acceptor sites, *i.e.*, other hydroxyl groups. Therefore, these silanol groups are primarily located inside the channels of the molecular sieve and are presumably associated with structural imperfections or local defects.[27]   Due to its large breadth, the band around 3530 cm^{-1} can be assigned to hydrogen-bonded hydroxyl groups[28] which are formed on defect sites resulting from structural imperfections, or by dealumination.[29] Hydroxyl groups directly related to the tetrahedral incorporation of Al^{3+} into MCM-41 were not observed. Note that bridging hydroxyl groups (SiOHAl groups) are typically characterized by a sharp band in the region between 3650 and 3550 cm^{-1}.[30]

To visualize the orientation and density of the hydroxyl groups on the internal surface of MCM-41, an atomistic simulation was carried out.[31] The hydroxyl groups inside the pores of siliceous MCM-41 with a pore diameter of 30 Å and a concentration of 1.2 mmol/g OH-groups is shown in Figure 2. At this concentration, it can be clearly seen that the internal surface is not completely covered with hydroxyl groups. Moreover, most of the internal OH groups are isolated and somewhat tilted towards other oxygen atoms, while only a small number of hydroxyl groups form strongly hydrogen-bonded hydroxyl groups. A detailed study of the type and concentration of hydroxyl groups present on MCM-41 as function of the Si/Al ratio and of the pore size is reported in ref. 32.

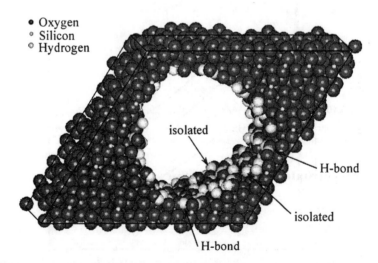

Figure 2. Orientation and density of OH groups on the internal surface of MCM-41 with a pore diameter of 30 Å and a OH group concentration of 1.2 mmol/g. Reprinted with permission from *Micropor. Mesopor. Mater.* **1999**, *27*, 321. Copyright 1999 Elsevier Science.

3.1.2 Acid Sites Generated by Impregnation of MCM-41 with $H_3PW_{12}O_{40}$

The type and density of acid sites generated by the impregnation of MCM-41 with $H_3PW_{12}O_{40}$ was studied by adsorption of pyridine (0.1 mbar) and water (0.9 mbar) at 423 K. The IR spectra of siliceous MCM-41 and Pt/HPW(30)/MCM-41 are compared in Figure 3.

After adsorption of pyridine on siliceous MCM-41, only bands assigned to hydrogen-bonded pyridine (1448 and 1597 cm^{-1})[30] were observed, while on Pt/HPW(30)/MCM-41, Brønsted-type acid sites, indicated by the bands at 1542 and 1614 cm^{-1}, were additionally present. After the co-adsorption of 0.9 mbar H_2O on Pt/HPW(30)/MCM-41, the concentration of Brønsted type acid sites increased by about 75%, while the concentration of Lewis type acid sites was not affected by the presence of water vapor.

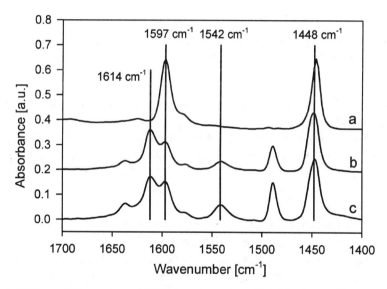

Figure 3. Difference IR spectra at 423 K after adsorption of (**a**) 0.1 mbar pyridine on siliceous MCM-41; (**b**) 0.1 mbar pyridine on Pt/HPW(30)/MCM-41; and (**c**) 0.1 mbar pyridine and 0.9 mbar H_2O on Pt/HPW(30)/MCM-41.

3.1.3 Chemical State of the Metallic Sites

Changes in the XANES during the adsorption of 1010 ppm NO at 573 K on the pre-reduced Pt/MCM-41 catalyst are shown in Figure 4 (the spectra were subtracted from that of reduced Pt/MCM-41). During exposure to NO, the intensity of the peak above the absorption edge strongly increased as a function of time. Conceptually, XANES probes the density of states above the Fermi level and, thus, can be used to investigate oxidation/reduction processes of the metal

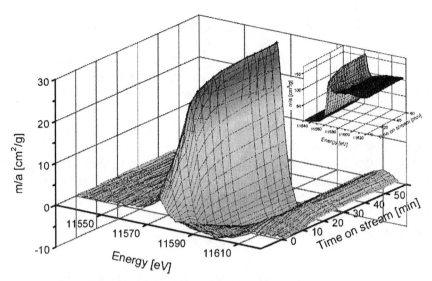

Figure 4. Changes in the XANES region during the adsorption of 1010 ppm NO at 573 K on pre-reduced Pt/MCM-41 (Raw data shown on insert).

component.[33] The changes observed clearly indicate that the metal is oxidized during adsorption of NO.

The NO conversion and oxidation state of Pt in Pt/MCM-41 (SiO$_2$) during the reduction of NO with propene are shown in Figure 5. Under reaction conditions, a reducing agent (C$_3$H$_6$) and O$_2$ are additionally present, therefore it is essential to study the chemical state of the catalyst *in situ* while the reaction is being carried out.

As discussed in the next section, the conversions of NO and C$_3$H$_6$ over Pt/MCM-41 start simultaneously and increase with increasing temperature until the hydrocarbon conversion reaches ~100 %. The XANES observed during this reaction indicates that Pt is only slightly oxidized up to the temperature of maximum NO conversion (453 K). As conversion decreases at higher temperatures, platinum is oxidized. This clearly indicates that reduced Pt particles are significantly more active for the C$_3$H$_6$-NO-O$_2$ reaction compared to platinum oxide particles. At propene conversion levels below 100%, propene is available over the whole catalyst bed. At higher temperatures, propene is additionally consumed via direct oxidation and, therefore, a fraction of residual oxygen remains on the surface. Due to the lack of reducing agent, the formation of Pt-O species was observed at this temperature. With increasing temperature, the rate of propene oxidation increases, which leads to a further formation of Pt-O species and to a decrease in NO$_x$ conversion. At temperatures above 623 K, the rate for direct oxidation of C$_3$H$_6$ is so high that further NO reduction is not observed.

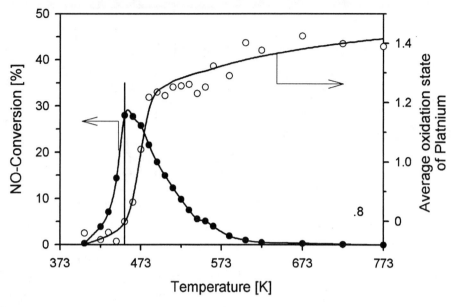

Figure 5. Conversion of NO (●) and oxidation state of Pt in Pt/MCM-41 (SiO$_2$) (○) during the C$_3$H$_6$-NO$_x$-O$_2$ reaction.

3.1.4   Catalytic Activity

The conversion of NO and C$_3$H$_6$ as a function of temperature is shown in Figure 6. The maximum NO conversion and selectivity for N$_2$ formation are summarized in Table 2. NO and C$_3$H$_6$ conversion start at the same temperature, and both rates increase with increasing temperature until total conversion of the

Table 2. NO conversion and N$_2$ selectivity of Pt/HPW/MCM-41 catalysts.

Sample	NO conversion	Temperature at maximum	N$_2$ selectivity
	%	conversion/K	%
Pt/MCM-41	62	483	35
Pt/HPW(15)/MCM-41	60	493	35
Pt/HPW(30)/MCM-41	50	503	40
Pt/HPW(60)/MCM-41	47	513	45
HPW(30)/MCM-41	30	593	32
Pt/SiO$_2$	29	513	29

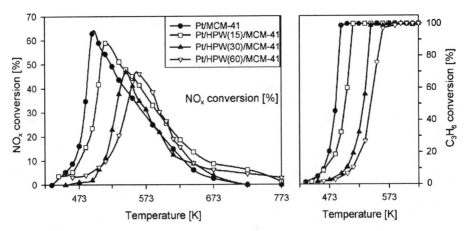

Figure 6. Conversion of NO and C_3H_6 over Pt/HPW/MCM-41 as a function of temperature.

reducing agent is reached between 480 and 520 K. At the temperature of maximum NO conversion, and a further temperature increase leads to a decrease in NO conversion, while the conversion of C_3H_6 remains at 100%. The highest NO conversion was observed for Pt/MCM-41, and decreases with $H_3PW_{12}O_{40}$ loading of the catalysts, while selectivity for N_2 formation increases from 35% (Pt/MCM-41) up to 45% (Pt/HPW(60)/MCM-41) on the tungstophosphoric acid-containing catalysts. Also, the temperature at which the maximum NO conversion was reached increased with the $H_3PW_{12}O_{40}$ loading of the catalysts. The effect of adding 2.5 vol% water vapor into the gas stream on the NO conversion at 573 K is shown in Figure 7. At this temperature, NO conversion without water

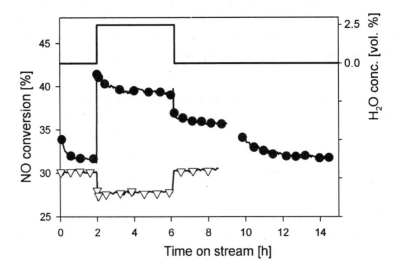

Figure 7. Changes in the activity of Pt/MCM-41 ($\triangledown$) and Pt+HPW/MCM-41 ($\bullet$) during a stepwise change of water vapor concentration between 0 and 2.5 vol% at 573 K.

is about 30% over both catalysts. After stepwise addition of 2.5 vol% water vapor into the reactant gas stream, the NO conversion decreased immediately to 28% over Pt/MCM-41, while for Pt/HPW(30)/MCM-41 it increased initially to 42% and reached a constant level of 39% within one hour onstream. After removal of the water vapor from the feed, the activity of Pt/MCM-41 immediately returned to its initial value (30%). On Pt/HPW(30)/ MCM-41, the NO conversion reached a constant level of 36%, while the initial NO conversion level was only restored on the Pt/HPW(30)/MCM-41 catalyst after heating to 773 K in He for one hour.

The influence of water vapor concentration on the activity at 573 K for catalysts loaded with increasing concentrations of $H_3PW_{12}O_{40}$ is shown in Figure 8. At this temperature, the activity and selectivity of the catalysts under reaction conditions without water vapor present is independent of the $H_3PW_{12}O_{40}$ loading.

On all tungstophosphoric acid-containing catalysts, an increase in NO conversion in the presence of water vapor was observed, while for Pt/MCM-41 the activity continuously decreased. On the catalyst with the highest $H_3PW_{12}O_{40}$ loading, the NO conversion increases from 30 to 40% (*i.e.*, a conversion level 25 rel% higher compared to water-free reaction conditions) at water vapor concentrations above 2 vol%.

NO conversion as a function of C_3H_6 concentration over Pt/HPW(30)/MCM-41 at 573 K with and without water vapor present is shown in Figure 9 (NO$_x$ concentration 1010 ppm). Independent of the presence of water

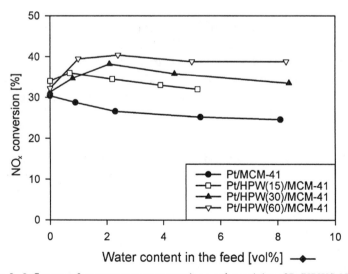

Figure 8. Influence of water vapor concentration on the activity of Pt/HPW/MCM-41.

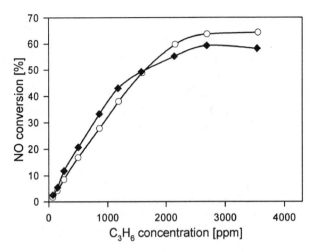

Figure 9. NO_x conversion of Pt/HPW(30)/MCM-41 in the absence (○) and presence (◆) of H_2O vapor (NO_x concentration 1010 ppm).

vapor, the NO conversion increases almost linearly up to a C_3H_6 concentration of 2100 ppm. Further increases in the concentration of the hydrocarbon did not result in higher NO conversion. At C_3H_6 concentrations below 1550 ppm, NO conversion is higher in the presence of water vapor, while above this concentration the catalyst is more active under water-free reaction conditions. At the highest propene concentration, the NO conversion was 64% under water-free reaction conditions compared to 58% in the presence of water vapor.

3.1.5   Stability of the Catalysts under Reaction Conditions

The number of accessible Pt atoms on the Pt/HPW/MCM-41catalysts before and after reaction for 24 hr at 573 K in the presence of 2.5 vol% water vapor, determined b y CO adsorption, is summarized in Table 3.   While f or Pt/MCM-41, the H/Pt ratio was not affected by the reaction, on catalysts loaded with $H_3PW_{12}O_{40}$ a significant reduction in the number of accessible metal atoms was observed after reaction.

Table 3. H/Pt ratio of catalysts before and after reaction for 24 hr at 2.5 vol% water vapor.

Sample	H/Pt ratio before reaction	H/Pt ratio after reaction
Pt/MCM-41	0.63	0.6
Pt/HPW(15)/MCM-41	0.16	0.09
Pt/HPW(30)/MCM-41	0.31	0.12
Pt/HPW(60)/MCM-41	0.1	0.06

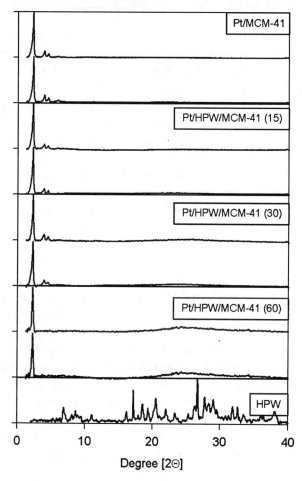

Figure 10. Comparison of the XRD patterns, with normalized intensity, before (top) and after reaction (bottom trace) for 24 hr in the presence of 2.5 vol% H_2O.

The XRD patterns of the catalysts before and after reaction at 573 K for 24 hr in the presence of 2.5 vol% water vapor are compared in Figure 10. All catalysts showed the typical XRD pattern of MCM-41 type materials with a very strong (100) and additional (110), (200) and (210) peaks.[34] From the position of the (100) reflection, a unit cell size of 39.4 Å was determined for the MCM-41 support. With increasing $H_3PW_{12}O_{40}$ loading of the catalysts, the intensity of the (110), (200) and (210) reflections decreased. On Pt/HPW(60)/MCM-41, where only the (100) reflection was observed, an additional broad peak in the region $15° < 2\theta < 35°$ was present. However, even on Pt/HPW(60)/MCM-41, reflections from crystalline $H_3PW_{12}O_{40}$ were not observed, as shown by the comparison with the XRD pattern of $H_3PW_{12}O_{40} \cdot 6H_2O$.

4. DISCUSSION

IR spectra clearly reveal that the surface hydroxyl groups on MCM-41 type materials are only weakly acidic. Bridging hydroxyl groups (Brønsted acid sites) were not observed after incorporation of Al into MCM-41. After loading with $H_3PW_{12}O_{40}$, a small concentration of Brønsted type acid sites was detected, which increased significantly after the co-adsorption of water. Also, the number of accessible Pt atoms is affected by the loading of $H_3PW_{12}O_{40}$.

It was reported that partially hydrated $H_3PW_{12}O_{40}$ supported on MCM-41 type materials forms Keggin-type units with an approximate diameter of 12 Å.[17] As shown in Figure 11, the pore diameter of the mesoporous MCM-41 type support is sufficiently large to allow deposition of Pt and $H_3PW_{12}O_{40}$ clusters, even at a loading of 60 wt%, without blocking the accessibility of both sites to the reactants. From geometrical considerations, an average distance of 760 Å and 12 Å was determined between the Pt and the HPW particles, respectively. Also, the absence of XRD reflections from crystalline $H_3PW_{12}O_{40} \cdot 6H_2O$ in the Pt/HPW/MCM-41 catalysts confirmed the presence of small clusters of tungstophosphoric acid inside the pores. Nevertheless, the number of accessible metal atoms decreased markedly on the $H_3PW_{12}O_{40}$-containing catalysts, especially after 24 hrs reaction time, which indicates that the tungstophosphoric acid partially migrates onto the surface of the metal particles during the reaction.

On $H_3PW_{12}O_{40}$-containing catalysts, NO conversion was significantly enhanced in the presence of water vapor, while on Pt/MCM-41, similar to most other catalysts reported,[9-11] the presence of water vapor led to a decrease in NO conversion. It is important to note that after adding water vapor to the reactant gas mixture, the activity immediately changed for the Pt/MCM-41 catalysts, while on Pt/HPW(30)/MCM-41 an induction period of about 30 min was observed, and the

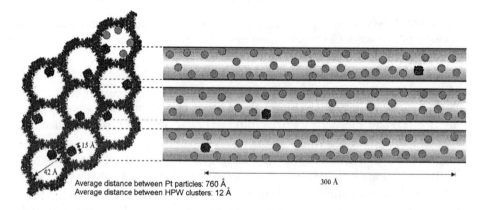

Average distance between Pt particles: 760 Å
Average distance between HPW clusters: 12 Å

300 Å

Figure 11. Dimensions of the Pt and $H_3PW_{12}O_{40}$ clusters in MCM-41 with a pore diameter of 42 Å, containing 1.6 wt% Pt and 60 wt% $H_3PW_{12}O_{40}$. The size of the clusters was chosen according to EXAFS results for Pt and to literature data for $H_3PW_{12}O_{40}$.[17]

final conversion level was reached slowly. Similar behavior was observed after removing the water vapor from the gas stream, where heating the Pt/HPW(30)/MCM-41 catalysts in He at 773 K for 1 hr was necessary to reach the initial conversion level (see Figure 7). The presence of the induction period indicates that the structural properties of the $H_3PW_{12}O_{40}$ component are affected by water vapor. In contrast, changes in the sorption properties, such as a preferential adsorption of one of the reactants in the presence of water vapor, should result in an immediate change in activity, as observed on Pt/MCM-41.

The experiments reveal that the metal and the tungstophosphoric acid are in intimate contact, which seems to be essential to achieve the additional activity in the presence of water vapor. Note that the role of Pt in the catalytic reduction of NO in an oxidative atmosphere depends on the type of reducing agent used.[35] For reactions using unsaturated hydrocarbons as the reducing agent (*e.g.*, C_3H_6), it was proposed that NO reduction occurs only on the Pt particles and that under reaction conditions the Pt surface is mostly covered with carbonaceous species. Under reaction conditions, the Pt was found be in a reduced state (see Figure 5).

In an *in situ* IR study, we identified –NCO species as reactive intermediates on the surface of Pt/MCM-41 during the reduction of NO with C_3H_6.[36] The -NCO reacts with adsorbed NO/NO_2 and forms N_2 and N_2O by the following mechanism:

$$NO + CO \rightarrow \text{-NCO} + O$$
$$\text{-NCO} + NO \rightarrow N_2 + CO_2$$
$$\text{-NCO} + NO \rightarrow N_2O + CO$$
$$\text{-NCO} + NO_2 \rightarrow N_2O + CO_2$$
$$\text{-NCO} + H_xC_yO_z \rightarrow \text{-CN} + H_xC_yO_{z+1}$$

XANES indicates that reduced metal particles are significantly more active for NO_x reduction, therefore rapid removal of O atoms from the metal surface is essential to achieve a high NO reduction rate.

McCormick *et al.*[37,38] reported that, after adsorption of NO on hydrated SiO_2-supported tungstophosphoric acid, protonated NOH^+ species were formed, which almost completely decomposed into N_2 and N_2O (~50 % selectivity to N_2) in the presence of O_2 and H_2O around 700 K. The authors proposed a combined adsorption/decomposition mechanism, where tungstophosphoric acid adsorbs NO into the bulk structure, *i.e.*:

$$H_3PW_{12}O_{40} \cdot 6H_2O + 3NO \rightarrow H_3PW_{12}O_{40} \cdot 3NO + 6H_2O$$

Subsequent decomposition of NO into N_2 and O_2 was proposed.

The reactions reported here were carried out at a much lower temperature (573 K), where NO conversion over HPW(30)/MCM-41 was only 8%. Supported tungstophosphoric acid, in general known to be a very active catalyst for oxidation reactions,[39] led to an increase in C_3H_6 conversion and a decrease in NO_x conversion in the presence of water vapor. Therefore, the activity increase in the presence of water observed on Pt/HPW/MCM-41 cannot be explained by suppression of direct oxidation of the hydrocarbon, as proposed for Co/ZSM-5.[40] Also, experiments indicate that an additive contribution of $H_3PW_{12}O_{40}$ to the activity of the metal component is not the main reaction route contributing to the increase in the overall activity observed in the presence of water vapor.

Therefore, we speculate that the improved activity of Pt/HPW/MCM-41 in the presence of water vapor results from a reaction occurring at the interface between the metal and acid sites. The conversion of NO as a function of the propene concentration indicates that the hydrocarbon is partially adsorbed on the tungstophosphoric acid. Thus, we further speculate that $-C_xH_y$ species, formed by the adsorption of C_3H_6 on Brønsted-type acid sites, react with $-NO$ or $-NCO$ adsorbed on the metal sites. The formation of highly acidic Brønsted-type acid sites on the hydrated tungstophosphoric acid (see Figure 3) generates additional sorption sites, which increase the local concentration of reducing agent on the perimeter between the acid and the metal clusters and thus give rise to higher activity in the presence of water vapor. With increasing C_3H_6/NO ratios, the local concentration of C_3H_6 approaches a level where it starts to displace NO from the metal sites and, therefore, the activity does not increase further. In the presence of water vapor, this effect is more pronounced because the concentration of Brønsted type acid sites is higher. Consequently, Pt/HPW/MCM-41 catalysts are more active in the presence of water vapor at lower C_3H_6 concentrations. In the presence of water vapor, however, the critical concentration of reducing agent was reached at a lower C_3H_6/NO ratio, therefore the Pt/HPW/MCM-41 catalysts are more active when water vapor is not present at high concentrations of reducing agent.

5.   CONCLUSIONS

Loading of Pt/MCM-41 catalysts with $H_3PW_{12}O_{40}$ leads to an increase in the selectivity for N_2 formation and to a slight decrease in the catalytic activity for reduction of NO with propene. The decrease in the number of accessible Pt atoms indicates that the metal atoms are partially covered by the tungstophosphoric acid, which lowers the activity. The main product of NO reduction over Pt/MCM-41 with propene is N_2O, which is characteristic of Pt group metals supported on oxides.[41,42]

In contrast to all other catalysts studied so far, Pt/HPW/MCM-41 catalysts showed a significant increase in NO conversion in the presence of water vapor, while for Pt/MCM-41 a small suppression of NO conversion was found. The improved activity of the Pt/HPW/MCM-41 catalysts is a ttributed to the generation of Brønsted-type acid sites on hydrated $H_3PW_{12}O_{40}$, which act as additional sorption sites for the reducing agent. These sites, only present on hydrated $H_3PW_{12}O_{40}$, enhance the local concentration of C_3H_6 on the surface of $H_3PW_{12}O_{40}$-containing catalysts in the presence of water vapor.

The advantages of using MCM-41 type materials as supports result from the high specific surface areas of these materials ($\sim$1000 m^2/g) and from their regular pore sizes compared to other oxide supports. This allows a high loading with metallic and acidic clusters without experiencing severe limitations in the accessibilities of these sites for the reactants. Therefore, transport limitations can be avoided and higher space velocities of the reactants over the catalysts can be achieved, which lead directly to higher activity of the catalysts under industrial conditions. Additionally, the catalysts will not be susceptible to a blocking of the pores by reaction products or small particles present in the gas stream. The spatial separation of the active clusters (see Figure 11) contributes to the high stability of the catalyst under reaction conditions. The potential to combine different functionalities, such as a NO_x reduction and a N_2O decomposition function or a total oxidation function for VOC within the same material, will open up several possibilities for the successful application of MCM-41-based catalysts in the field of exhaust gas treatment.

ACKNOWLEDGEMENTS

Research was financially supported by the "Fonds zur Förderung der Wissenschaftlichen Forschung" (Project P10874 CHE) and the "Hochschuljubiläumsfonds der Österreichischen Nationalbank " (Project 7119).

REFERENCES

1.   Seinfeld, J. H. *Science* **1989**, *243*, 745.
2.   Armor, J. N. *Catal. Today* **1977**, *38*, 163.
3.   Bosch, H.; Janssen, F. *Catal. Today* **1988**, *2*, 369.
4.   Armor, J. N. *Catal. Today* **1995**, *26*, 99.
5.   Iwamoto, M. *Stud. Surf. Sci. Catal.* **2000**, *130*, 23.
6.   Iwamoto, M.; Furukawa, H.; Mine, Y.; Uemura, F.; Mikuriya, S.; Kagawa, J. *J. Chem. Soc., Chem. Comm.* **1986**, 1271.
7.   Li, Y.; Armor, *J. Appl. Catal. B* **1992**, *1*, L31.
8.   Desai, A. J.; Kovalchuk, V. I.; Lombardo, E. A.; d'Itri, J. L. *J. Catal.* **1999**, *184*, 396.
9.   Iwamoto, M.; Mizuno, N.; Yahiro, H. *Proc. 10th Int. Congr. Catal.*, Elsevier: Budapest, **1992**, 1285.

10.   Chen, H. Y.; Sachtler, W. M. H. *Catal. Lett.* **1998**, *50*, 125.
11.   Sumiya, S.; Saito, M.; He, H.; Feng, Q. C.;Takezawa, N.; Yoshida, K. *Catal. Lett.* **1998**, *50*, 87.
12.   Feng, X.; Hall, W. K. *J. Catal.* **1997**, *166*, 368.
13.   Joyner, R. W.; Stockenhuber, M. *Catal. Lett.* **1997**, *45*, 15.
14.   Iwamoto, M.; Yahiro, H.; Shin, H. K.; Watanabe, M.; Guo, J.; Konno, M.; Chikahisa, T. Murayama, T. *Appl. Catal. B* **1995**, *5*, L1.
15.   Hamada, H.; Kintaichi, Y.; Sasaki, M.; Ito, Y. *Appl. Catal.* **1991**, *75*, L1.
16.   Burch, R.; Millington, P. J. *Catal. Today* **1995**, *26*, 185.
17.   Kozhevnikov, I. V.; Sinnema, A.; Jansen, R. J. J.; Pamin, K.; van Bekkum, H. *Catal. Lett.* **1995**, *30*, 241.
18.   Cheng, C. F.; Park, D. H.; Klinowski, J. *J. Chem. Soc., Faraday Trans.* **1997**, *93*, 193.
19.   Schießer, W.; Vinek, H.; Jentys, A. *Catal. Lett.* **1998**, *56*, 189.
20.   Jentys, A.; Schießer, W.; Vinek, H. *J. Chem. Soc., Chem. Comm.* **1999**, 335.
21.   Jentys, A.; Rumplmayr, G.; Lercher, J. A. *Appl. Catal.* **1989**, *53*, 299.
22.   Corma, A. Fornes, V. Navarro, M. T.; Perez-Pariente, J. *J. Catal.* **1994**, *148*, 569.
23.   Borade, R. B.; Clearfield, A. *J. Catal.* **1995**, *31*, 267.
24.   Kubelkowa, L.; Kotrla, J. Florian, J. *J. Phys. Chem.* **1995**, *99*, 10285.
25.   Jentys, A.; Pham, N. H.; Vinek, H. *J. Chem. Soc., Faraday Trans.* **1996**, *92*, 3287.
26.   Gallei, E.; Eisenbach, D. *J. Catal.* **1995**, *37*, 474.
27.   Jacobs, P. A.; van Ballmoos, R. *J. Phys. Chem.* **1982**, *86*, 3050.
28.   Wollery, G. L.; Alemany, L. B.; Dessau, R. M.; Chester, A. W. *Zeolites* **1986**, *6*, 14.
29.   Dessau, R. M.; Schmitt, K. D.; Kerr, G. T.; Wollery, G. L.; Alemany, L. B. *J. Catal.* **1987**, *104*, 484.
30.   Ward, J. *J. Catal.* **1967**, *9*, 225.
31.   Kleestorfer, J.; Vinek, H.; Jentys, A. *J. Mol. Catal. A: Chem.* **2001**, *166*, 53.
32.   Jentys, A.; Kleestorfer, K.; Vinek, H. *Micropor. Mesopor. Mater.* **1999**, *27*, 321.
33.   Englisch, M.; Lercher, J. A.; Haller, G. L. in *Series on Synchrotron Radiation Techniques and Applications*, Vol. 2, ed. Y. Iwasawa. World Scientific: Tokyo, **1996**, 276.
34.   Schüth, F. *Ber. Bunsenges. Phys. Chem.* **1995**, *99*, 1315.
35.   Burch, R.; Sullivan, J. A.; Watling, T. C. *Catal. Today* **1998**, *42*, 13.
36.   Schießer, W.; Vinek, H.; Jentys, A.; *Appl. Catal. B* **2001**, *33, 263*.
37.   Herring, A. M.; McCormick, R. L. *J. Phys. Chem. B* **1998**, *102*, 3175.
38.   McCormick, R. L.; Boonrueng, S. K.; Herring, A. M. *Catal. Today* **1998**, *42*, 145.
39.   Cavani, F. *Catal. Today* **1998**, *41*, 73.
40.   Maisuls, S. E.; Seshan, K.; Feast, S.; Lercher, J. A. *Appl. Catal. B* **2001**, *29*, 69.
41.   Amiridis, M. D.; Zhang, T.; Farrauto, R. J. *Appl. Catal. B* **1996**, *10*, 203.
42.   Cant, N. W.; Angove, D. E.; Chambers, D. C.; *Appl. Catal. B* **1998**, *17*, 63.

Chapter 10

POLYMERIZATION WITH MESOPOROUS SILICATES

Keisuke Tajima, Takuzo Aida
Department of Chemistry and Biotechnology, Graduate School of Engineering, The University of Tokyo

Keywords: mesoporous silica, polymerization, template polymerization, composite material

Abstract: This short review focuses on precision polymerization using mesoporous silicate materials for the formation of controlled macromolecular architectures. Synthesis and modification of mesoporous silicates are overviewed briefly in relation to the development of designer polymerization catalysts, and then the following points are highlighted: catalytic activity, control of primary and higher-order structures of polymers, and fabrication of silica/functional polymer nanocomposite materials with emphasis on confinement effects on polymer properties.

1.   INTRODUCTION

Imagine if we could produce a polymer with a defect-free molecular structure we want, or if we could bend polymer chains to form shapes we like. Although we have our own hands, they are too large to directly manipulate such molecular-level events. Therefore, we have to seek excellent catalysts and processors that allow production and processing of polymeric materials with desired primary and higher-order structures. Development of molecular catalysts with finely-tuned ligands has enabled us to obtain polymers with controlled primary structures. However, such catalysts are designed to exhibit high performance in homogeneous, non-constrained media, and the polymerization simply follows the statistics of numerous elementary reaction steps, leading to the formation of poorly controlled higher-order structures. We therefore have to perform post-processing of polymers to fabricate ordered materials with desired

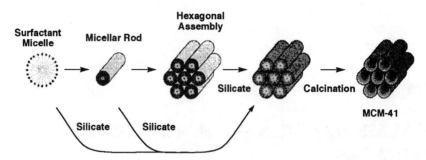

Figure 1. Schematic representation of "liquid crystal-templating (LCT)" mechanism for the synthesis of mesoporous silicate MCM-41. Adapted from *J. Am. Chem. Soc.* **1992**, *114*, 10834. Copyright 1992 American Chemical Society.

properties. On the other hand, if we can design nanoscopic reactors or processing machines, then we may be able to fabricate high-performance polymeric materials without post-processing, since polymerization occurs in constrained environments, leading to the formation of polymeric materials with long-range molecular order.[1]

Recent developments of mesoporous silicate materials provide serious candidates for the confined spaces, which are attractive as nanoscopic reactors for controlled macromolecular synthesis. Since the first report on mesoporous silicate MCM-41 in 1992,[2] various mesoporous silicate materials have been developed. These inorganic materials are prepared using rod-like or spherical micelles as templates, and have an ordered hexagonal or cubic array of uniformly sized nanoscopic channels with pore diameters varying from 15 to 300 Å (Figure 1). Compared with crystalline microporous zeolites with pore diameters of 3-10 Å, mesoporous materials have advantages as catalysts and/or host materials for inclusion chemistry.[3] For example, they can incorporate relatively large molecules and may also align them along the channels. Furthermore, modification of their structural features and chemical properties is much easier than for microporous zeolites, due to larger pore diameters and the amorphous nature of the silicate framework. Extensive studies have also been performed on the incorporation of catalytically active species into mesoporous silicates. Because of these interesting characteristics, the use of mesoporous silicates as reaction media for macromolecular synthesis now attracts great attention. For polymerization with mesoporous silicates, the following possibilities can be envisaged:

1. Easy access of monomers to active sites: Due to the large surface areas of mesoporous silicates, active sites for polymerization can be uniformly dispersed on the silicate surface, thereby allowing monomer molecules easy access to the growing ends of the polymer chains attached to the active sites.

2. Control of primary structures: In nanoscopic channels, conformational changes of included polymer chains may be restricted, such that the elementary reactions of polymerization proceed in a controlled manner. Polymerizations within such confined nanoscopic channels are expected to show different regioselectivity and/or stereoselectivity than those in non-constrained media. Furthermore, some bimolecular side reactions such as recombination and disproportionation, typical of free-radical polymerization, can be suppressed, so that one may control molecular weight and even molecular weight distribution of polymers.

3. Control of higher-order structures: Due to their regular arrangement of nanoscopic channels, mesoporous silicates may serve as potential inorganic nano-molds for the fabrication of polymeric materials with controlled higher-order structures and multi-level molecular ordering.

4. Confinement effects on included polymers: In nanoscopic channels, included polymer chains may be isolated and arranged such that they eventually exhibit unique chemical and physical properties different from those of polymers in bulk phase.

The following sections provide an overview of recent developments in mesoporous materials with potential catalytic activity, macromolecular synthesis with mesoporous silicates focusing on control of primary and higher-order macromolecular architectures, and fabrication of silica/functional polymer nanocomposite materials.

2.   DESIGN OF MESOPOROUS SILICATES AS POTENTIAL NANOREACTORS FOR POLYMERIZATION

There are extensive reviews on mesoporous silicate materials functionalized by incorporation of non-transition and transition metal ions,[5] which may be elaborated into designer mesoporous catalysts for controlled macromolecular synthesis.

2.1   General Features of Synthesis of Mesoporous Silicates

The mesoporous silicate MCM-41, consisting of regularly arranged nanoscopic channels with a narrow channel-diameter distribution, was synthesized in 1992,[2] using a hexagonal assembly of rod micelles composed of long-chain alkylammonium salts as templates under basic conditions. This achievement triggered a variety of synthetic studies on mesoporous silicate materials. Not only basic (pH~11) but also acidic conditions (pH~2) are applicable, allowing rapid formation of mesoporous silicates under rather mild conditions. Examples of amphiphilic molecules include cationic, anionic and non-ionic surfactants with long alkyl chains, where the diameter of the silicate channel can be varied in the range 20-40 Å by changing the length of the hydrophobic alkyl chain. Addition of hydrophobic guest molecules such as 1,3,5-trimethylbenzene to the hydrothermal reaction system causes swelling of the template micelles and results in the formation of large silicate channels up to ~100 Å. Micelles of amphiphilic block copolymers having hydrophilic and hydrophobic segments can also be utilized as templates, for the synthesis of mesoporous silicates with much larger channels (40-300 Å).[6]

At concentrations of amphiphile close to the critical micelle concentration (cmc), powder-like mesoporous silicates are generally obtained. Although micelles are dynamic under such dilute conditions, interaction of the surfactants with charged silicate intermediates facilitates the formation of mesostructures as thermodynamically favored products. Recently, utilization of condensed phases of lyotropic liquid crystals of amphiphilic molecules as templates has been found to be effective for the fabrication of well-defined mesostructures such as monoliths or films. More recently, an interesting synthetic approach was proposed, involving evaporation of the solvent from a surfactant/silica sol condensation system to allow rapid formation of mesoporous silicate films.[7]

Mesoporous silicates generally adopt hexagonal or cubic phases, depending on the type of template structure and hydrothermal condensation conditions. Two-dimensional (2-D) hexagonal architectures consisting of unidirectional channels are anisotropic and may align included polymer chains in a certain direction. On the other hand, cubic and 3-D hexagonal architectures consisting of interconnected channels may be advantageous for diffusion of guest molecules.

2.2   Mesoporous Silicates Functionalized with Organic Modifiers

Surface modification of mesoporous silicates with organic functionalities has been investigated for tuning chemical and physical properties of the silicate channels. For example, treatment of mesoporous silicates with organosilicon derivatives such as $RSi(OR')_3$ and $RSiCl_3$ allows incorporation of various organic functionalities on the silica surface. Use of $(MeO)_3SiCH_2CH_2CH_2SH$ as an organic modifier for MCM-41 results in a mesoporous material functionalized with a monolayer of thioalkyl functionalities, which serves as an effective adsorbent for heavy metal ions (Figure 2).[9] $(MeO)_3SiCH_2CH_2CH_2NH_2$ as a

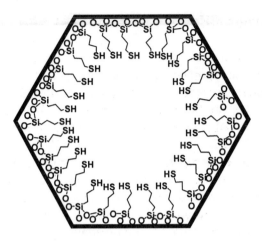

Figure 2. Schematic representation of an MCM-41 channel modified with $(MeO)_3SiCH_2CH_2CH_2SH$.

modifier has the potential to incorporate catalytically active metal clusters in the silicate channels.[10] Treatment of MCM-41 with Me_3SiCl results in trimethylsilylation of the surface silanol functionalities to give a mesoporous silicate with hydrophobic channels. This material preferentially includes hydrophobic guest molecules within the channels. More recently, periodic mesoporous organosilicates (PMO's) have been synthesized using organosilicates as silica sources for the sol-gel based hydrothermal condensation.[11] Compared with the post-modification approach described above, this direct method enables much more efficient incorporation of organic functionalities into the silicate structure (Figure 3).

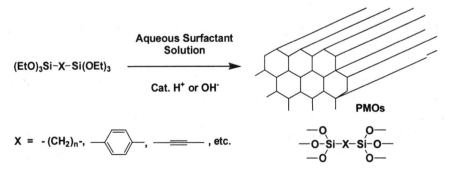

Figure 3. Synthetic scheme for periodic mesoporous organosilicates (PMO's).

2.3   Mesoporous Silicates Functionalized with Metal Ions

Incorporation of metal ions on the silica surface or into the silicate framework creates mesoporous materials with potential catalytic activity. Classical examples include hydrothermal synthesis of mesoporous metallosilicates by the combined use of tetraalkylorthosilicates ($Si(OR)_4$) and metal alcoholates ($M(OR)_n$), where a variety of non-transition and transition metal ions such as Al,[12] Ti,[13] Zr,[14] Sn,[15] V,[16] B,[17] Cr,[18] Mn,[19] Mo,[20] W,[21] Fe,[22] Cu,[23] Zn[24] and Ga[25] can be incorporated. More recently, utilization of a long alkyl-chain amphiphilic molecule bearing a tris(bipyridine)ruthenium(II) complex as the hydrophilic head group for the hydrothermal synthesis was found to give, after calcination, a mesoporous silicate bearing catalytically active Ru(II) species.[26] One-pot synthesis of a titanium-containing MCM-41 (Ti-MCM-41) was achieved by addition of titanocene dichloride to a liquid crystal synthetic gel and subsequent calcination.

Post-modification approaches are also effective for the incorporation of metal ions such as Al,[28] Ti,[29] Mn,[30] Cr,[31] Mo,[32] Ru,[33] Pd[34] and so forth[35] onto the silica surface. The synthetic approach includes adsorption or ion exchange by exposure of mesoporous silicates to solutions of metal salts or metal clusters. Alternatively, direct exchange of ionic template surfactants in mesostructured silica, before calcination, with cationic metal complexes has also been reported.[36] On the other hand, utilization of immobilized organic ligands on the silicate surface for the chelation of metal ions is effective for the fabrication of mesoporous silicate materials having metal complex functionalities.[37] Examples include *in situ* formation of a manganese(III) Schiff-base complex on the silica surface of MCM-41[38] and a rhodium bis(phosphine) complex on the surface of SBA-15.[39]

3.   POLYMERIZATION WITH MESOPOROUS SILICATES

3.1   Catalytic Activity and Control of Primary Structures of Polymers

In polymerization with metal complexes, organic ligands around the active sites may affect the chain growth steps significantly, by changing, for example, the lifetime of the growing polymer chain or the stereochemical course of the reaction. Inorganic supports for the catalysts may also provide certain environments around the active sites which affect the rate of polymerization, probability of side reactions such as chain transfer and termination, and stereochemistry of chain growth.

Metallocene catalysts, in combination with co-catalysts such as methylalumoxane (MAO), are known to bring about polymerization of olefins. Supporting metallocene catalysts on inorganic materials often results in prolongation of the lifetime of the catalyst. Furthermore, supported metallocene catalysts require lower amounts of co-catalyst than homogeneous catalysts for initiation of polymerization. Rahiala and coworkers reported the polymerization of ethylene using zirconocene dichloride supported on commercial Grace silica, mesoporous silica MCM-41, and aluminum-containing MCM-41 (Al-MCM-41).[40] Among these inorganic supports, Al-MCM-41 gave the highest catalytic activity (1.58×10^4 kg PE mol Zr^{-1} hr^{-1}). Nitrogen

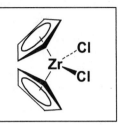

adsorption/desorption isotherms and powder pattern X-ray diffraction (XRD) studies showed no substantial decrease in surface area (~1000 m²/g) or channel diameter of MCM-41 upon immobilization of zirconocene on the silica surface.

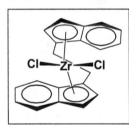

Woo and coworkers reported a novel supported catalyst by treatment of MCM-41 with MAO, followed by dichloro[*rac*-ethylenebis(indenyl)]zirconium, an efficient catalyst for the isotactic polymerization of propylene (Figure 4).[41] The polymerization of propylene with this supported catalyst system gave a polypropylene (PP) with an isotactic pentad fraction (mmmm) of 83% (T_m = 131.5 °C), higher than that of the polymer obtained

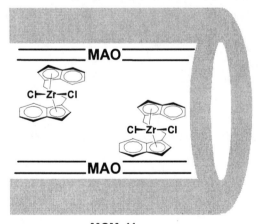

MCM-41

Figure 4. An illustration of zirconocene supported on MCM-41 pretreated with methylalumoxane (MAO).

with the same zirconocene catalyst under homogeneous conditions (78%; T_m = 129.8 °C). Pretreatment of MCM-41 with MAO is believed to consume the surface silanol functionalities and suppress protonolysis of the active centers. The supported catalyst requires a smaller amount of MAO (Al/Zr = 800) than the homogeneous catalytic system (Al/Zr = 8000) to initiate the polymerization. Kaminsky and Woo *et al.* reported the formation of syndiotactic PP with MCM-41-supported dichloro[Me$_2$C(Cp)(Flu)]zirconium,[42] where the syndiotacticity and T_m of the PP are higher than those for the PP obtained with catalyst or with the homogeneous or an amorphous SiO$_2$-supported system.

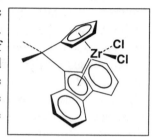

O'Hare and coworkers also investigated the polymerization of propylene with dichloro- [*rac*-ethylenebis(indenyl)]zirconium supported on MCM-41 and an inorganic clay with a 2-D sheet-like structure.[43] They showed that the catalytic activity of the former system (1.2 x 10³ kg PP mol^{-1} Zr^{-1} hr^{-1}) is lower than that of the latter system probably due to a diffusion control in the MCM-41 channels. However, polymerization with the former catalyst system under more sophisticated conditions was found to give a PP with a higher T_m (141 °C) than the clay-supported system (T_m = 135 °C). An EXAFS study of the supported catalyst suggested an interaction between the Zr-Cl moieties and the MAO-MCM-41 adduct.[44]

MAO is known to be a mixture of oligomers with different numbers of repeating units, and the active species responsible for the activation of metallocenes is not clear. Soga and coworkers succeeded in fractionation of the MAO oligomers by size using the nanoscopic channels of MCM-41 with different channel diameters (Figure 5). They also investigated the polymerization of

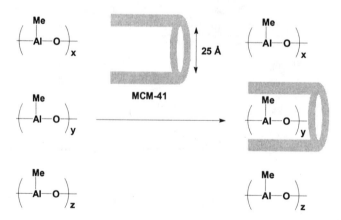

Figure 5. Fractionation of methylalumoxane (MAO) oligomers with MCM-41.

ethylene with Cp_2ZrCl_2 in conjunction with isolated MAO fractions,[45] and found that a MAO fraction separated using a silicate channel with a diameter of 25 Å serves as the most efficient co-catalyst. A similar result was reported for the polymerization of propylene with dichloro- [*rac*-ethylenebis(indenyl)]zirconium in combination with fractionated MAO oligomers.[46]

Weckhuysen and coworkers utilized calcined $Cr(acac)_3$-mounted MCM-41 (Cr-MCM-41) as a catalyst for the polymerization of ethylene.[47] Nitrogen adsorption/desorption isotherms and XRD studies showed that Cr-MCM-41 has a hexagonal structure with a large surface area. According to electron spin resonance (ESR) spectroscopy, the chromium ions are uniformly dispersed on the MCM-41 surface when the loading of Cr^{3+} is lower than 1 wt%, where the formation of inactive Cr_2O_3 clusters is negligible. In particular, Cr-MCM-41 prepared with Al-MCM-41 containing 1 wt% $Cr(acac)_3$ was the most active supported catalyst, with a polymerization rate of 7.3 x 10^2 kg PE mol Cr^{-1} hr^{-1}.

Looveren and coworkers reported that an MCM-41/MAO complex, prepared by the *in situ* hydrolysis of Me_3Al on the surface of MCM-41, is capable of initiating the co-oligomerization of ethylene and propylene when combined with dichloro[*rac*-ethylenebis(indenyl)]zirconium as catalyst.[48] In this system, the channel size of MCM-41 was claimed to affect the molecular weight of the resulting co-oligomer. When MCM-41 prepared with a C_{16} surfactant was used as the support, a co-oligomer with M_n of 1280 was obtained. On the other hand, use of MCM-41 prepared with a shorter-chain C_{14} surfactant as the support results in the formation of a co-oligomer with a higher M_n (1730). In sharp contrast, no polymerization took place when an MCM-41/MAO adduct, prepared by post-synthesis loading of MAO onto MCM-41, was used. A composite material prepared by hydrolysis of Me_3Al on the surface of amorphous silica showed a much lower activity than that with MCM-41, suggesting that the nanoscopic silicate channels in MCM-41 may isolate the individual active sites and suppress the bimolecular deactivation of chain growth.

Aida and coworkers utilized an aluminum-containing MCM-41 (Al-MCM-41) for free-radical polymerization of methylmethacrylate (MMA) with benzoyl peroxide (BPO) as initiator.[49] The polymer molecular weight with Al-MCM-41 is much higher than that formed in solution under otherwise identical conditions and can be controlled over a wide range (M_n = 1.3-3.0 x 10^5) by changing the initial molar ratio of MMA to BPO, although the molecular weight distribution is not narrow (M_w/M_n = 2.5). An electron paramagnetic resonance (EPR) study showed the formation of long-lived propagating polymer radicals within the nanoscopic channels, where about 25% of the initial intensity of the EPR signal remained even after a month (Figure 6). On the other hand, the

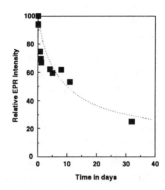

Figure 6. An electron paramagnetic resonance (EPR) trace of free radical polymerization of methyl methacrylate (MMA) by 2,2′-azobisisobutyronitrile (AIBN) within the silicate channels of MCM-41. Reproduced with permission from *Macromol. Rapid. Commun.* **1997**, *18*, 991. Copyright 1997 Wiley Publishing.

stereoregularity of the polymer is similar to that obtained in solution, indicating that the silicate channels of MCM-41 are not small enough to affect the stereochemical course of chain growth.

Al-MCM-41 is much more acidic than MCM-41. Aida and Kageyama *et al.* found that Al-MCM-41 in the presence of an alcohol can bring about ring-opening polymerization of lactones to give polyesters with narrow molecular weight distributions (MWD).[50] By changing the molar ratio [monomer]/[alcohol], the polymer molecular weight can be controlled over a rather wide range. Sequential polymerization of two different lactones with the Al-MCM-41/alcohol system resulted in the formation of a block copolymer with narrow MWD. The polymerization rate was enhanced when the Al content of Al-MCM-41 was increased.[51] In contrast, no polymerization took place using pure silicate MCM-41 or modified Al-MCM-41 with methylated silanol functionalities. Furthermore, use of the microporous aluminosilicate zeolite-Y with much smaller channels (8 Å) instead of Al-MCM-41 resulted in no polymerization. These observations suggest that the Lewis-acidic aluminum moieties and Brønsted-acidic silanol functionalities in Al-MCM-41, as well as the large diameter (27 Å) of the aluminosilicate channel, are essential for polymerization. The polymerization mechanism shown in Figure 7 has been postulated, in which the monomer is activated by cooperation of the Lewis-acidic site and the Brønsted-acidic functionality on the aluminosilicate surface. Titanium-containing MCM-41 (Ti-MCM-41), in conjunction with an alcohol, also initiates the ring-opening polymerization of lactones.[51] The polymerization proceeds rather sluggishly but gives a polymer containing a fraction of extremely high molecular weight.

In connection with the above observations, ring-opening polymerization of lactide was found to proceed in a tin-containing mesoporous silica (Sn-HMS), prepared by direct hydrothermal synthesis with tetraethylorthosilicate (TEOS) and

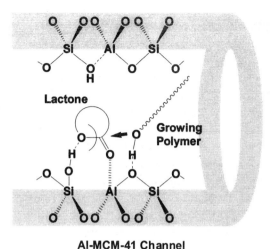

Al-MCM-41 Channel

Figure 7. Proposed mechanism for the activation of lactone on the interior aluminosilicate surface of Al-MCM-41 for ring-opening polymerization with alcohol.

tin isopropoxide.[15] The activity of Sn-HMS for the polymerization is superior to that of an amorphous silica with post-synthesis-loaded Sn moieties, probably because of the large surface area and suitable dispersion of active metal centers in Sn-HMS. This catalytic system gives a polyester with a narrow MWD, as indicated by the ratio M_w/M_n of 1.1, suggesting that the confined space of the silicate channel suppresses the "back-biting" reaction of the propagating polymer and intermolecular transesterification, both of which are very likely to occur in the ring-opening polymerization of cyclic esters.

Condensation of pyrrole and benzaldehyde to form a 4:4 cyclic co-oligomer (porphyrinogen) was found to occur selectively within the mesoporous aluminosilicate channels of FSM-16.[52] The reaction is likely catalyzed by the acidic sites on the silica surface. Oxidation of the resulting product gives tetraphenylporphyrin (TPP), a useful dye molecule, whose yield depends on the channel size of FSM-16. For example, use of FSM-16 with a channel diameter of 28 Å resulted in the formation of TPP in excellent yield (38%), higher than those with channels of 20 Å (18%) and 34 Å (28%) in diameter. These results indicate that the nanoscopic channels affect the probability of chain cyclization in the condensation of pyrrole and benzaldehyde.

Cationic polymerization of cyclohexyl vinyl ether within MCM-41 channels has been reported,[53] where the polymer molecular weight is controlled by the monomer-to-initiator mole ratio, although the MWD is not narrow (M_w/M_n = 1.4-2.2).

3.2   Control of Higher-Order Structures of Polymers

The chemical and physical properties of polymeric materials depend strongly on their secondary and ternary structures as well as primary structures such as molecular weight, stereoregularity, comonomer sequence, and so forth. Polymerization with regularly arranged nanoscopic channels of mesoporous silicates as nanoflasks is also interesting, since it can form polymeric materials with well-defined short- and long-range orderings of polymer chains.

Recently, Aida and Kageyama *et al.* reported that a titanocene complex supported by a particular type of mesoporous silicate called mesoporous silica fiber (MSF, channel diameter = 27 Å),[54] in conjunction with MAO, produced crystalline nanofibers of ultrahigh molecular weight (M_n = 6,200,000) linear polyethylene.[55] The polymer thus obtained appears to be a cocoon-like white mass with a low bulk density. Scanning electron microscopy (SEM) shows that the fibers are 30-50 nm in diameter (Figure 8), with a crystalline density close to the theoretical upper limit (1.0 g cm^{-3}) for densely packed polyethylene. Unlike ordinary high-density polyethylene consisting of folded-chain crystals, the polyethylene fibers consist exclusively of extended-chain crystals, according to small angle x-ray scattering (SAXS) and differential scanning calorimetry (DSC). For the formation of such crystalline nanofibers of polyethylene, an "extrusion polymerization" mechanism, which mimics biological formation of natural fibers such as crystalline cellulose fibers, has been postulated, where the polymer chains formed at the activated titanocene sites within individual mesopores are extruded into the solvent phase and assembled to form the extended-chain crystals (Figure 9). Thus, the nanoscopic channels in this particular case most likely serve as a

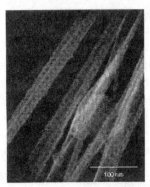

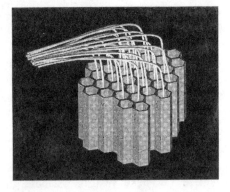

Figure 8. SEM micrograph of crystalline nanofibers of ultrahight molecular weight, linear polyethylene, formed by polymerization with alcohol. Adapted from *Science* **1999**, *28*, 2113. Copyright 1999 American Association for the Advancement of Science.

Figure 9. Proposed mechanism for the formation of fibrous polyethylene with a mesoporous silica fiber (MSF)-supported titanocene. Adapted from *Science* **1999**, *28*, 2113. Copyright 1999 American Association for the Advancement of Science.

template which suppresses the kinetically favored chain-folding process, since the channel diameter (27 Å) is almost one order of magnitude smaller than the lamellar thickness (270 Å) of ordinary folded-chain polyethylene crystals. Since most studies on precision polymerization have focused on the control of primary structures of polymers, this finding, which enables control of a ternary structure, is unique and may indicate a high potential of aligned mesoporous channels for the fabrication of novel polymer materials from specialty as well as commodity monomers.

Along this line, a novel hybrid thin film of polyethylene and silica (Figure 10) was synthesized[56] by polymerization of ethylene on a mesoporous silica film[57] doped with titanocene. In relation to these studies, polymerization of ethylene with a Cr^{3+}-doped MCM-41 (Cr-MCM-41) also gave polyethylene nanofibers with a length of several microns and a diameter of 50-100 nm, covering the outer surface of the MCM-41 particle.[47] On the other hand, isotactic polymerization of propylene with MCM-41-supported dichloro[*rac*-ethylenebis(indenyl)]zirconium was reported to give spherulite-like polymeric particles with an approximate diameter of 10 μm, as observed by SEM.[43] Use of a homogeneous catalyst or a clay-supported catalyst resulted in the formation of polypropylene with only a porous open structure, which appears to be composed entirely of sub-particles around 1 μm in size.

Mallouk and Ozin *et al.* utilized MCM-41 as a mold for the fabrication of a fibrous phenolformaldehyde resin.[58] Here, an acid-catalyzed polyaddition/condensation of phenol and formaldehyde was carried out within the silicate channels of MCM-41, then the silicate framework was destroyed by HF,

Figure 10. SEM micrograph of a polyethylene/silica/mica hybrid thin film, formed by polymerization of ethylene on a mesoporous silica film-supported titanocene. Reproduced with permission from *J. Polym. Sci. Part A: Polym. Chem.* **2000**, *38S*, 4821. Copyright 2000 Wiley Publishing.

Figure 11. TEM micrograph of polymeric mesofibres, formed by polyaddition/condensation of phenol and fomaldehyde within the silicate channels of MCM-41. Reprinted with permission from *J. Mater. Chem.* **1998**, *8*,13. Copyright 1998 Royal Society of Chemistry.

leaving poly(phenolformaldehyde) mesofibers with a diameter of approximately 20 Å and an aspect ratio higher than 10^3 (Figure 11). In contrast, when the polymerization was conducted in the presence of the non-porous silica called Cab-O-Sil, a non-fibrous, agglomerated polymeric mass resulted. Hyeon and coworkers reported the fabrication of a polypyrrole/poly(methylmethacrylate) coaxial nanocable using the silicate channels of mesoporous SBA-15.[59] The first step of the synthesis involves free-radical polymerization of methylmethacrylate (MMA) with benzoyl peroxide within the silicate channels of SBA-15. The pore diameter decreased from 75 to 62 Å, suggesting that the inner surface of SBA-15 became coated with the produced polymer (PMMA). Subsequent loading of pyrrole into the PMMA-coated channels, followed by oxidative polymerization of the included pyrrole with $FeCl_3$, resulted in the formation of a polypyrrole inner domain (Figure 12). The coaxial polypyrrole/PMMA wire was isolated by treatment of the resulting silica/polymer composite material with HF. Atomic force microscopy (AFM) showed the unidirectionally oriented morphology of the isolated wire.

Ordered mesoporous carbons have been synthesized using mesoporous silica as a template.[60] A carbon precursor such as sucrose is infiltrated into the nanoscopic channels of MCM-48 or SBA-15, and the resulting adduct is heated to induce polymerization. Subsequent pyrolysis of the resulting polymeric precursor results in the formation of a carbon/silica composite material. The silica template is washed out with a NaOH solution to leave a pure carbon material, whose N_2 adsorption/desorption isotherm shows a large surface area due to its mesoporous structure. TEM observation of the material shows an ordered hexagonal porous structure (Figure 13), whose XRD pattern is almost identical to that of the templating mesoporous silica. Thus, the carbon material is an exact negative replica of the templating mesoporous silica.

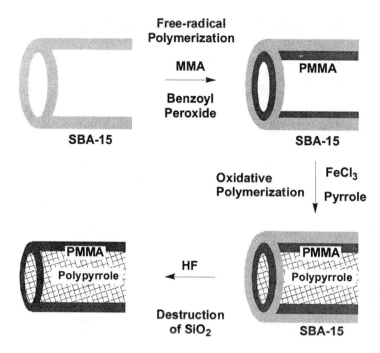

Figure 12. Schematic representation of the formation of a polypyrrole/poly(methylmethacrylate) coaxial nanocable with the channels of mesoporous silica SBA-15.

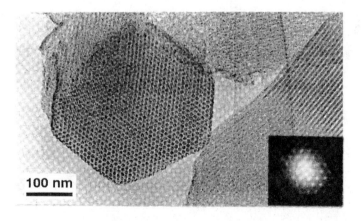

Figure 13. TEM micrograph of an ordered mesoporous carbon, prepared by pyrolysis of sucrose within the channels of mesoporous silica SBA-15. Reprinted with permission from *J. Am. Chem. Soc.* **2000**, *122*, 10712. Copyright 2000 American Chemical Society.

In relation to the synthesis of organic polymers, formation of metal clusters templated with mesoporous silicates has also attracted great attention, and several examples of the fabrication of metallic nanowires within nanoscopic silicate channels have been reported. Ryoo and coworkers reported the synthesis of Pt nanowires inside the channels of mesoporous silicate materials.[61] A detailed TEM study showed preferential growth of the Pt nanowires along the (110) axis within the MCM-41 channels. The Pt nanowires exhibit high thermal stability in comparison to those without a silicate coating. Bimetallic nanowires of Pt-Rh and Pt-Pd were also synthesized by impregnation of a mesoporous organic/inorganic hybrid (HMM-1) with a mixture of precursor metal salts, followed by exposure of the resulting composite material to UV light.[62] Yang and coworkers reported the synthesis of Ag nanowires with mesoporous silica SBA-15.[63] The nanoscopic channels of SBA-15 were impregnated with $AgNO_3$ and the resulting composite material was heated to induce thermal decomposition of included $AgNO_3$ to give the Ag nanowires. TEM observation showed that the Ag nanowires are of uniform diameter (50-60 Å) with a high aspect ratio from 100 to 1000 (Figure 14). Stucky and coworkers also reported the template synthesis of Pt, Au and Ag nanowires with mesoporous silica SBA-15.[64] Upon treatment with HF to destroy the silica wall, the included nanowires with a diameter of 7 Å and a length of 50-1000 nm were isolated. On the other hand, Pd nanowires were synthesized by low temperature chemical vapor infiltration (CVI) of an organometallic Pd precursor into the nanoscopic channels of some mesoporous silicates.[65]

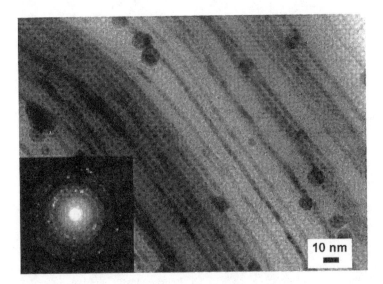

Figure 14. TEM micrograph of Ag nanowires within the channels of mesoporous silica SBA-15. Reprinted with permission from *Chem. Commun.* **2000**, 1063. Copyright 2000 Royal Society of Chemistry.

3.3 Fabrication of Silica/Functional Polymer Nanocomposite Materials

Inclusion polymerization in ordered nanoscopic silicate channels can produce aligned polymer nanodomains, segregated by the silica wall. If a polymer chain has bulky substituents, each silica channel can accommodate only a limited number of polymer chains. Thus, one may investigate the chemical and physical properties of a segregated nanodomain of polymer strands or even those of a single polymer chain. Furthermore, one can also expect some anisotropic properties for such polymer/silica composite materials.

Bein and coworkers reported the synthesis of a graphite-type conducting carbon wire by the polymerization of acrylonitrile within the nanoscopic channels of MCM-41 and subsequent pyrolysis at 1000 °C (Figure 15).[66] The resulting graphite/MCM-41 composite material showed a microwave conductivity of ~0.1 S cm⁻¹. A polyaniline wire encapsulated within the nanoscopic silicate channels was also synthesized by incorporation of aniline and its subsequent oxidative polymerization. However, the resulting composite material showed a microwave conductivity of only 0.0014 S cm⁻¹.[67] Variation of the channel diameter may make it possible to tune the conducting properties. Similarly, incorporation of 1,4-diiodo-1,3-butadiyne and 1-iodo-1,3,5-hexatriyne into the silicate channels of MCM-41, followed by UV irradiation, gave oligoyne compounds within the channels,[68] which show enhanced chemical and thermal stability.

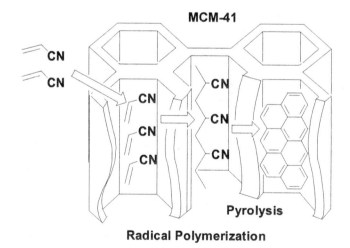

Figure 15. An illustration of free radical polymerization of acrylonitrile within the silicate channels of MCM-41, and subsequent pyrolysis to form a conducting carbon nanowire.

Thermal properties of polymer chains such as the glass transition temperature are affected by spatial confinement with silicate channels.[69] Moller and coworkers have investigated the polymerization of methylmethacrylate within MCM-41 and MCM-48,[70] where N_2 adsorption/desorption isotherm profiles of the resulting composite materials, together with their scanning and transmission electron micrographs, showed formation of a polymer exclusively inside the mesopores. According to differential scanning calorimetry, the included polymer (PMMA) did not show any glass transition behavior, while ordinary PMMA has a glass transition at 111°C. This observation suggests a possible interaction between the inner surface of MCM-41 or MCM-48 and PMMA, which may suppress long-range molecular motion of polymer strands.

O'Hare and Ozin *et al.* independently reported ring-opening polymerization of [1]silaferrocenophane within the nanoscopic channels of mesoporous silicates.[71,72] When MCM-41 was exposed to a solution or a vapor of [1]ferrocenophane, ring-opening of the monomer took place to give an adduct with MCM-41.[71] Upon heating at 140 °C, ring-opening polymerization of residual monomer molecules took place, to give a poly(ferrocenyl-silane)/MCM-41 composite material (Figure 16).[72] Pyrolysis of this composite material at 900 °C under N_2 resulted in decomposition of the included polymer to give superparamagnetic iron nanoparticles with an average diameter of 20 ± 5 Å.

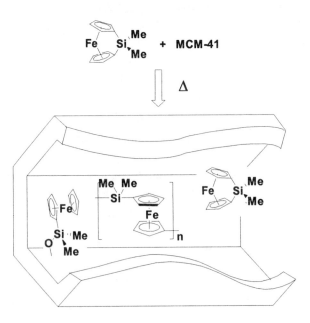

Figure 16. An illustration of ring-opening polymerization of [1]silaferrocenophane within the silicate channels of MCM-41.

Nano-sized semiconducting materials show unique optical and electronic properties, referred to as "quantum size effects". Synthesis of mesoporous silicates containing semiconductors inside the channels is generally performed by incorporation of small precursor molecules and subsequent polymerization. Nanoparticles of Fe_2O_3 were synthesized within the silicate channels of MCM-41 by post-synthesis loading of iron(III) nitrate and subsequent oxidation. This composite material displays a wider band gap than bulk Fe_2O_3 due to the "quantum size effect".[73] Ozin and coworkers reported the formation of silicon nanoclusters by chemical vapor deposition (CVD) of silicon precursors inside the oriented hexagonal channels of a surfactant-containing silicate film.[74] The resulting composite material showed a yellow-orange photoluminescence, characteristic of nano-sized silicon clusters. Silicon nanowires were also synthesized within large channels of a mesoporous silicate prepared using an amphiphilic block copolymer micelle as template.[75] In this case, the mesoporous silica was exposed to a carbon dioxide supercritical fluid solution of diphenylsilane, to allow complete (~95%) and rapid filling of the silicate channels with the Si precursor. The resulting composite material showed absorption and photoluminescence spectra characteristic of (100)-oriented silicon nanowires. Of interest, its fluorescence excitation spectrum is blue-shifted in comparison with ordinary silicon nanowires without a silicate coating, suggesting a quantum confinement effect. In relation to this observation, infiltration of MCM-41 with a soluble gallium precursor solution, followed by exposure to ammonia, was reported to produce a photoluminescent composite material of MCM-41 containing gallium nitride (GaN) in the silicate channels.[76] A significant blue shift of the photoluminescence indicated the presence of quantum-confined GaN clusters inside the silicate channels.

More recently, synthesis of mesoporous silicate materials including conjugated polymer chains in their channels was reported. The synthetic strategy includes the formation of silicate hexagonal structures using micelles of diacetylenic surfactants as templates, and subsequent polymerization of the included surfactants. Recently, Aida and Tajima have reported the synthesis of stick-like silicate materials with diameters of 15-25 μm, using ammonium ion-terminated diacetylenic surfactants (Figure 17). Subsequent heating resulted in the formation of photoluminescent red-colored materials with aligned nanodomains of polydiacetylenes within the silicate channels.[77] Upon photoexcitation at 450 nm, the silicate microsticks emitted a yellowish-green luminescence centered at 550 nm, which is considerably red-shifted from that of the polymer extracted from the silicate channels. This observation indicates a possible confinement effect of the silicate channel on the effective conjugation length of the included polymer.

Brinker and coworkers reported the synthesis of mesostructured silica films with various mesoscopic orders using non-ionic amphiphilic diacetylene

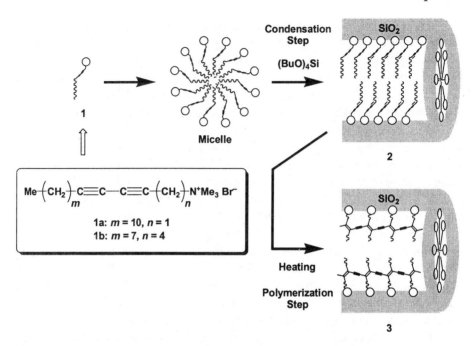

Figure 17. Fabrication of micrometer-scale photoluminescent silicate sticks **3** by "sol-gel based *in situ* polymerization" of diacetylenic surfactant monomers **1**. Reprinted with permission from *Angew. Chem. Int. Ed.* **2001**, *40*, 3803. Copyright 2001 Wiley Publishing.

monomers as surfactants.[78] When the silica films with hexagonal and lamellar phases were exposed to UV light, polymerization of the included diacetylenes took place to form blue-colored polydiacetylene/silica composite materials. Under ideal conditions, the silica film with a cubic phase did not show any color change associated with polymerization, suggesting the importance of appropriate orientation of the included monomers. Interestingly, the polymerized films exhibited a color change from blue to red upon exposure to certain solvents (solvatochromism), heating (thermochromism), and mechanical abrasion (mechanochromism). In relation to these studies, a conjugated polymer/ mesoporous silica composite material was prepared by post-synthesis loading of poly[2-methoxy-5-(2'-ethylhexyloxy)-1,4-phenylenevinylene] (MEH-PPV) into the channels of a mesoporous silicate glass.[79] Since the silicate glass was prepared in a magnetic field, the nanoscopic channels in this material were unidirectionally aligned. Therefore, the guest polymer chains included in the silicate channels should also be aligned in a single direction.[80] Accordingly, excitation of this composite material with polarized light resulted in fluorescence emission with anisotropic character, due to the unidirectional orientation of the chromophoric polymer chains along the nanoscopic channels (Figure 18). This observation suggests the potential of such nanoscopically aligned luminescent materials for optoelectronic devices.

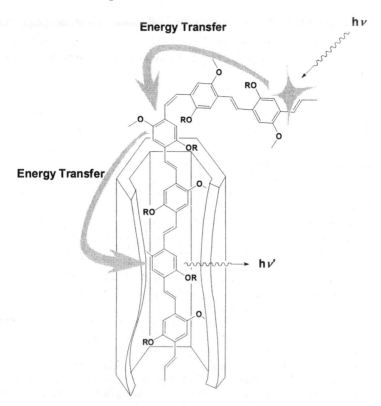

Figure 18. Proposed photochemical events for photoluminescence from a mesoporous silica/poly(phenylene vinylene) composite material.

4. PERSPECTIVES

Recent developments in the synthesis of mesoporous silicates have triggered extensive studies on a new class of supported catalysts for controlled polymerization and macromolecular architectures. Mesoporous silicates with transition metal ions in the silicate framework are quite interesting, if the metal centers themselves can catalyze polymerization without doping. Template-assisted polymerization, where the morphology of the channel is translated into the product shape, is also interesting. These studies may lead to the development of high-performance polymeric materials and inorganic/organic nanocomposite materials for electronic and optical devices. Further developments of mesoporous silicates with long-range structural order such as mesoporous silica films, monoliths and fibers represent the next step toward new composite materials. For example, micropatterning of mesostructured and mesoporous silicate materials has been reported,[81] which is applicable to waveguide and lasing materials. Furthermore, from a viewpoint of designer materials, one may also consider fabrication of more complex silica/polymer composite materials containing hetero-conjunct polymer nanodomains of different chemical and physical properties.

REFERENCES

1. Tajima, K.; Aida, T. *Chem. Commun.* **2000**, 2399.
2. Beck, J.S.; Vartuli, J. C.; Roth, W. J.; Leonowicz, M. E.; Kresge, C. T.; Schmitt, K. D.; Chu, C. T. -W.; Olson, D. H.; Sheppard, E. W.; McCullen, S. B.; Higgins, J. B.; Schlenker, J. L.; *J. Am. Chem. Soc.* **1992**, *114*, 10834.
3. Moller, K.; Bein, T. *Chem. Mater.* **1998**, *10*, 2950.
4. Johnson, B. F. G. ; Raynor, S. A.; Shephard, D. S.; Mashmeyer, T.; Thomas, J. M.; Sankar, G.; Bromley, G.; Oldroyd, R.; Gladden L.; Mantle, M. D. *Chem. Commun.* **1999**, 1167.
5. Sayari, A. *Chem. Mater.* **1996**, *8*, 1840; Corma, A. *Chem. Rev.* **1997**, *97*, 2373; Ying, J. Y.; Mehnert C. P.; Wong, M. S. *Angew. Chem. Int. Ed.* **1999**, *38*, 56; Ciesla U.; Schüth, F. *Micropor. Mesopor. Mater.* **1999**, *27*, 131.
6. Göltner, C. G.; Berton, B.; Krämer, B; Antonietti, M. *Chem. Commun.* **1998**, *2287*; Göltner, C. G.; Henke, S.; Weissenberger, M. C.; Antonietti, M. *Angew. Chem. Int. Ed.* **1998**, *37*, 613.
7. Attard, G. S.; Glyde, J. C.;Göltner, C. G. *Nature* **1995**, *378*, 366.
8. Lu, Y.; Ganguli, R.; Drewien, C. A.; Anderson, M. T.; Brinker, C. J.; Gong, W.; Guo, Y.; Soyez, H.; Dunn, B.; Huang, M. H.; Zink, J. I. *Nature* **1997**, *389*, 364; Brinker, C. J.; Lu, Y.; Sellinger, A.; Fan, H. *Adv. Mater.* **1999**, *11*, 579.
9. Feng, X.; Fryxell, G. E.; Wang, L. -Q.; Kim, A. Y.; Liu, J.; Kemner, K. M. *Science* **1997**, *276*, 923; Liu, J.; Feng, X.; Fryxell, G. E.; Wang, L. -Q.; Kim, A. Y.; Gong, M. *Adv. Mater.* **1998**, *10*, 161.
10. Shephard, D. S.; Zhou, W.; Maschmeyer, T.; Matters, J. M.; Roper, C. L.; Persons, S.; Roper, C. L.; Parsons, S.; Johnson, B. F. G.; Duer M. J. *Angew. Chem. Int. Ed.* **1998**, *37*, 2719.
11. Inagaki, S.; Guan, S.; Fukushima, Y.; Oshuna T.; Terasaki, O. *J. Am. Chem. Soc.* **1999**, *121*, 9611; Ishii, C. Y.; Asefa, T.; Coombs, N.; MacLachlan M. J.; Ozin, G. A. *Chem. Commun.* **1999**, 2539; Asefa, T. A.; Ishii, C. Y.; MacLachlan M. J.; Ozin, G. A. *J. Mater. Chem.* **2000**, *10*, 1751; MacLachlan, M. J.; Asefa T.; Ozin, G. A. *Chem. Eur. J.* **2000**, *6*, 2507; Lu, Y.; Fan, H.; Doke, N.; Loy, D. A.; Assink, R. A.; LaVan D. A.; Brinker C. A. *J. Am. Chem. Soc.* **2000**, *122*, 5258.
12. Janicke, M. T.; Landry, C. C.; Christiansen, S. C.; Kumar, D.; Stucky G. D.; Chmelka, B. F. *J. Am. Chem. Soc.* **1998**, *120*, 6940.
13. Corma, A.; Navarro, M. T.; Pariente, J. P. *J. Chem. Soc., Chem. Comm.* **1994**, 147; Corma, A.; Kan, Q.; Rey, F.; Chem. Commun. **1998**, 579; Camblor, M. A.; Corma, A.; Esteve, P.; Martínez, A.; Valencia, S. *J. Chem. Soc., Chem. Comm.* **1997**, 795; Corma, A.; Jordá, J. L.; Navarro, M. T.; Rey, F. *Chem. Commun.* **1998**, 1899; Corma, A.; Domine, M.; Gaona, J. A.; Jordá, J. L.; Navarro, M. T.; Rey, F.; Pérez-Pariente, J.; Tsuji, J.; McCulloch, B.; Nemeth, N. T. *Chem. Commun.* **1998**, 2211; Prakash, A. M.; Sung-Suh H. M.; Kevan, L. *J. Phys. Chem. B.* **1998**, *102*, 857.
14. Jones, D. J.; Jiménez-Jimenez, J.; Jimenez-López, J.; Maireles-Torres, P.; Olivera-Pastor, P.; Rodriguez-Castellón, E.; Rozière, J. *J. Chem. Soc., Chem. Comm.* **1997**, 431.

15. Abdel-Fattah, T. M.; Pinnavaia, T. J. *J. Chem. Soc., Chem. Comm.* **1996**, 665.

16. Reddy, J. S.; Sayari, A. *J. Chem. Soc., Chem. Comm.* **1995**, 2231; Neumann, R.; Khenkin, A. M. *J. Chem. Soc., Chem. Comm.* **1996**, 2643.

17. Sayari, A.; Danumah, C.; Moudrakovski, I. L. *Chem. Mater.* **1995**, *7*, 813; Sayari, A.; Moudrakovski, I. L.; Danumah, C.; Ratcliffe, C. I.; Ripmeester, J. A.; Preston, K. F.; *J. Phys. Chem.*, **1995**, *99*, 16373; Chatterjee., M.; Iwasaki, T.; Hayashi, T.; Onodera, Y.; Ebina, T.; Nagase, T. *Chem. Mater.* **1999**, *11*, 1368.

18. Ulagappan, N.; Rao, C. N. R. *J. Chem. Soc., Chem. Comm.* **1996**, 1047.

19. Zhao, D.; Goldfarb, D. *J. Chem. Soc., Chem. Comm.* **1995**, 875; Zhang, J.; Goldfarb, D. *J. Am. Chem. Soc.* **2000**, *122*, 7034.

20. Zhang, W.; Wang, J.; Tanev, P. T.; Pinnavaia, T. J. *J. Chem. Soc., Chem. Comm.* **1996**, 979.

21. Zhang, W.; Suo, J.; Zhang, X.; Li, S. *J. Chem. Soc., Chem. Comm.* **1998**, 241.

22. Yuan, Z. Y.; Liu, S. Q.; Chen, T. H.; Wang, J. Z.; Li, H. X. *J. Chem. Soc., Chem. Comm.* **1995**, 973.

23. Karakassides, M. A.; Fournaris, K. G.; Travlos, A.; Petridis, D. *Adv. Mater.* **1998**, *10*, 483

24. Hartmann, M.; Racouchot, S.; Bischof, C. *J. Chem. Soc., Chem. Comm.* **1997**, 2367.

25. Tuel, A.; Gontier, S. *Chem. Mater.* **1996**, *8*, 114.

26. Jervis, H. B.; Raimondi, M. E.; Raja, R.; Machmeyer, T.; Seddon, J. M.; Bruce, D. W. *Chem. Commun.* **1999**, 2031.

27. Raimondi, M. E.; Marchese, L.; Gianotti, E.; Maschmeyer, T.; Seddon, J. M.; Coluccia, S. *Chem. Commun.* **1999**, 87.

28. Mokaya, R.; Jones, W. *J. Chem. Soc., Chem. Comm.* **1997**, 2185; Anwander, R.; Palm, C.; Gerstberger, G.; Groeger, O.; Engleheart, G. *Chem. Commun.* **1998**, 1811; Mokaya, R. *Angew. Chem. Int. Ed.* **1999**, *38*, 2930.

29. Oldroyd, R. D.; Thomas, J. M.; Sankar, G. *J. Chem. Soc., Chem. Comm.* **1997**, 2025; Krijnen, S.; Abbenhuis, H. C. L.; Hanssen, R. W. J. M.; van Hooff, J. H. C.; van Santen, R. A. *Angew. Chem. Int. Ed.* **1998**, *37*, 356.

30. Eswaramoorthy, M.; Neeraj, Rao, C. N. R. *Chem. Commun.* **1998**, 615.

31. Rao, R. R.; Weckhuysen, B. M.; Schoonheydt, R. A.; *Chem. Commun.* **1999**, 445; Zhu, Z.; Hartmann, M.;Maes, M.; Czernuszewicz, R. S.; Kevan, L. *J. Phys. Chem. B.* **2000**, *104*, 4690.

32. Ferreira, P.; Gonçalves, I. S.; Kühn, F. E.; Pillinger, M.; Rocha, J.; Thursfield, A.; Xue, W. -M.; Zhang, G. *J. Mater. Chem.* **2000**, *10*, 1395; Shannon, I. J.; Maschmeyer, T.; Oldroyd, R. D.; Sankar, G.; Thomas, J. M.; Pernot, H.; Balikdjian, J. -P.; Che, M. *J. Chem. Soc., Farad. Trans.* **1998**, *94*, 1495.

33. Zhou, W.; Thomas, J. M.; Shephard, D. S.; Johnson, B. F. G.; Ozkaya, D.; Maschmeyer, T.; Bell, R. G.; Ge. Q. *Science* **1998**, *280*, 705.

34. Mehnert, C. P.; Ying, J. Y.; *J. Chem. Soc., Chem. Comm.* **1997**, 2215; Mehnert, C. P.; Weaver, D. W.; Ying, J. Y. *J. Am. Chem. Soc.* **1998**, *120*, 12289.

35. Ryoo, R.; Jun, S.; Kim, J. M.; *J. Chem. Soc., Chem. Comm.* **1997**, 2225; Anwander, R.; Runte, O.; Eppiner, J.; Gerstberger, G.; Herdweck, E.; Spiegler, M. *J. Chem. Soc., Dalton Trans.* **1998**, 847.

36. Yonemitsu, M.; Tanaka, Y.; Iwamoto, M. *Chem. Mater.* **1997**, *9*, 2679; Badiei, A.-R. Bonneviot, L. *Inorg. Chem.* **1998**, *37*, 4142.

37. Liu, C.-J.; Li, S.-G.; Pang, W.-Q.; Che, C.-M. *J. Chem. Soc., Chem. Comm.* **1997**, 65; Rao, Y. V. S.; De Vos, D. E.; Bein, T.; Jacobs, P. A. *J. Chem. Soc., Chem. Comm.* **1997**, 355.

38. Sutra, P.; Brunel, D. *J. Chem. Soc., Chem. Comm.* **1996**, 2485.

39. Crudden, C. M.; Allen, D.; Mikoluk, M. D.; Sun, J. *Chem. Commun.* **2001**, 1154.

40.   Rahiala, H.; Beurroies, I.; Eklund, T.; Hakala, K.; Gougeon, R.; Trens, P.; Rosenholm, J. B. *J. Catal.* **1999**, *188*, 14.
41.   Ko, Y.-S.; Han, T.-K.; Park, J.-W.; Woo, S.-I. *Macromol. Rapid Commun.* **1996**, *17*, 749.
42.   Kaminsky, W.; Strübel, C.; Lechert, H.; Genske, D.; Woo, S.-I. *Macromol. Rapid Commun.* **2000**, *21*, 909.
43.   Tudor, J.; O'Hare, D. *J. Chem. Soc., Chem. Comm.* **1997**, 603.
44.   O'Brien, S.; Tudor, J.; Maschmeyer, T.; O'Hare, D. *J. Chem. Soc., Chem. Comm.* **1997**, 1905.
45.   Sano, T.; Doi, K.; Hagimoto, H.; Wang, Z.; Uozumi, T.; Soga, K. *Chem. Commun.* **1999**, 733.
46.   Sano, T.; Hagimoto, H.; Jin, J.; Oumi, Y.; Uozumi, T.; Soga, K. *Macromol. Rapid Commun.* **2000**, *21*, 1191.
47.   Weckhuysen, B. M.; Rao, R. R.; Pelgrims, J.; Schoonheydt, R. A.; Bodart, P.; Debras, G.; Collart, O.; Van der Voort, P.; Vansant, E. F. *Chem. Eur. J.* **2000**, *6*, 2960.
48.   van Looveren, L. K.; Geysen, D. F.; Vercruysse, K. A.; Wouters, B. H.; Grobet, P. J.; Jacobs, P. A. *Angew. Chem. Int. Ed.* **1998**, *37*, 517.
49.   Ng, S. M.; Ogino, S.; Aida, T. *Macromol. Rapid. Commun.* **1997**, *18*, 991.
50.   Kageyama, K.; Ogino, S.; Aida, T.; Tatsumi, T. *Macromol.* **1988**, *31*, 4069; Kageyama, K.; Tatsumi, T.; Aida, T. *Polym. J.* **1999**, *31*, 1005.
51.   Kageyama, K.; Tatsumi, T.; Aida, T. *Polym. J.* **1999**, *31*, 1005.
52.   Shinoda, T.; Izumi, Y.; Onaka, M. *J. Chem. Soc., Chem. Comm.* **1995**, 1801.
53.   Spange, S.; Graeser, A.; Rehak, P.; Jäger, C.; Schulz, M. *Macromol. Rapid Commun.* **2000**, *21*, 146.
54.   Huo, Q.; Zhao, D.; Feng, J.; Weston, K.; Buratto, S. K.; Stucky, G. D.; Schacht, S.; Schüth, F. *Adv. Mater.* **1997**, *9*, 974.
55.   Kageyama, K.; Tamazawa, J.; Aida, T. *Science* **1999**, *285*, 2113.
56.   Tajima, K.; Ogawa, G.; Aida, T. *J. Polym. Sci. Part A: Polym. Chem.* **2000**, *38S*, 4821.
57.   Tolbert, S. H.; Schäffer, T. E.; Feng, J.; Hansma, P. K.; Stucky, G. D. *Chem. Mater.* **1997**, *9*, 1962.
58.   Johnson, S. A.; Khushalani, D.; Coombs, N.; Mallouk, T. E.; Ozin, G. A. *J. Mater. Chem.* **1998**, *8*, 13.
59.   Jang,, J.; Lim, B.; Lee, J.; Hyeon, T. *Chem. Commun.* **2001**, 83.
60.   Ryoo, R.; Joo, S. H.; Jun. S. *J. Phys. Chem. B* **1999**, *103*, 7743; Jun, S.; Hoo, H.; Ryoo, R.; Kruk, M.; Jaroniec, M.; Liu, Z.; Ohsuna, T.; Terasaki, O. *J. Am. Chem. Soc.* **2000**, *122*, 10712; Ryoo, R.; Joo, S. H.; Kruk, M.; Jaroniec, M. *Adv. Mater.* **2001**, *13*, 677; Shin, H. J.; Ryoo, R.; Kruk, M.; Jaroniec, M. *Chem. Commun.* **2001**, 349; Yoon, S. B.; Kim, J. Y.; Yu, J.-S. *Chem. Commun.* **2001**, 559.
61.   Ko, C. H.; Ryoo, R. *J. Chem. Soc., Chem. Comm.* **1996**, 2467; Ryoo, R.; Ko, C. H.; Park, I. S. *J. Chem. Soc., Chem. Comm.* **1999**, 1413; Liu, Z.; Sakamoto, Y.; Ohsuna, K.; Hiraga, K.; Terasaki, O.; Ko, C. H.; Shin, H. J.; Ryoo, R. *Angew. Chem. Int. Ed.* **2000**, *39*, 3107; Shin, H. J.; Ko, C. H.; Ryoo, R. *J. Mater. Chem.* **2001**, *11*, 260.
62.   Fukuoka, A.; Sakamoto, Y.; Guan, S.; Inagaki, S.; Sugimoto, N.; Fukushima, Y.; Hirahara, K.; Iijima, S.; Ichikawa, M. *J. Am. Chem. Soc.* **2001**, *123*, 3373.
63.   Huang, M. H.; Choudrey, A.; Yang, P. *Chem. Commun.* **2000**, 1063.
64.   Han, Y.-J.; Kim, J.-M.; Stucky, G. D. *Chem. Mater.* **2000**, *12*, 2068.
65.   Lee, K.-B.; Lee, S.-M.; Cheon, J. *Adv. Mater.* **2001**, *13*, 517.
66.   Wu, C.-G.; Bein, T. *Science* **1994**, *266*, 1013.
67.   Wu, C.-G.; Bein, T. *Science* **1994**, *264*, 1757; Wu, C.-G.; Bein, T. *Chem. Mater.* **1994**, *6*, 1109.
68.   Hlavat'y, J.; Rathousk'y, J.; Zukal, A.; Kavan, L. *Carbon* **2001**, *39*, 53.
69.   Frisch, H.; Mark, J. E. *Chem. Mater.* **1996**, *8*, 1735.

70. Moller, K.; Bein, T.; Fischer, R. *Chem. Mater.* **1998**, *10*, 1841.
71. O'Brien, S.; Tudor, J.; Barlow, S.; Drewitt, M. J.; Heyes, M. J.; O'Hare, D. *J. Chem. Soc., Chem. Comm.* **1997**, 641; O'Brien, S.; Keates, J. M.; Barlow, S.; Drewett, M. J.; Payne, B. R.; O'Hare, D. *Chem. Mater.* **1998**, *10*, 4088.
72. MacLachlan, M. J.; Aroca, P.; Coombs. N.; Manners, I.; Ozin, G. A. *Adv. Mater.* **1998**, *10*, 144; MacLachlan, M. J.; Ginzburg, M.; Coombs, N.; Nandyala, P.; Raju, P.; Greedan, J. E.; Ozin, G. A.; Manner, I. *J. Am. Chem. Soc.* **2000**, *122*, 3878.
73. Abe, T.; Tachibana, Y.; Uematsu, T.; Iwamoto, M. *J. Chem. Soc., Chem. Comm.* **1995**, 1617.
74. Dag, Ö.; Ozin, G. A.; Yang, H.; Reber, C.; Bussière, G. *Adv. Mater.* **1999**, *11*, 474.
75. Coleman, N. R. B.; Morris, M. A.; Spalding, T. R.; Holmes, J. D. *J. Am. Chem. Soc.* **2001**, *123*, 187; Coleman, N. R. B.; O'Sullivan, N.; Ryan, K. M.; Crowley, T. A.; Morris, M. A.; Spalding, T. R.; Steytler, D. C.; Holmes, J. D. *J. Am. Chem. Soc.* **2001**, *123*, 7010.
76. Winkler, H.; Birkner, A.; Hagen, V.; Wolf, I.; Schmechel, R.; von Seggern, H.; Fischer, R. A. *Adv. Mater.* **1999**, *11*, 1444.
77. Tajima, K.; Aida, T. *Angew. Chem. Int. Ed.* **2001**, *40*, 3803.
78. Lu, Y.; Yang, Y.; Sellinger, A.; Lu, A.; Huang, J.; Fan, H.; Haddad, R.; Lopez, G.; Burns, A. R.; Saski, D. Y.; Shelnutt, J.; Brinker, C. J. *Nature* **2001**, *410*, 913.
79. Wu, J.; Gross, A. F.; Tolber, S. H. *J. Phys. Chem. B* **1999**, *103*, 2374; Nguyen, T. -C.; Wu, J.; Doan, V.; Schwartz, B. J.; Tolber, S. H. *Science* **2000**, *288*, 652; Schwartz, B. J.; Nguyen, T.-Q.; Wu, J.; Tolbert, S. H. *Synthetic Metals* **2001**, *116*, 35.
80. Tolbert, S. H.; Firouzi, A.; Stucky, G. D.; Chmelka, B. F. *Science* **1997**, *278*, 264.
81. Yang, H.; Coombs, N.; Ozin, G. A. *Adv. Mater.* **1997**, *9*, 811; Trau, M.; Yao, N.; Kim, E.; Xia, Y.; Whitesides, G. M.; Aksay, I. A. *Nature* **1997**, *390*, 674.

Chapter 11

DESIGNING POROUS SOLIDS OVER MULTIPLE PORE SIZE REGIMES

Andreas Stein, Rick C. Schroden
Department of Chemistry, University of Minnesota, Minneapolis, MN 55455

Keywords: microporous, mesoporous, macroporous, bimodal pores, hierarchical structures, templating, colloidal crystal, silicates, catalysts

Abstract: This chapter highlights some of the recent synthetic approaches to inorganic solids (especially silicates) with designed porosity on multiple levels, covering pore size ranges from a few Ångstroms (micropores) through two to tens (mesopores) and hundreds (macropores) of nanometers. To achieve controlled porosity in these size regimes, molecular structure directors, supramolecular arrays of surfactants, block copolymers, emulsions, colloidal crystals, bacteria and porous carbon have been used as templates. Hierarchical porosity with bimodal or higher pore size distributions can be attained by employing combinations of these methods in multi-template processes.

1.   INTRODUCTION

Porous solids, such as zeolites and amorphous silica, have proven highly successful in a variety of commercial applications, including catalysis, ion exchange, chromatography and sorption. Porosity provides the materials with high specific surface areas, which may increase interaction of the solids with substrate molecules or guest species. Reactive surface groups are often inherently present (*e.g.*, surface silanols on silicates), or they may be added to the support (*e.g.*,

Nanostructured Catalysts, edited by S. Scott *et al.*
Kluwer Academic/Plenum Publishers, 2003

metal clusters). In zeolites, where pores are uniform in size and shape, selective transport or reactivity is possible. A desire to extend typical applications for zeolites to large guest molecules has led to the development of new synthetic techniques that produce porous solids with larger void dimensions than those attainable in zeolites. The increased pore diameters facilitate penetration of reactant molecules and reduce the resistance to diffusion of reactants or products.

Recent computer simulations have demonstrated that diffusion and reactivity of guests in porous catalysts depends not only on pore size ranges and distributions but also on the pore architecture.[1] In order to combine the selective properties of micropores with easier mass transport, it is necessary to design solids with multiple pore size regimes. In those systems, the smaller pores provide the largest fraction of the surface area. However, they also pose the greatest diffusional resistance. Diffusion limitations are minimized by making micropores more accessible and by keeping diffusion paths in micropores relatively short. In model studies, structures containing micropores in close contact with macropores exhibited superior performance, *e.g.*, for coke combustion under long residence times.[1] While such simulations provide a theoretical basis for designing more efficient catalysts, synthetic developments in the last few years have made it possible to prepare catalyst structures with controlled porosity and interconnectivity over multiple pore size ranges.

Traditionally, zeolites with additional meso- or macropores have been obtained by pelletization processes. Textural porosity is created between adjacent crystallites or grains. However, the structure of textural porosity is not uniform. Bimodal pore structures involving zeolites are also prepared by supporting zeolite crystallites on membranes[2] or by forming zeolite composites with other porous matrices.[3] The methods presented in this chapter aim at greater control of the pore structure at each size level. Several approaches to inorganic solids with hierarchical porosity, based on single and multiple templating processes, as well as solid-conversion reactions are featured. Here, the term "templating" is used in a broad sense and may encompass mechanisms of structure direction,[2,4] space filling, spatial patterning and "true" templating.[5] The pore size definitions are based on IUPAC conventions, where micropores range from 0.2-2 nm, mesopores from 2-50 nm, and macropores from 50 nm upwards.[6]

2.   SURFACTANT TEMPLATING

A wide range of mesoporous materials has been synthesized by surfactant templating techniques, a topic which has recently been reviewed.[7] The syntheses involve cooperative interactions of sol-gel precursors and surfactants, resulting in mesostructures of the inorganic phase around periodic (or sometimes nonperiodic) arrays of surfactant micelles. Removal of the surfactant opens up uniform channels, whose diameters can range from two to tens of nanometers, depending on the choice of surfactant and reaction conditions. The resulting products have large pore volumes and surface areas, often exceeding 1 mL/g and 1000 m²/g, respectively.

Functionalized mesoporous solids have been considered for a wide range of heterogeneous catalysis reactions.[8] It has been noted that mesoporous catalysts differ significantly in many respects from their functionalized, amorphous silica counterparts.[9] In several investigations, confinement of the catalyst in the mesoporous solid improved the activity compared to attachment to amorphous or nonporous silica, either due to enhanced selectivity in a sterically homogeneous environment or due to higher catalyst turnover brought about by stabilization of the catalyst within the channels. In other instances, the performance of the mesoporous catalyst was worse than for a catalyst attached to a non-porous support, due to limited accessibility of the active sites in the mesopores.[10] In the latter studies, the pores were typically smaller than 4.0 nm, and improved performance might be expected with larger mesoporous hosts. On the other hand, enlarging the channels reduces host-guest selectivity. This limitation may be overcome by introducing secondary microporosity into the walls.

The popular class of mesoporous silicates prepared by templating with cationic surfactants (M41S) possesses walls with typical thicknesses around 1 nm—too narrow to introduce structured microporosity. However, more recently introduced preparations involving triblock copolymers or non-ionic surfactants result in thicker-walled structures, which may be designed with bimodal pore structures.

One such material is SBA-15, a structure synthesized from tetraethoxysilane (TEOS) as the silica source under acidic conditions, using the commercially available ethylene oxide/propylene oxide triblock copolymer $EO_{20}PO_{70}EO_{20}$ (Pluronic P123, BASF) as the surfactant.[11,12] The product contains an array of hexagonally ordered mesopores (*p6mm* symmetry). Wall thicknesses (ca. 2-6 nm) and pore sizes (4.6-30 nm) can be adjusted, for example by the use of additives (*e.g.*, 1,3,5-trimethylbenzene) or by the choice of the synthesis temperature (35-80°C) and time. The morphology of SBA-15 is controllable by addition of cosolvents, cosurfactants, or electrolytes to produce fibers, hard spheres, doughnuts, rope-structures, gyroids and discoids.[13] The template can be removed by extraction with ethanol or by heating at 140°C.[12] It has now been demonstrated that the hexagonally arranged mesopores are interconnected by smaller pores which appear to be less ordered and to have a distribution of sizes ranging from micropores to approximately 3.4 nm.[14,15] It has been proposed that poly(ethylene oxide) chains of the triblock copolymer template penetrate the silica walls during the synthesis, producing the secondary porosity upon calcination.[15] In addition, low-molecular-weight impurities in the commercial triblock copolymer may have templated some disordered domains. The secondary porosity is maintained up to 900°C, but disappears at 1000°C, forming a channel structure resembling that of the hexagonal mesoporous sieve MCM-41. Interconnection of mesopores in SBA-15 by complementary micropores was further demonstrated by forming platinum[14] and carbon[16] inverse replicas of the pore structure. The carbon material, for example, consisted of nanorods that were interconnected by spacers, which separated the rods by a distance of 3 nm. Secondary microporosity has also been noted in mesoporous silicates templated by poly(butadiene-*b*-ethylene oxide) block-copolymers.[17]

Another type of bimodal mesoporous system has been obtained by addition of sodium salts to a synthesis mixture for mesoporous silica that was templated by non-ionic surfactants (the oleyl polyethylene oxide, Brij 97) under neutral conditions.[18] Samples of the resulting MSU-1 structure contained two pore size maxima (*e.g.*, 5.2/3.4 nm with NaCl and 6.2/3.8 nm with Na_2SO_4). Transmission electron micrographs (TEM) showed regions of hexagonally ordered and worm-like disordered pores, which were interpreted as a structure with two interconnected pore systems. Secondary microporosity has also been attained by delaminating the layered zeolite MCM-22 to produce a partially crystalline micro-/mesoporous material.[19]

3.   EMULSION TEMPLATING

Whereas surfactant templating techniques depend on cooperative interactions between the surfactant and the inorganic precursor, other techniques rely on phase separation between a liquid or solid template and a sol-gel or other suitable fluid precursor, which surrounds regions of the template. In one such technique, the emulsion templating process, a solid is grown around liquid colloidal droplets that may be stabilized by a surfactant.[20,21] If desired, narrow size dispersity of emulsion droplets can be obtained by shearing and/or fractionation, resulting in a typical droplet polydispersity of *ca.* 10%. Sol-gel precursors such as tetramethoxysilane (TMOS) are added to the emulsion. To induce ordering of droplets, the mixture is concentrated to at least 50% volume fraction of droplets. This process facilitates interconnection between pores after gelation and drying. During drying, the porous ceramic product undergoes up to 50% shrinkage. The emulsion templating technique permits the preparation of macroporous solids, including silicates and other oxides. The walls surrounding the macropores of emulsion-templated materials are thick enough that the controlled introduction of secondary pores would be feasible.

4. TEMPLATING USING SOLID MOLDS

A variety of small particles are suitable as molds for templating mesoporous and macroporous structures. These include porous carbon, colloidal particles, and even bacteria. If the particles themselves are porous, they may be filled with fluid sol-gel precursors, which are subsequently polymerized. Alternately, the particles can be aggregated (*e.g.*, as colloidal crystals) and interstitial spaces filled with fluid precursors, including sol-gel precursors and pre-formed nanocrystals. In most cases, removal of the template, either by calcination or dissolution, is necessary to obtain a porous product.

4.1 Porous Carbon Molds

Carbon Black Pearls with an average particle diameter of 12 nm were used as templates to prepare micrometer-sized ZSM-5 crystals containing mesopores throughout the structure.[22] The spheres were first impregnated to incipient wetness with a zeolite precursor gel containing all ingredients except the silicate source. A small excess of TEOS was then added. The material was aged at room temperature and hydrothermally treated. The mesopore templates (Carbon Black Pearls) and the micropore structure directing agent (tetrapropylammonium hydroxide, TPAOH) were removed by calcination at 550°C, producing mesoporous ZSM-5 crystals with a mesopore volume of 1 mL/g and a micropore volume of 0.09 mL/g.

Similarly, a meso-/microporous titanium silicate-1 (TS-1) catalyst was prepared from tetraethylorthotitanate and TEOS precursors by templating with Carbon Black Pearls and using TPAOH as a structure director.[23] This material exhibited improved activity in epoxidation of cyclohexene compared to conventional microporous TS-1, but did not change the product distribution. For the epoxidation of oct-1-ene, no significant difference in activity was observed for meso-/microporous TS-1 compared to conventional microporous TS-1, suggesting that this reaction was not limited by diffusion through the micropores, unlike the cyclohexene reaction.

4.2 Supercellular Templating with Bacteria

Structures containing both meso- and macropores or micro- and macropores have been prepared by templating with bacteria.[24,25] *Bacillus subtilis* bacteria consist of decimeter-long, multicellular threads (0.5 μm in diameter), which are coaligned to form a pseudohexagonal array. The dried bacteria were dipped into a silica sol or a synthesis mixture for MCM-41 (cetyltrimethylammonium bromide (CTAB), TEOS, NaOH, water).[24] The bacterial fibers were infiltrated by the silica precursors within a few minutes. Drying compacted the composite. After calcination at 600°C, ordered macroporous fibers with *ca.* 0.5 μm diameter channels were obtained. In the case of the MCM-41 mixture, the products contained mesoporous walls (50-200 nm thick) in addition to the macropores. It is interesting to note that, unlike silica sols with negative surface charges, positively charged colloids (including titania and alumina) were not easily infiltrated into the bacterial template.

To obtain macroporous zeolite, nanocrystalline silicalite particles were first synthesized by methods that restricted the particle sizes to *ca.* 50 nm.[25] TPAOH was employed as the structure director. Bacterial assemblies consisting of pseudohexagonally packed, multicellular filaments were infiltrated with a sol of the silicalite nanoparticles.[25] After drying and calcination at 600°C, hollow channels formed, which were surrounded by *ca.* 100 nm thick walls of coalesced zeolite nanoparticles.

4.3   Latex Sphere Templating

Uniform, spherical void spaces can be obtained with polymer latex spheres as templates. Such spheres can be readily synthesized with narrow size dispersities (*ca.* 3% for spheres with diameters of a few hundred nanometers, larger for spheres outside of this range—tens of nanometers to micrometers). In applications where macropores need not be arranged periodically, one can combine latex sphere suspensions with sol-gel mixtures, allow the inorganic phase to condense, and remove the template by calcination to create an open macroporous structure. In this case, secondary porosity is controlled by the sol-gel chemistry. Latex sphere templating methods are summarized in Figure 1.

For example, Antonietti *et al.* used spherical polymer latex particles in the size range from 20 - 200 nm as templates for porous silica.[29] The polymer surfaces were functionalized with carboxylate, sulfate, sulfonate or poly(ethylene oxide) to facilitate compatibility between the template and the silicate system. The porous silica was prepared by adding TMOS to a latex suspension at pH 2, followed by calcination at 450°C to produce a continuous foam-like structure. Bimodal pore size distributions were obtained by dual templating with latex fillers and block-copolymer surfactants. The bimodal product contained 100 nm macropores arising from the latex spheres and 6 nm mesopores (hexagonal regions as well as branches), attributed to the block copolymer templates. Under the acidic conditions used, cationic lattices demixed during templating. Better interaction between the template and the silica network was achieved with poly(ethylene glycol)-modified spheres or with spheres containing high surface

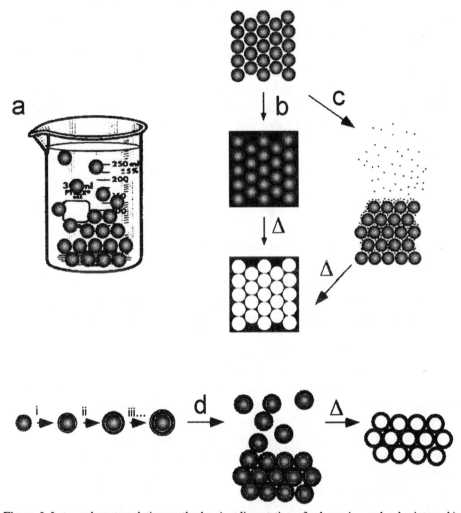

Figure 1. Latex sphere templating methods: a) sedimentation of spheres in a sol-gel mixture; b) infiltration of colloidal crystals with sol-gel precursors, followed by calcination for template removal and curing; c) infiltration of colloidal crystals with nanocrystals, followed by calcination; d) assembly of core-shell structures. These are prepared by (i) priming colloidal PS particles with a polyelectrolyte layer; (ii) depositing silica or zeolite nanoparticles; (iii) depositing alternating layers of polyelectrolyte and silica or zeolite until a desired wall thickness is achieved, followed by centrifugation and calcination. Figure 1d was adapted from references 26-28.

concentrations of sulfate or sulfonate groups. Sample porosities were reported to be between 71 and 83%. Wall thicknesses were controlled by varying the ratio of latex spheres to silica. With the latex suspension templating approach, macropores were not ordered nor fully interconnected, except at very high latex sphere content.

A variation of this method combines sedimentation and aggregation processes to prepare macroporous amorphous silica with a high density of macropores surrounded by 34-57 nm thick walls (Figure 1a).[30] A mixture of TEOS and polystyrene (PS) spheres dispersed in ethanol was stirred for 20 min at room temperature. Then stirring was stopped to permit settling of the spheres. It·was proposed that a layer of the precursor formed on the beads before the beads agglomerated as they settled. A relatively high TEOS/PS ratio was required in order to achieve sufficient coverage of the spheres to maintain a well-formed structure. The resulting structures were stable up to about 900-1000°C before they collapsed. Addition of cetyltrimethylammonium to the synthesis gel with a composition identical to that used for conventional MCM-48 samples[31] resulted in macrostructured silica with a cubic arrangement of mesopores in the walls.[32] Whereas relatively easy scale-up is possible by this method, the macropores are generally not periodically structured.

A versatile approach toward more ordered macroporous solids is based on colloidal crystal templates. This method has been used to synthesize three-dimensionally ordered macroporous (3DOM) metal oxides, metals, polymers, semiconductors and hybrid materials.[33] The general procedure involves the formation of colloidal crystals from uniformly-sized polymer or silica spheres (Figure 2, top left). The interstitial spaces in the colloidal crystal are then infiltrated with a fluid that can be converted into a solid framework. After removal of the template, a three-dimensional system of periodically arranged, usually interconnected macropores is obtained. Due to the flexibility of this method, it has been possible to introduce secondary porosity into the wall structures, particularly for silica-based systems.

The spheres may be packed by a variety of methods,[34] the simplest ones being sedimentation methods[35,36] or pressing a PS nanosphere powder into a pellet.[37] In the latter method, ordered domains are limited to a few micrometers, compared to hundreds of micrometers by sedimentation methods. Approaches to control the packing have recently been reviewed.[34] With single sphere sizes, the packing structure of spheres, and therefore of the voids in the final 3DOM product, is typically face-centered cubic.[38] Hexagonal packing of macropores in a 3DOM silica system has also been observed by TEM images and diffraction patterns.[39] More complex AB_2 and AB_3 superlattice structures can be obtained from binary sphere mixtures with sphere diameter ratios of 0.58 and 0.62, respectively.[40] These complex structures have been investigated as templates to produce macroporous silicates with dual macropore structure.[37]

Figure 2. Top left: Scanning electron micrograph (SEM) of a PS colloidal crystal. Top right: SEM of 3DOM silica showing a close-packed layer of macropores which are interconnected through smaller windows. The white and grey regions are the silica walls, the dark spots are the windows. Bottom right: Low magnification TEM image of 3DOM silica functionalized with polyoxometalate clusters. The dark regions are silica walls. Bottom left: High magnification image of the walls in the same material, with the clusters appearing as dark spots.

Several approaches have been taken to assemble relatively well-ordered macroporous structures with multiple porosity:

- Infiltrating colloidal crystals with sol-gel precursors which are subsequently transformed into micro- or mesoporous walls;
- Filling colloidal crystals with pre-assembled, porous nanocrystals;
- Forming spherical core-shell structures (polymer core, micro-/mesoporous nanocrystalline shell) which are subsequently packed.

These approaches will be detailed below.

4.3.1 Infiltration of Colloidal Crystals with Sol-Gel Precursors

Velev *et al.* described the first approach that employed colloidal crystals to template layers of silica with ordered arrays of macropores.[35] PS spheres containing positive (amidine) or negative (sulfate) surface charges were filtered from a suspension to form colloidal crystals on a filter membrane. The sphere surfaces were modified with CTAB surfactant to aid silica polymerization. Any excess surfactant was washed away. The voids were then filled with an aqueous silicate solution prepared by base hydrolysis of TMOS. Due to the strong adsorption of the CTAB surfactant on the spheres, the surfactant did not act as a secondary template for mesopores. In the absence of CTAB, the aqueous silicate solution did not appear to fill the voids in the colloidal crystal efficiently, and the structure collapsed after calcination. The porosity of the ordered macroporous silica samples was estimated to be 78 vol%. Any secondary porosity, if present, was not described.

Holland *et al.* used solvent-free or ethanolic TEOS or TMOS precursors to infiltrate PS colloidal crystals (Figure 1b).[38] A hardened silicate/PS composite formed overnight, even without the addition of any surfactant, acid or base catalyst. The PS template was removed by calcination or solvent extraction, producing macroporous silica (Figure 2, top right). In this reaction, moisture from the atmosphere provided water for hydrolysis, and surface carboxylate groups produced during emulsion polymerization of the spheres were thought to catalyse hydrolysis and condensation reactions. More uniform penetration of the colloidal crystal was attained with less viscous precursors. However, the viscosity and polymerization rate of the alkoxide could be adjusted by dilution of the alkoxide precursor. The walls of the macroporous silica were X-ray amorphous with a wide distribution of non-templated mesopores from 2-40 nm. Dilution of the silicate precursor yielded thinner walls (from ca. 90 nm with undiluted TEOS, to *ca.* 54 nm with a 55:45 TEOS:ethanol mixture by weight), and larger windows between voids. Very high dilution levels resulted in macropore structures that were not well-packed, presumably due to excessive swelling of the polymer spheres.

Macroporous silica with walls containing more uniform mesopores was prepared by combining colloidal crystal templating with surfactant templating.[38] An aqueous solution of cetyltrimethylammonium hydroxide, TEOS and TPAOH was stirred at 70°C for 1 hr, filtered, and passed through a PS colloidal crystal template. After removal of both surfactant and PS by calcination, a structure with a bimodal distribution was obtained: macropores with a diameter of a few hundred nanometers at the locations of the original PS spheres, and mesopores (< 4 nm) within the walls. The surface area of the macroporous product increased from *ca.* 200 m^2/g in the absence of a surfactant template to over 1300 m^2/g with the surfactant. The BJH pore volume doubled to 0.80 mL/g and the pore size distribution tightened significantly with a median mesopore diameter of 2.3 nm.

Other surfactants can also be employed in dual templating processes. For example, Luo *et al.* infiltrated a PS colloidal crystal with a silica precursor gel containing TEOS, solvent and the amphiphilic triblock copolymer $EO_{20}PO_{70}EO_{20}$ (Pluronic P123) under acidic conditions.[37] The relatively thin walls (up to *ca.* 80 nm) in the macroporous product contained mesopores with a tight pore size distribution around 7.7 nm (surface area: 915 m^2/g).

Yin *et al.* combined PS sphere templating (203 nm spheres) with surfactant copolymer templating ($EO_{106}PO_{70}EO_{20}$, BASF) to produce macropores (120 nm) and mesopores (4-5 nm, 8 nm spacing) within the walls after calcination.[39,41] TEM images revealed that the mesopores provided sets of connecting paths between adjacent macropores. As in the previous systems, additional interconnections were generated by the larger windows created from the close-packing of spheres.

Another dual templating method has been employed to synthesize macroporous zeolites.[42] In this case, the macropore structure was again controlled by templating with PS spheres, and the micropore structure was influenced by tetrapropylammonium hydroxide as a structure-directing agent. Two challenges were faced in this synthesis: (1) The growth of zeolite crystals within the colloidal crystal template had to be limited to avoid loss of a uniform macropore structure; this was addressed by employing conditions that favor formation of nanocrystals. (2) Typical zeolite syntheses involve hydrothermal processing temperatures that exceed the glass transition temperature of the polymer spheres. Phase separation or sphere melting was avoided by employing a solid-state transformation (a "dry conversion method")[43] that converted preformed macroporous amorphous silica to silicalite. TEOS and an aqueous solution of TPAOH were mixed with PS colloidal crystals and allowed to react overnight, forming a solid composite. This composite was heated in a Teflon-lined autoclave under autogenous pressure at 130 °C for 40 hr to transform the amorphous silica into silicalite. Based on nitrogen sorption data, the silicalite walls around the 250 nm diameter macropores were approximately 50% crystalline. The periodic order of the template was not fully maintained during the synthesis, however, mercury porosimetry measurements demonstrated that the macropores were interconnected. The walls comprised of silicalite were very thin (average: 113 nm), resulting in short diffusion paths for potential guest species. As a result, one might expect improved reaction efficiencies and reduced blocking of channels. This material combines the advantages of facile transport of guest species through the macropores with the selectivity of the zeolitic micropores.

4.3.2 Infiltration of Colloidal Crystals with Nanocrystals

The aforementioned difficulties concerning the incompatibility of a polymer sphere template with required sol-gel processing conditions can be overcome by employing preformed nanocrystals as precursors and infiltrating colloidal crystals with the nanocrystals (Figure 1c). This technique is applicable when nanocrystals can be readily synthesized and when they can be attached to each other by sintering or other bond formation processes. The diameter ratio of nanocrystals to polymer spheres must be sufficiently small to permit thorough penetration of the precursors into the colloidal crystal template. The technique is most suitable for thin layers of colloidal crystals. It has been applied to structures without secondary porosity in the walls (*e.g.*, titania),[44,45] as well as to the hierarchical structures presented below.

For example, Huang *et al.* prepared nanocrystalline silicalite (30-80 nm grains) using TPAOH as a structure directing agent.[46] After calcination at 550°C, the precursor nanocrystals were suspended in ethanol and then infiltrated into an ordered array of 300 nm PS spheres. The silicalite crystallites interacted via surface silanol groups, possibly aided by ethanol. Calcination resulted in further condensation of the macroporous zeolite skeleton, which exhibited a final macropore size of 160 nm. A similar approach was used by Wang *et al.* for ZSM-5 and silicalite nanocrystals, although they employed aqueous suspensions of the zeolites (*ca.* 4 wt% solids) and colloidal crystals consisting of relatively large spheres (2600 ± 100 nm).[47] In this system, little shrinkage was observed, and void diameters were similar to the sphere sizes. In addition to zeolite micropores and templated macropores, this sample contained intercrystalline voids between agglomerated zeolite grains.

4.3.3 Assembly of Core-Shell Structures

When 3DOM structures are prepared by colloidal crystal templating, their wall thicknesses may be controlled to some extent by adjusting precursor concentrations, as noted earlier. Walls can be thinned by etching procedures or enhanced by surface grafting methods. Thermal processing/sintering also controls grain sizes in the walls, in particular when the walls consist of crystalline grains.[48] In all these methods, the wall thicknesses can be varied only within a limited range. Thicker walls are obtained by employing dilute latex sphere suspensions (section 4.3, above), but the product order remains limited.

These limitations have been addressed by a method based on assembly of core-shell structures.[26] TPA-silicalite-1 nanoparticles (*ca.* 50 nm diameters) were assembled on the surface of monodisperse PS spheres (640 nm diameter) with polycationic poly(diallyldimethylammonium) chloride interlayers (Figure 1d).[27,28] By alternate deposition of the negatively charged zeolite and positively charged polyelectrolyte layers, the thickness of the resulting core-shell structure was controlled. Calcination resulted in fusion of the zeolite nanoparticles by condensation of surface silanols. Uncalcined core-shell particles were used as building blocks for colloidal crystallization via centrifugation and sintering. The resulting structures contained continuous networks of relatively ordered macropores that were *ca.* 20% smaller in diameter than the original spheres. A wall thickness of 200 nm was reported, *i.e.*, thicker than in the sedimentation-aggregation process presented earlier.[30] The wall thickness was approximately twice that of the controllable thickness of zeolite and polyelectrolyte layers around the polystyrene cores. Hierarchical porosity was provided by the microporosity of silicalite and by the random mesoporosity between the zeolite grains.

5. SOLID CONVERSION OF PREFORMED SHAPES

A goal in the synthesis of hierarchical porous solids is to achieve a level of intricacy widely found in natural systems. As an example, diatomite is the elaborately structured product of silicification of single-celled algae. The structure of diatomite contains macropores in the size range from submicrometer to tens of micrometers. Since it is a cheap raw material, it has been employed to synthesize macroporous zeolites.[49,50]

In one method,[49] silicalite nanocrystals (80 nm) were affixed to the surface of the diatoms as seeds. The seeded diatoms were then treated hydrothermally (175°C, 3 d) in a synthesis mixture for silicalite, based on tetrapropylammonium bromide (TPABr) as the structure directing agent. Most silica was derived from the diatoms, although approximately 10% silica originated from added TEOS. Zeolite crystallites filled some of the macropore space in the diatoms, producing interparticle mesoporosity in addition to the microporosity. The zeolitized diatomite sample had a relatively low zeolite content, *ca.* 5% (w/w).

In an alternate approach,[50] a mixture containing TPABr, $NaAlO_2$ and aqueous NaOH was kneaded with diatomite. Upon heating, the siliceous diatomite surface was converted to ZSM-5. Further calcination resulted in macroporous ZSM-5 crystals. An NH_4^+-ion exchanged sample was found to have high gas permeability and was proposed to be suitable for use as a methanol-conversion catalyst.

6.   THE NEXT LENGTH SCALE: SPATIAL PATTERNING

The porous structures described so far may be suitable as supports, catalysts, *etc.*, for bulk processes. It is, however, possible to combine many of the above synthetic methods with stamping techniques that employ elastomeric stamps to achieve surface patterns of the porous solids with micrometer-size or larger features. The resulting structures may be suitable for device applications, such as patterned microreactors on a chip. Typically, polydimethylsiloxane (PDMS) stamps or micromolds are employed for the microcontact printing or micromolding processes.[51] These are readily prepared with the desired patterns.

Mesoscopic or mesoporous silica has been patterned either on self-assembled monolayers (SAMs)[52] or within the confinements of capillaries provided by a PDMS stamp.[53] In the first approach,[52] a hexadecanethiol pattern was prepared on a gold surface. When this substrate was immersed in an appropriate surfactant-silicate gel mixture (TEOS, CTACl, acidic aqueous conditions), disk- and ribbon-shaped particles of a mesoporous silicate formed interconnected layers predominantly on the SAM patterned lines on the gold surface. The silicate was attached to the patterned surface via surfactant-alkanethiol interfaces.

The second approach (growth of mesoscopic and mesoporous silica within microcapillaries) was carried out with a similar precursor gel (TEOS, CTACl, HCl, H_2O).[53] In this case, an electric field was applied which served multiple functions. It induced electro-osmotic fluid flow within the capillaries, thereby increasing the filling of channels with the reaction precursors. It also led to alignment of the surfactant micelles, which formed hexagonally packed arrays of tubes that were aligned parallel to the substrate and along the capillary axis. In addition, the electric field caused Joule heating, thereby increasing the silica polymerization rate. The hexagonal phase was maintained upon calcination in spite of structural shrinkage.

Yang *et al.* combined PDMS micromolding, polymer sphere templating, and cooperative assembly of amphiphilic triblock copolymers with sol-gel precursors to create patterned oxides (silica, niobia, titania) with multiple pore size regimes.[54] Patterned solids were obtained by compressing sol-gel mixtures with PDMS stamps, and allowing the materials to harden. Alternately, a PDMS stamp fabricated with microchannels (typically tens to hundreds of micrometers) was placed on a substrate, and the channels were filled with latex spheres. The remaining void spaces were then filled with a sol-gel precursor, which was allowed to polymerize. If additional mesoporosity was desired, block copolymer templates could also be included. For example, with $EO_{106}PO_{70}EO_{106}$ (Pluronic F127) a cubic mesophase was formed, and with $EO_{20}PO_{70}EO_{20}$ (Pluronic P123) a hexagonal mesophase. Removal of the stamps, followed by calcination, produced porous materials with multiple length scales on the order of 10, 100 and 1000 nm.

Huang *et al.* applied a similar method to fabricate micropatterned silicalite films.[46] An ethanolic sol droplet of 30-80 nm silicalite nanocrystals was placed on a silicon wafer, followed by application of a PDMS stamp and solvent removal. A micropatterned silicalite film resulted after calcination at 550°C. Ethanolic sols were preferred over aqueous sols, since ethanol wetted both the substrate and the nanoparticles well. Patterns with feature sizes down to 200 nm and areas of nearly a square centimeter were achieved.

7. ADDING FUNCTIONALITY

While pure silicate supports provide suitable physical features, such as high surface area, tunable porosity and thermal and mechanical stability, their surface reactivity is limited to the silanol groups that are usually present. However, the surface chemistry may be readily modified by attaching appropriate functional groups, either during the synthesis of the porous silicate or subsequently by surface grafting or impregnation. Hybrid inorganic-organic mesoporous silicates have recently been reviewed.[10,55,56] Similar functionalization techniques can be employed with 3DOM silicates and hierarchically structured supports.

3DOM silicates with organic functional groups, including vinyl ($—HC=CH_2$), cyanoethyl ($—C_2H_4CN$) and mercaptopropyl ($—C_3H_6SH$), have been prepared by "one-pot" syntheses.[38] Typically, the trialkoxyorganosilane precursors containing these functional groups were diluted with TEOS or TMOS to increase the degree of linking in the resulting network, and dissolved in alcohol to improve penetration of the colloidal crystal template. To preserve the organic functional groups, the PS template was selectively removed by extraction with a 1:1 (v/v) mixture of THF and acetone. Organic functional groups in 3DOM materials may serve a variety of applications. The thiol end of the mercaptopropyl group has an affinity for heavy metals like mercury, which could make these materials useful in environmental cleanup.[57,58] Functional groups may also serve as anchors for large guest molecules, such as enzymes, proteins and other biological molecules. Guests that are too large to be anchored within zeolites or mesoporous sieves would have ample room on a 3DOM support. Furthermore, the large pores may allow more efficient guest transport through the structure.

Whereas in the above experiments the secondary porosity of the walls was not specifically controlled, dual-templating techniques have allowed the preparation of hierarchical meso-/macroporous structures with organic functionalities directly incorporated during the synthesis.[59] Mesopore templating involved CTAB micelles and co-condensation of a mixture of TEOS with the dye-functionalized precursor 3-(2,4-dinitrophenylaminopropyl)triethoxysilane under acidic conditions. The precursor gel was aged for four days before infiltration into a PS colloidal crystal. Solvent was allowed to evaporate and the material was aged for two more days to increase the degree of condensation. The PS latex was extracted with toluene, and CTAB with HCl/EtOH. The yellow chromophore remained linked to the mesostructural silica network.

A recent development in the field of hybrid materials has been the synthesis of mesoporous silicates with wall structures consisting of covalently bonded silica-organic networks.[60-62] Surfactant-templated syntheses of these materials use a precursor that has two trialkoxysilyl groups connected by an organic bridge, such as 1,2-bis(trimethoxysilyl)ethane or -ethene. This technique permits stoichiometric incorporation of organic groups in silicate networks. The introduction of suitable functional groups in the walls may allow further tunability of the mechanical, surface chemical or physical properties of the hybrid composite. This technique can also be extended to colloidal crystal templating or dual templating approaches. Ordered 3DOM structures are obtained by similar techniques used for pure silica systems, although template removal by solvent extraction is not always complete.[63]

Inorganic clusters such as polyoxometalates (POMs) have been integrated in 3DOM silica structures via organic linkages, both by direct-synthesis and grafting methods. For example, lacunary γ-decatungstosilicate clusters (γ-$SiW_{10}O_{36}$) were incorporated into the wall structures of 3DOM silica by a direct synthesis approach to introduce the redox active properties of the POM to the 3DOM material (Figure 2, bottom).[64] The clusters were reacted with TEOS, with or without addition of 1,2-bis(triethoxysilyl)ethane as a linking group, and the mixture was molded by a polymer colloidal crystal. The linking groups connected the clusters covalently to the silica support. When the polymer spheres were removed by extraction, the intact POM clusters remained attached to the hybrid 3DOM structures and were highly dispersed throughout the walls (Figure 2, bottom left). The materials were demonstrated to exhibit catalytic activity for the epoxidation of cyclooctene with an anhydrous H_2O_2/t-BuOH solution at room temperature.

Johnson *et al.* studied the influence of pore structure on the activity of various silica samples that were modified with transition metal-substituted polyoxometalates (TMSP).[65] Clusters of the type $[Co^{II}(H_2O)PW_{11}O_{39}]^{5-}$ and $[SiW_9O_{37}\{Co^{II}(H_2O)\}_3]^{10-}$ were chemically anchored to functionalized macroporous (350-450 nm pores), mesoporous (30-60 Å pores), and non-porous silica surfaces modified by prior treatment of the hydroxylated surfaces with $(EtO)_3Si(CH_2)_3NH_2$. While the mesoporous silica support had the highest surface area, partially restricted access of the clusters (*ca.* 1 nm in diameter) to the mesopore channels (*ca.* 2.8 nm in diameter) limited the cluster loading in these samples. Attachment of a TMSP cluster to a propyl amine ligand (*ca.* 0.8 nm) extending into the pore channel would decrease the pore opening to less than half of its original diameter, thus effectively blocking other clusters from entering that channel. The more open macroporous structure was able to support more TMSP clusters per amine anchoring group. Because the clusters were attached datively to the surface, they were retained in catalytic reactions involving the epoxidation of cyclohexene to cyclohexene oxide, with comparable values of reaction rate and conversion on all three supports.

Additional functionality in 3DOM structures arises from their optical properties.[66] These materials have a stunning appearance over many length scales. They often appear opalescent and colored due to Bragg diffraction of light by the periodic structure with feature sizes comparable to the wavelength of visible light (*i.e.*, a sub-micrometer diffraction grating). For example, macroporous silica with an average pore spacing of 250 nm appears bright blue in reflected light and yellow when illuminated from the rear. Striking color changes in the visible spectrum are produced when the voids of 3DOM solids are filled with solvents. The color of macroporous silica in the above example changes to a deep green in reflected light and a pale violet in transmitted light when the pores are filled with methanol. The wavelength of these colors is linearly related to the refractive index of the fluid occupying the macropores and can be tuned by modifying the refractive index of the wall material (through compositional or structural changes), as well as the size and shaping of the pores.

8. CONCLUSION AND FUTURE OUTLOOK

This chapter has demonstrated the power of templating methods to design intricate porous structures with feature sizes ranging from the molecular level to visible length scales. The focus has been on silica-based materials, because at this stage, the flexibility in structure is greatest with silicate systems. Other compositions can be synthesized by appropriate techniques at each of the above size levels, though—especially for smaller features—syntheses are typically more difficult and specialized to relatively few compositions.

One of the major motivations in developing hierarchical pore structures has been the desire to facilitate transport (penetration, flow, diffusion) of ever larger guest molecules in porous hosts, while maintaining the advantages of small-pore structures: selectivity for interactions, capability of separation, and high surface areas. Materials with multiple pore size distributions combine the benefits of each pore size regime. In addition, they can exhibit features that reach beyond these goals, including the interesting optical effects mentioned above, and potential changes in product distributions of catalytic reactions (compared to bulk reactions)[67] achievable by the small sizes of nanocrystalline zeolite grains formed within a colloidal crystal matrix. Thus, the most promising developments may arise from combining multiple properties of these complex structures. For example, in mesoporous 3DOM structures, the macropores are large enough to stabilize large guests (such as biological molecules), while the smaller, functionalized pores may be used to slowly deliver a reagent that can interact with the guest species. As another example, the large voids might be filled with conductive polymers that transmit signals arising from interactions with the support material and guest molecules. In this design one would combine sensing and signal transmission function within a single nanostructured material. A further example of multifunctional nanostructures would be the combination of host-guest selectivity within the walls of a macroporous material with the potential photonic crystal properties of the periodic macroporous solid. By taking advantage of the interactions between guest molecules and the different host structures, one can chemically design new materials with optimized properties for sensing, filtration, thermal insulation, selective catalysis, or for new electronic (low-dielectric-constant), optical, or magnetic materials. While the presently achievable complexity still lags behind many natural structures, great strides have been made in past years which permit us to become more creative in the design of artificial porous solids.

ACKNOWLEDGMENTS

Portions of some of the work described here were funded by 3M, Dupont, the David & Lucile Packard Foundation, the McKnight Foundation, the NSF (DMR-9701507) and the MRSEC Program of the NSF under Award Number DMR-9809364. The authors thank Dr. Christopher F. Blanford and Mr. Sergey Sokolov for providing electron microscopy images.

REFERENCES

1.   El-Nafaty, U. A.; Mann, R. *Chem. Eng. Sci.* **1999**, *54*, 3475-3484.
2.   Bein, T. *Chem. Mater.* **1996**, *8*, 1636-1653.
3.   Komarneni, S.; Katsuki, H.; Furuta, S. *J. Mater. Chem.* **1998**, *8*, 2327-2329.
4.   Davis, M. E. *Chem. Eur. J.* **1997**, *3*, 1745-1750.
5.   Davis, M. E.; Lobo, R. F. *Chem. Mater.* **1992**, *4*, 756-768.
6.   Sing, K. S. W.; Everett, D. H.; Haul, R. A. W.; Moscou, L.; Pierotti, R. A.; Rouquérol, J.; Siemieniewska, T. *Pure Appl. Chem.* **1985**, *57*, 603-619.
7.   Stein, A. in *The Role of Surfactants and Amphiphiles in the Synthesis of Porous Inorganic Solids*; Texter, J., Ed.; Marcel Dekker: New York, 2001, pp 819-851.
8.   Clark, J. H.; Macquarrie, D. J. *Chem. Commun.* **1998**, 853-860.
9.   Macquarrie, D. J.; Jackson, D. B. *Chem. Commun.* **1997**, 1781-1782.
10.   Stein, A.; Melde, B. J.; Schroden, R. C. *Adv. Mater.* **2000**, *12*, 1403-1419.
11.   Zhao, D.; Huo, Q.; Feng, J.; Chmelka, B. F.; Stucky, G. D. *J. Am. Chem. Soc.* **1998**, *120*, 6024-6036.

12. Zhao, D.; Feng, J.; Huo, Q.; Melosh, N.; Fredrickson, G. H.; Chmelka, B. F.; Stucky, G. D. *Science* **1998**, *279*, 548-552.
13. Zhao, D.; Sun, J.; Li, Q.; Stucky, G. D. *Chem. Mater.* **2000**, *12*, 275-279.
14. Ryoo, R.; Ko, C. H.; Kruk, M.; Antochshuk, V.; Jaroniec, M. *J. Phys. Chem. B* **2000**, *104*, 11465.
15. Kruk, M.; Jaroniec, M.; Ko, C. H.; Ryoo, R. *Chem. Mater.* **2000**, *12*, 1961-1968.
16. Jun, S.; Joo, S. H.; Ryoo, R.; Kruk, M.; Jaroniec, M.; Liu, Z.; Ohsuna, T.; Terasaki, O. *J. Am. Chem. Soc.* **2000**, *122*, 10712-10713.
17. Göltner, C. G.; Berton, B.; Krämer, E.; Antonietti, M. *Adv. Mater.* **1999**, *11*, 395-398.
18. Bagshaw, S. A. *Chem. Commun.* **1999**, 1785-1786.
19. Corma, A.; Fornes, V.; Pergher, S. B.; Maesen, T. L. M.; Buglass, J. G. *Nature* **1998**, *396*, 353-356.
20. Imhof, A.; Pine, D. J. *Nature* **1997**, *389*, 948-951.
21. Imhof, A.; Pine, D. J. *Adv. Mater.* **1998**, *10*, 697-700.
22. Jacobsen, C. J. H.; Madsen, C.; Houzvicka, J.; Schmidt, I.; Carlsson, A. *J. Am. Chem. Soc.* **2000**, *122*, 7116-7117.
23. Schmidt, I.; Krogh, A.; Wienberg, K.; Carlsson, A.; Brorson, M.; Jacobsen, C. J. H. *Chem. Commun.* **2000**, 2157-2158.
24. Davis, S. A.; Burkett, S. L.; Mendelson, N. H.; Mann, S. *Nature* **1997**, *385*, 420-423.
25. Zhang, B.; Davis, S. A.; Mendelson, N. H.; Mann, S. *Chem. Commun.* **2000**, 781-782.
26. Rhodes, K. H.; Davis, S. A.; Caruso, F.; Zhang, B.; Mann, S. *Chem. Mater.* **2000**, *12*, 2832-2834.
27. Caruso, F.; Caruso, R. A.; Möhwald, H. *Science* **1998**, *282*, 1111-1114.
28. Caruso, F.; Caruso, R. A.; Möhwald, H. *Chem. Mater.* **1999**, *11*, 3309-3314.
29. Antonietti, M.; Berton, B.; Göltner, C.; Hentze, H. P. *Adv. Mater.* **1998**, *10*, 154-159.
30. Vaudreuil, S.; Bousmina, M.; Kaliaguine, S.; Bonneviot, L. *Micropor. Mesopor. Mater.* **2001**, *44*, 249.
31. Vartuli, J. C.; Schmitt, K. D.; Kresge, C. T.; Roth, W. J.; Leonowicz, M. E.; McCullen, S. B.; Hellring, S. D.; Beck, J. S.; Schlenker, J. L.; Olson, D. H.; Sheppard, E. W. *Chem. Mater.* **1994**, *6*, 2317-2326.
32. Danumah, C.; Vaudreuil, S.; Bonneviot, L.; Bousmina, M.; Giasson, S.; Kaliaguine, S. *Micropor. Mesopor. Mater.* **2001**, *44*, 241.
33. Stein, A. *Micropor. Mesopor. Mater.* **2001**, *44*, 227.
34. Xia, Y.; Gates, B.; Yin, Y.; Lu, Y. *Adv. Mater.* **2000**, *12*, 693-713.
35. Velev, O. D.; Jede, T. A.; Lobo, R. F.; Lenhoff, A. M. *Nature* **1997**, *389*, 447-448.
36. Holland, B. T.; Blanford, C. F.; Stein, A. *Science* **1998**, *281*, 538-540.
37. Luo, Q.; Li, L.; Yang, B.; Zhao, D. *Chem. Lett.* **2000**, 378-379.
38. Holland, B. T.; Blanford, C. F.; Do, T.; Stein, A. *Chem. Mater.* **1999**, *11*, 795-805.
39. Yin, J. S.; Wang, Z. L. *Appl. Phys. Lett.* **1999**, *74*, 2629-2631.
40. Pusey, P. N.; Poon, W. C. K.; Ilett, S. M.; Bartlett, P. *J. Phys.: Condens. Matter* **1994**, *6*, A29-A36.
41. Yin, Y. S.; Wang, Z. L. *Microsc. Microanal.* **1999**, *5 (Suppl. 2: Proceedings)*, 818-819.
42. Holland, B. T.; Abrams, L.; Stein, A. *J. Am. Chem. Soc.* **1999**, *121*, 4308-4309.
43. Shimizu, S.; Kiyozumi, Y.; Mizukami, F. *Chem. Lett.* **1996**, 403-404.
44. Subramania, G.; Constant, K.; Biswas, R.; Sigalas, M. M.; Ho, K. M. *Appl. Phys. Lett.* **1999**, *74*, 3933-3935.
45. Subramania, G.; Manoharan, V. N.; Thorne, J. D.; Pine, D. J. *Adv. Mater.* **1999**, *11*, 1261.
46. Huang, L.; Wang, Z.; Sun, J.; Miao, L.; Li, Q.; Yan, Y.; Zhao, D. *J. Am. Chem. Soc.* **2000**, *122*, 3530-3531.
47. Wang, Y. J.; Tang, Y.; Ni, Z.; Hua, W. M.; Yang, W. L.; Wang, X. D.; Tao, W. C.; Gao, Z. *Chem. Lett.* **2000**, 510-511.

48.   Blanford, C. F.; Yan, H.; Schroden, R. C.; Al-Daous, M.; Stein, A. *Adv. Mater.* **2001**, *13*, 26.

49.   Anderson, M. W.; Holmes, S. M.; Hanif, N.; Cundy, C. S. *Angew. Chem. Int. Ed.* **2000**, *39*, 2707-2710.

50.   Inui, S.; Kimura, S.; Nonaka, S. *Japanese Patent JKXXAF JP 04187515 A2 19920706*; Chuo Silica K. K.: Japan, 1992.

51.   Zhao, X. M.; Xia, Y. N.; Whitesides, G. M. *J. Mater. Chem.* **1997**, *7*, 1069-1074.

52.   Yang, H.; Coombs, N.; Ozin, G. A. *Adv. Mater.* **1997**, *9*, 811-814.

53.   Trau, M.; Yao, N.; Kim, E.; Xia, Y.; Whitesides, G. M.; Aksay, I. *Nature* **1997**, *390*, 674-676.

54.   Yang, P.; Deng, T.; Zhao, D.; Feng, P.; Pine, D.; Chmelka, B. F.; Whitesides, G. M.; Stucky, G. D. *Science* **1998**, *282*, 2244-2246.

55.   Moller, K.; Bein, T. *Chem. Mater.* **1998**, *10*, 2950-2963.

56.   Ozin, G. A.; Chomski, E.; Khushalani, D.; MacLachlan, M. J. *Curr. Opin. Coll. Surf. Sci.* **1998**, *3*, 181-193.

57.   Feng, X.; Fryxell, G. E.; Wang, L. Q.; Kim, A. Y.; Liu, J.; Kemner, K. M. *Science* **1997**, *276*, 923-926.

58.   Mercier, L.; Pinnavaia, T. J. *Adv. Mater.* **1997**, *9*, 500-503.

59.   Lebeau, B.; Fowler, C. E.; Mann, S.; Farcet, C.; Charleux, B.; Sanchez, C. *J. Mater. Chem.* **2000**, *10*, 2105-2108.

60.   Melde, B. J.; Holland, B. T.; Blanford, C. F.; Stein, A. *Chem. Mater.* **1999**, *11*, 3302-3308.

61.   Inagaki, S.; Guan, S.; Fukushima, Y.; Ohsuma, T.; Terasaki, O. *J. Am. Chem. Soc.* **1999**, *121*, 9611-9614.

62.   Asefa, T.; MacLachlan, M. J.; Coombs, N.; Ozin, G. A. *Nature* **1999**, *402*, 867-871.

63.   Melde, B. J.; Skugrud, K.; Stein, A. unpublished work.

64.   Schroden, R. C.; Blanford, C. F.; Melde, B. J.; Johnson, B. J. S.; Stein, A. *Chem. Mater.* **2001**, *13*, 4314.

65.   Johnson, B. J. S.; Stein, A. *Inorg. Chem.* **2001**, *40*, 801.

66.   Blanford, C. F.; Schroden, R. C.; Al-Daous, M.; Stein, A. *Adv. Mater.* **2001**, *13*, 26.

67.   van der Pol, A. J. H. P.; Verduyn, A. J.; van Hooff, J. H. C. *Appl. Catal.* **1992**, *92*, 113.

Chapter 12

STRATEGIES FOR THE CONTROL OF POROSITY AROUND ORGANIC ACTIVE SITES IN INORGANIC MATRICES

Christopher W. Jones
School of Chemical Engineering, Georgia Institute of Technology, Atlanta, GA 30332, USA

Keywords: catalysis, silica, organic functionalization, inorganic/organic hybrid, OFMS, shape-selective, molecular imprinting

Abstract: The functionalization of inert supports with catalytic sites has been utilized for the preparation of solid catalysts for decades. Recently, there has been a surge in the use of silane chemistry to functionalize the surfaces of porous silica materials to create high surface area, catalytically active solids. Silica-organic hybrid materials have been prepared with acidic, basic, oxidizing and reducing catalytic sites, with most attention directed at creating new or unique molecular sites on the silica surface. Whereas the molecular structure of the active site can be of paramount importance in homogeneous catalysts, in heterogeneous catalysts both the structure of the active site and the porosity around the active site are of extreme importance, since reactant and product diffusion can strongly influence catalytic function. Here, strategies for the control of porosity around organic active sites in silica hosts are discussed.

1. INTRODUCTION

The molecular functionalization of silica solids with organic species has proven to be a promising technique for the creation of catalytic materials.[1] In particular, the decoration of silica surfaces via reaction with organosilanes is a

Nanostructured Catalysts, edited by S. Scott *et al.*
Kluwer Academic/Plenum Publishers, 2003

versatile method for incorporating moieties with acidic, basic or oxidation/reduction catalytic functions. Indeed, chemists have developed an array of techniques for the synthesis of discrete molecular active sites that are covalently bound to silica and other solid surfaces.

The molecular architecture of the active site is, however, only part of the battle in the rational design of "molecular" solid catalysts. Control of the porosity around the active center gives the chemist another variable that can be manipulated to influence catalytic properties. For example, zeolites are well-known as a class of catalysts where the size, shape and connectivity of the pores can have a dramatic influence on catalytic properties.[2] These microporous, crystalline aluminosilicates have pores that are roughly the size of small molecules (3-8 Å in diameter) and hence the size and shape of the pores can profoundly influence the rate of reactant and product transport through the catalyst as well as dictate the manner in which a reactant approaches and binds to an active site.[3]

Here we discuss techniques for the synthesis of porous silica materials with controlled porosity and internal organic catalytic sites. Two distinct classes of materials are discussed: highly ordered, crystalline silica materials with catalytic organic groups randomly distributed throughout the internal pores of the solid (organic-functionalized molecular sieves: OFMS's),[4] and amorphous, disordered silica materials with well-defined, multifunctional organic moieties distributed throughout their porosity (molecularly-imprinted silicas).[5]

2.   ORGANIC-FUNCTIONALIZED MOLECULAR SIEVES (OFMS's)

2.1   Preparation

OFMS's are crystalline, microporous silicates akin to zeolites. Whereas traditional zeolites substitute metal atoms for silicon atoms to generate catalytic sites, OFMS's contain intracrystalline organic species tethered to silicon atoms that can behave as catalytic centers. A schematic representation of acid sites in a traditional zeolite and in a sulfonic acid-functionalized OFMS is shown in Figure 1.

Crystalline silica molecular sieves are routinely synthesized hydrothermally using organic structure-directing agents (SDA's) to kinetically steer the syntheses to specific products.[6] The resulting silica solids are essentially non-porous, with the organic SDA's occluded within the micropores of the material. By combustion of the SDA at high temperatures in air (>500°C), the organic species are removed and significant microporosity is generated.

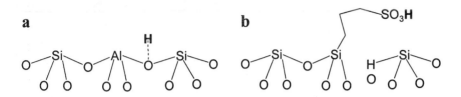

Figure 1. Schematic illustration of active sites in an acidic zeolite (**a**) and a sulfonic acid-functionalized OFMS (**b**). Acidic proton shown in bold.

To prepare crystalline silica molecular sieves with covalently tethered organic active sites as described in Figure 1, new synthetic methodologies are required, since high temperature combustion steps would result in the loss of both the SDA and the intended organic active site functionality. To this end, the applicability of solvent extraction was evaluated for the removal of the occluded SDA's from the micropores of the solid.

The ability to remove organic SDA's from crystalline high-silica molecular sieves depends on both the relative size of the SDA compared to the size of the micropores and the strength of the SDA-framework interaction.[7] Efficient extraction of the SDA without disruption of the crystalline framework occurs most readily in systems with relatively small SDAs that are weakly complexed with the silica framework.

Based on the above requirement and the additional prerequisite that the pore size must be large enough to contain both the SDA's and the intended tethered organic functionality, pure silica beta (*BEA topology) molecular sieves synthesized with tetraethylammonium fluoride SDAs were targeted as hosts for the first OFMS materials.[4] OFMS's with the *BEA topology have been prepared with several organic functional groups tethered within the micropores of the solid. In particular, OFMS's with tethered intracrystalline aromatics,[4,8] alkylthiols,[9] and alkylamines[10] have been directly synthesized using variations of the scheme described in Figure 2.

After extraction of the SDA, the tethered organic species are accessible to adsorbed species, allowing further modification of the organics or utilization of the species as catalytic sites. For example, the tethered aromatic groups (denoted PE for phenethyl) can be effectively sulfonated by SO_3 vapor to generate phenylsulfonic acid moieties.[4,8] Similarly, the mercaptopropyl groups can be oxidized to sulfonic acids using aqueous hydrogen peroxide[9] and the aminopropyl groups can be transformed into Schiff bases via reaction with aldehydes.[10]

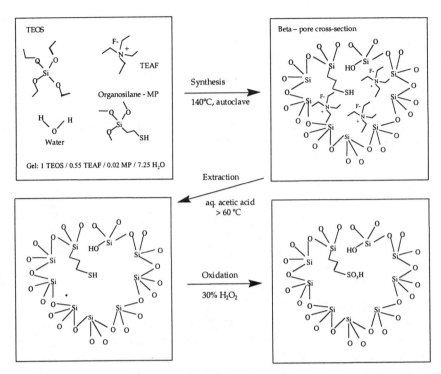

Figure 2. Schematic illustration of synthetic procedure for the preparation of OFMS's of the *BEA topology. Preparation of a propylsulfonic acid OFMS is illustrated.

2.2   Shape Selectivity in Catalysis

Crystalline molecular sieves such as zeolites are well-known for their shape-selective properties. Because these crystalline materials have a well-defined pore structure with pore openings roughly the size of small molecules (3-8 Å), they are able to discriminate between potential adsorbates/reactants of different sizes and shapes on the basis of molecular size or structure. With respect to catalysis, this shape-selectivity is commonly broken into three categories: reactant, product and transition state selectivity.[2,3,11]

A molecular sieve that illustrates reactant shape selectivity discriminates between multiple potential reactive molecules. For example, the zeolite CaA selectively dehydrates *n*-butanol in the presence of *sec*-butanol or isobutanol by largely excluding the bulky branched isomers from the zeolite pores that contain the vast majority of the acidic active sites.[12] An example of product shape selectivity in microporous molecular sieve catalysis is the selective formation of *p*-xylene during the alkylation of toluene over ZSM-5. In this case, all three isomers of xylene are formed within the pores the zeolite, but the bulky meta and ortho isomers are subsequently transformed to the faster diffusing para isomer that quickly escapes the zeolite pore system. In transition state selectivity, it is postulated that the transition state required for the formation of expected products that are not observed in a catalytic reaction cannot be accommodated within the pores of the molecular sieve, whereas the transition state for those that do form obviously can be accommodated. An example of a reaction where transition state shape selectivity is invoked to explain the observed catalytic behavior is in the transalkylation of biphenyl with polymethylbenzenes, where the 4 and 4,4' isomers are selectively produced.[13]

The shape-selective properties of OFMS's were evaluated using shape-selective adsorption or catalytic tests. The sulfonic acid-functionalized OFMS's were evaluated in the reactions of ethylene glycol with aldehydes and ketones to form cyclic acetals or ketals. Specifically, the OFMS's were tested in the conversion of 1-pyrenecarboxaldehyde (PYC) and cyclohexanone (HEX), as described below (Scheme 1).[4,9]

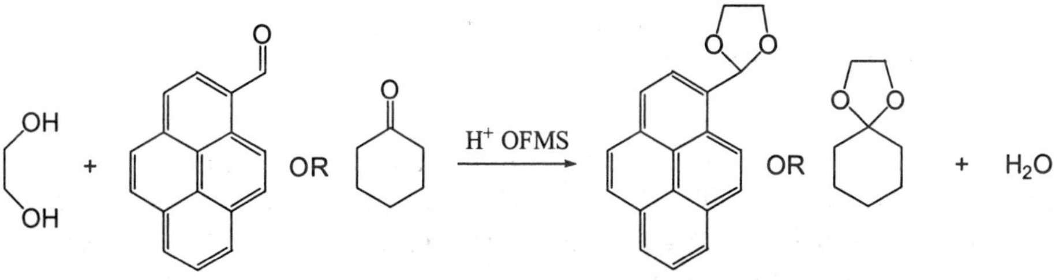

Scheme 1. Conversion of 1-pyrenecarboxaldehyde (PYC) and cyclohexanone (HEX).

This reaction is designed to be reactant shape selective, since 1-pyrenecarboxaldehyde is too large to enter the pores of molecular sieves with the *BEA topology, whereas cyclohexanone can easily diffuse into the structure. Table 1 contains data from several catalytic tests. These indicate that the propylsulfonic acid-functionalized OFMS is an effective shape-selective catalyst, effectively discriminating between the two reactants. This indicates that the vast majority of the organic functional groups are contained within the micropores of the OFMS and are not on the external surface. In contrast, the phenethylsulfonic acid-functionalized solid is not shape-selective unless a large base such as 2,4,6-tri-*t*-butylpyridine is added to the reaction. This bulky base cannot enter the OFMS pores and therefore only interacts with active sites on the external surface of the solid. However, even in the presence of this poison, the phenethylsulfonic acid OFMS is still a more active and less selective catalyst than the propylsulfonic acid OFMS.

The higher reactivity and lower selectivity of the phenylsulfonic acid OFMS warranted additional investigation.[9] The data in Table 1 indicate that all-silica beta molecular sieves that do not contain any organic functional groups are

Table 1. Initial rates and site-time yields for HEX and PYC conversion over acid catalysts. Adapted with permission from *Micropor. Mesopor. Mater.* **2001**, *42*, 21. Copyright 2001 Elsevier Science.

Catalyst	R_{HEX}[a]	R_{PYC}[a]	R_{HEX}/R_{PYC}	STY(min^{-1})[b]
PrSO₃H	0.26	0.0012	217	0.82
PrSO₃H[c]	0.18	0.00082	214	0.55
PrSO₃H	3.75	0.96	4	12
PrSO₃H[c]	3.6	0.19	19	11
Si-Beta-PrSO₃H	1.86	0.48	4	----
H-Al-Beta	15.8	0.3	53	12
H-Al-Beta[c]	15.1	0.26	58	12
TsOH	----	----	4	68

Reaction conditions: 60 °C, 0.55 mmol reactants, 16 mg OFMS catalyst, 1.5 g *o*-dichlorobenzene as solvent.
[a] mmol/g cat-min.
[b] for HEX conversion, assuming all sites are active and accessible.
[c] 12 mg 2,4,6-tri-*t*-butylpyridine added.
[d] from PQ corporation, Si/Al = 12.

also active and unselective catalysts after they are subjected to the SO₃ sulfonation procedure. The high activity and low selectivity observed in phenthylsulfonic acid OFMS and the sulfonated pure-silica material are therefore attributable to the formation of sulfonic acid sites not associated with organic moieties on the surface of the silica framework during the sulfonation step.[14] These highly accessible acid sites dominate the behavior of the phenylsulfonic acid OFMS, making it a non-selective catalyst.

The shape-selective properties of the propylamine-functionalized OFMS were probed using stoichiometric reactions. Samples of propylamine OFMS were reacted with excess 4-(dimethylamino)benzaldehyde or 4-dimethylamino-1-naphthaldehyde over 3Å sieves in methanol, Scheme 2.[10] Following the reaction,

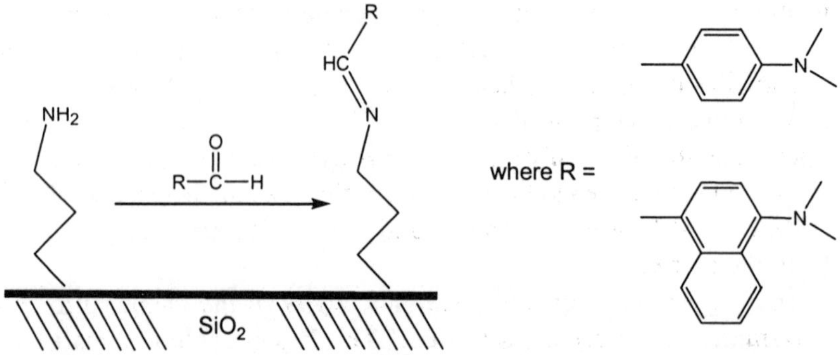

Scheme 2. Reactions of aldehydes with propylamine-functionalized silica.

the OFMS's were recovered, washed with methanol and dried under vacuum. Control materials, propylamine-functionalized amorphous silicates with a broad pore size distribution, were also subjected to reaction with aldehydes using the above technique. The resulting solids were analyzed by multiple spectroscopic and gravimetric methods.[10] Here, only the FT-Raman spectroscopic results will be presented.

The FT-Raman spectra of the benzaldehyde- and naphthaldehyde-treated amorphous silica control materials are shown in Figure 3. Over this non-shape-selective control material, it is evident that both the large naphthaldehyde and small benzaldehyde can easily react with the propylamine functional groups to generate Schiff bases. The amine N-H stretches in the Raman spectra (3310 cm^{-1}) are consumed and new C=N (1640 cm^{-1}) stretches appear. In contrast, when the same reactions are carried out over the OFMS's, only the smaller benzaldehyde reacts appreciably (see Figure 4). This indicates that the propylamine-functionalized OFMS is able to discriminate the subtle size difference between 4-(dimethylamino)benzaldehyde and 4-dimethylamino-1-naphthaldehyde. Thus, the vast majority of the organic functional groups in the OFMS reside within the shape-selective micropores of the solid.

These first examples of OFMS's containing acidic or basic organic moieties covalently bound within the pores of crystalline silicates with the zeolite beta topology (International Zeolite Association structure code: *BEA) illustrate one new route to controlling the porosity around organic sites in

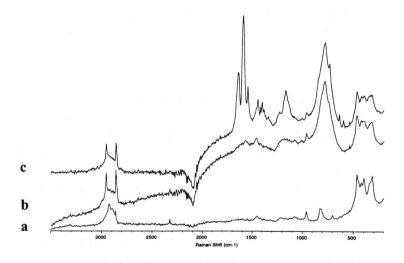

Figure 3. FT-Raman spectra of propylamine-functionalized amorphous silica (a) and the same silica reacted with 4-dimethylamino-1-naphthaldehyde (b) or 4-(dimethylamino)benzaldehyde (c). Reprinted with permission from *Micropor. Mesopor. Mater.* **1999**, *29*, 339. Copyright 1999 Elsevier Science.

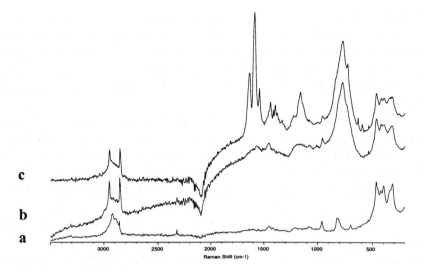

Figure 4. FT-Raman spectra of propylamine-functionalized OFMS (**a**) and OFMS reacted with 4-dimethylamino-1-naphthaldehyde (**b**) or 4-(dimethylamino)benzaldehyde (**c**). Reprinted with permission from *Micropor. Mesopor. Mater.* **1999**, *29*, 339. Copyright 1999 Elsevier Science.

inorganic/organic hybrid catalysts. By preparing a material where the organic groups are contained within the regular micropores of crystalline solids, the catalytic chemist can engineer the catalytic properties of the material by exerting control over the rate and the manner in which reactants and products interact with the active site. Since there are well over 100 different crystalline molecular sieve topologies,[15] one can imagine creating a whole family of shape-selective OFMS catalysts.

3.   MOLECULARLY-IMPRINTED SILICAS

The term molecular imprinting has been used to describe a number of different techniques for the selective functionalization of solid materials for molecular recognition applications such as catalysis and adsorption. A large volume of work on the molecular imprinting of organic polymers has been published (these will not be considered here), whereas a considerably smaller body of work exists on molecularly-imprinted inorganic oxides such as silica.[16] In nearly all cases, a templating moiety or "imprint" is used in an attempt to organize the inorganic silicates or organic species tethered to silica in a specific manner.[17] Before considering molecular imprinting as a technique for the control of porosity around organic active sites in inorganic matrices, we will briefly categorize the different types of investigations that have thus far been described as the "molecular imprinting of silica".[*]

3.1   Type 1: Imprinting to Generate Organic-free Porosity

In this type of imprinting, a specific templating molecule that interacts with silica is introduced during the silica particle synthesis. After condensation of the silica, the material is then calcined to remove any residual organic species (hence the term organic-free) and open the porosity that the organic species occupied. An example of this is the work of Maier *et al.*[18,19] and Pinel *et al.*[20] These groups describe the molecular imprinting of silica via the co-condensation of R^1-Si-$(OR^2)_3$ with TEOS in an acid catalysed sol-gel synthesis. The goal of this work is to produce siliceous materials with micropores that mimic the shape of the imprint molecule. A schematic of the idealized solid products produced in Maier's work is given in Figure 5.[19] It has been demonstrated that this methodology does not always produce selective catalysts, that is, the porosity created by the imprint is not highly shape-specific.[19-21] Materials of this type will not be considered further here, since the final materials do not contain significant organic species.

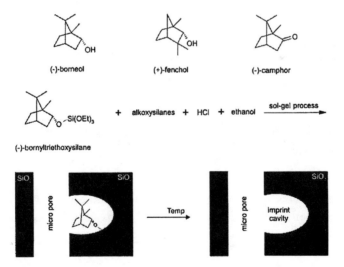

(-)-borneol   (+)-fenchol   (-)-camphor

(-)-bornyltriethoxysilane   +   alkoxysilanes   +   HCl   +   ethanol $\xrightarrow{\text{sol-gel process}}$

Figure 5. Schematic illustration of sol-gel synthesis of Type 1 molecularly-imprinted silica materials. Reprinted with permission from *Micropor. Mesopor. Mater.* **1999**, *29*, 389. Copyright 1999 Elsevier Science.

---

[*] The brief review here is not intended to be exhaustive. Rather, an overview of several different types of imprinting is presented.

3.2   Type 2: Imprinting to Spatially Position Organic Moieties on a Silica Surface

Here, an imprint is used to organize organic species spatially in existing porosity within a silica particle or on a silica particle's external surface. The goal of the imprint is to spatially position organic groups that are directly tethered to the surface. The imprint may serve to create additional porosity within the structure. This category can be further divided into imprinting via covalent interactions and imprinting via non-covalent interactions.

3.2.1   Non-covalent Interactions

Here, surface organic moieties are positioned during the synthesis of organic-functionalized silica materials via single or multiple non-covalent interactions[22,23] such as hydrogen bonding, ionic/Coulombic interactions or metal coordination. An example of this type of imprinting is the work of Dai *et al.*,[22] in which mesoporous silica is functionalized with $[Cu(aapts)_2S_2]^{2+}$, (where aapts is $H_2N(CH_2)_2NH(CH_2)_3Si(OCH_3)_3$ and S is water or methanol). The copper ion serves to organize the chelating diamine ligands and can be easily removed by protonation. A schematic of this approach is given in Figure 6.

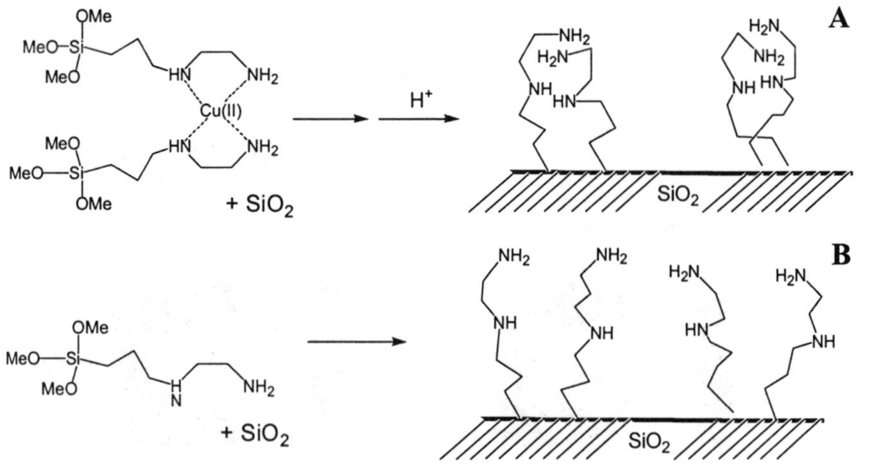

Figure 6. Schematic illustration of the use of metal coordination to organize organic species on the surface of silica. Imprinting **(A)** leaves organized amine sites on the surface after acidic deprotection. Synthesis in the absence of the metal imprint **(B)** results in random incorporation of amines.

3.2.2 Covalent Interactions

Imprinting through the use of covalent imprint-monomer interactions is more robust than using weaker non-covalent interactions to organize organic moieties. Wulff and coworkers spatially positioned amino groups on a silica surface using an imprint bound to monomers via hydrolyzable Schiff base linkages, as described in Figure 7.[24,25] Sasaki and coworkers subsequently adopted a similar approach.[26-27] In the Wulff and Sasaki investigations, the imprint-monomer functionalities were added to pre-formed silica materials. Hence, no additional porosity was created within the silica by the imprint.

Recently, Katz and Davis reported the preparation of a molecularly-imprinted silica using sol-gel techniques with spatially positioned amine groups contained in micropores templated by aromatic imprints. They utilized aminopropylsilanes covalently bound to the aromatic imprint via carbamate linkages [1; 2 (para); or 3 (1,3,5 positions) linkages with the resulting materials are denoted as one-, two- and three-point materials] to illustrate what appears to be a new, general technique for the preparation of novel catalytic materials.[5] The synthetic scheme developed by Katz and Davis is described in Figure 8. An imprint is hydrolyzed in the presence of TEOS and ethanol (which acts as a pore

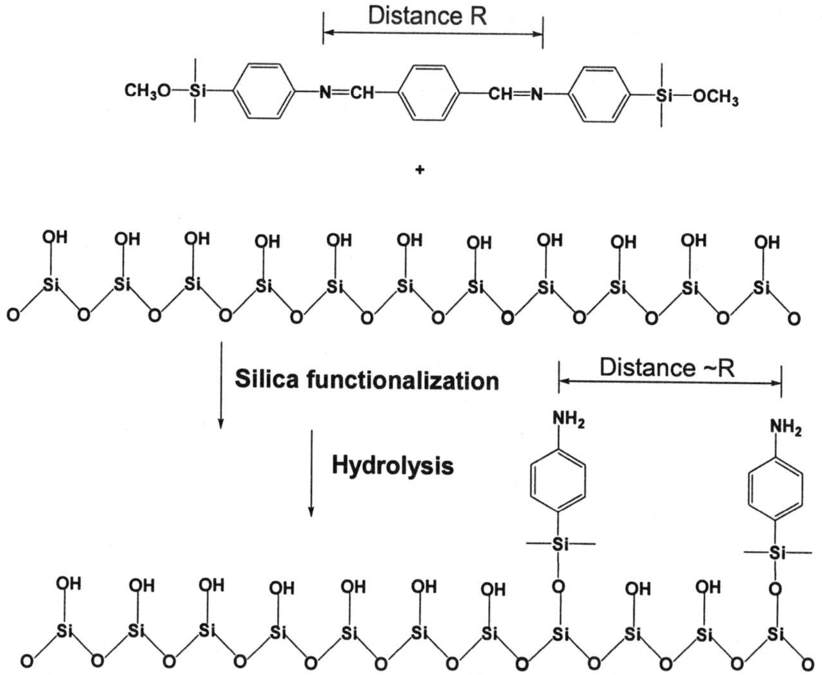

Figure 7. Schematic illustration of the use of organic ligand positioning on a silica surface via a covalently-bound imprinting molecule.

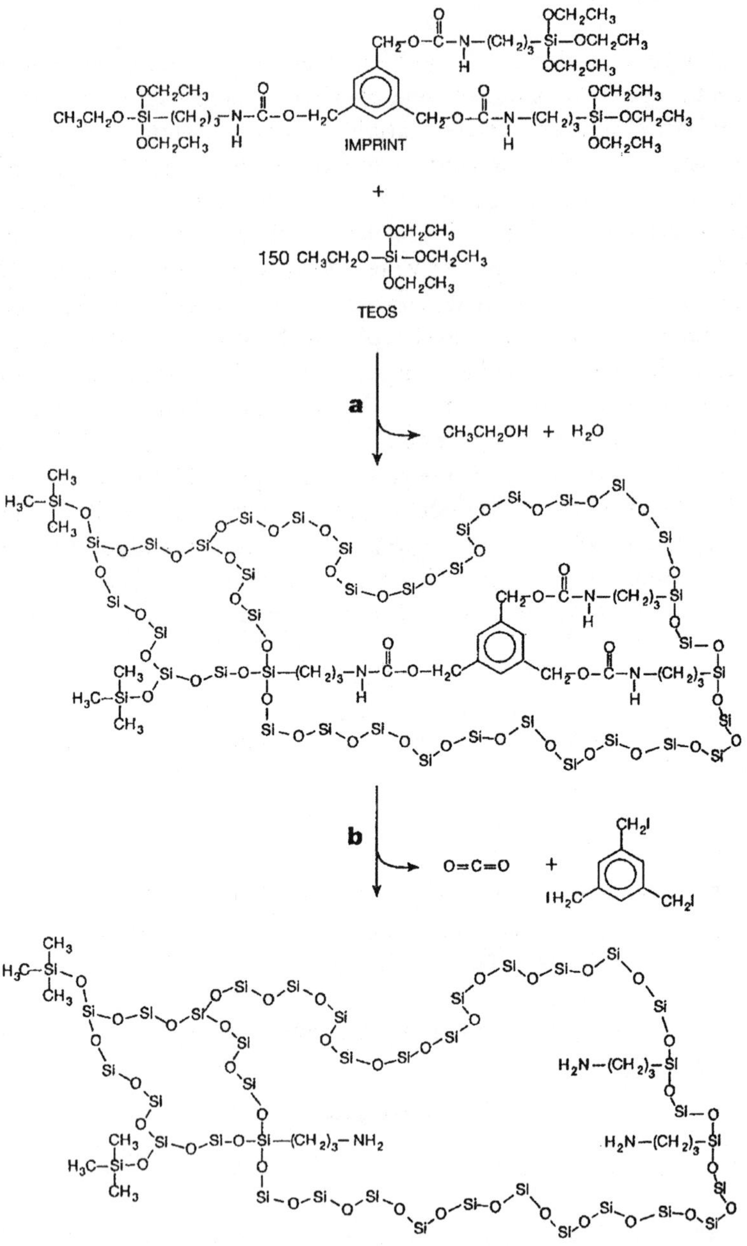

Figure 8. Synthetic procedure for the preparation of molecularly-imprinted polymers: **(a)** the sol-gel hydrolysis and consensation of alkoxysilanes catalysed by HCl (pH=2), and **(b)** deprotection via treatment with trimethylsilyl iodide in acetonitrile followed by washing with methanol and aqueous sodium bicarbonate. Reprinted with permission from *Nature* **2000**, *403*, 286. Copyright 2000 MacMillan Publishing.

filling agent) under acidic conditions to give a molecularly-imprinted silica material. The imprint is then removed via treatment with trimethylsilyl iodide, exposing spatially positioned amine groups around a nanocavity. Detailed characterization by solid state nuclear magnetic resonance (NMR) spectroscopy confirmed that the imprint species remained intact within the solid after synthesis and that the trimethylsilyl iodide treatment efficiently removed the imprint (73% of the aromatic imprint was removed in the one- and two-point materials).[5]

To characterize the accessibility of the amine functional groups as well as the proximity of the organic functional groups in the two- and three-point materials, a number of different experiments were carried out.[5] First, the deprotected materials were reacted with acetylacetone to form Schiff bases, which were clearly detected via UV-vis spectroscopy (316 nm), proving the accessibility of the sites to probe molecules. In another test, the one- and two-point materials were exposed to azelaoyl chloride, a probe molecule that has the correct dimensions to react with both amino groups in the two point material. In the one point material that contains isolated, single propylamine groups, only a single amide should be formed. As shown in Figure 9, infrared spectroscopy confirms

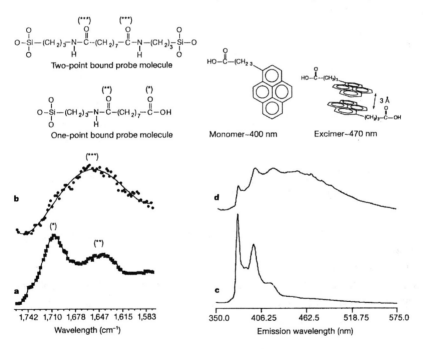

Figure 9. Infrared and fluorescence emission spectra from imprinted silicas in the presence of probe molecules: IR spectra of one-point (**a**) and two-point (**b**) materials contacted with azelaoyl chloride and fluorescence emission spectra of one-point (**c**) and three-point (**d**) materials contacted with pyrenebutyric acid. Reprinted with permission from *Nature* **2000**, *403*, 286. Copyright 2000 MacMillan Publishing.

that both acid chlorides react with amino groups in the two-point materials but only one reacts when exposed to the one-point material. In another adsorption test, the one- and three-point materials were exposed to pyrenebutyric acid and the resulting solids were studied using fluorescence emission spectroscopy. These data indicated that isolated pyrene groups exist in the one point material whereas excimers dominate the spectrum in the three-point material (see Figure 9).

Katz and Davis demonstrated that these materials are unique solid base catalysts.[5] The silicas were evaluated in the Knoevenagel condensation of malononitrile and isophthaldehyde, Scheme 3. They found that the unique environment of the silica micropores allows for size-selective reactions (akin to those that can occur in OFMS's) to take place. The two-point material inhibits addition of a second malononitrile moiety to the aromatic ring (overall rate 74 turnovers per site per hour). In contrast, randomly distributed propylamine groups on the surface of amorphous silica effectively produce the difunctionalized aromatic (367 turnovers per site per hour).[28] Hence, the small nanocavity around the amine groups in the two-point material does not allow for the formation of the doubly functionalized product that should readily be produced.

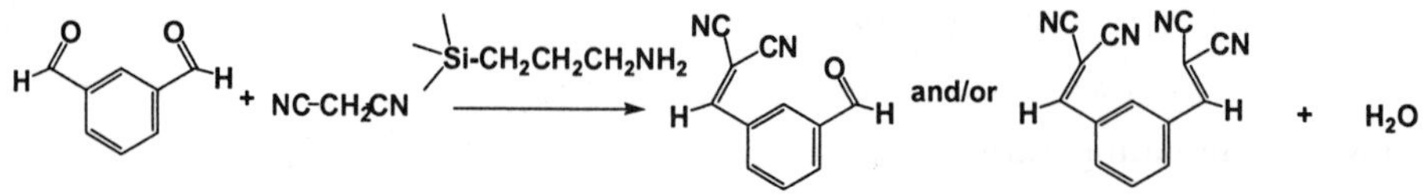

Scheme 3. Knoevenagel condensation.

4.   IMPRINTED SILICAS AND OFMS's - SIMILARITIES AND DIFFERENCES

OFMS's and molecularly imprinted silicas, while both catalytic inorganic/organic hybrids, are distinctly different materials. As such, the properties of the two classes of materials and the future outlook for the materials is worth further examination. Here, OFMS's will be compared with the most complex catalytic, molecularly-imprinted organic/inorganic hybrids described to date, the materials of Katz and Davis.

4.1   Silica Framework and Porosity

OFMS's are synthesized using techniques originally developed for the preparation of zeolites. Like zeolites, OFMS's are crystalline silicates, with every silicon and oxygen atom in the framework occupying crystallographically unique positions. As a result of this structure, the pores in OFMS's are of the same size and dimensions throughout the material. The organic moieties tethered to this framework (*e.g.,* the mercaptopropyl group in Figure 2) reside in this well-defined pore system. There is no experimental evidence that the attached organic residues create any additional space in the silicate.

In contrast, molecularly-imprinted (MI) silicas have a non-crystalline silica framework. Imprinted silicas can be prepared using sol-gel techniques that allow for some control over the pore size of the materials (as described in ref. 5) but because the materials are non-crystalline, there will likely always be some variation in the sizes of the pores. Whereas the organic species in OFMS's do not create additional porosity, MI silicas have the potential to create well-defined local porosity around the organic moieties. This will be discussed further in the next section.

4.2   Organic Species

OFMS's contain well-defined but simple organic species randomly distributed throughout the pores of the material, as described in Figure 10. To date, there is no experimental evidence to suggest that the organic groups reside in preferential locations within the pores of the silicate. As previously mentioned, a number of different organic species have been incorporated within the pores of OFMS's, including short chain aliphatic amines and thiols as well as small aromatics. Attempts to incorporate larger organic species within the pores of OFMS's result in a disruption of the crystal structure and multiphase, crystalline/amorphous mixtures are produced.[29] Hence, current techniques limit the synthesis of OFMS's to spatially small organic groups that can fit easily within the micropores of the crystalline silicate framework.

In MI silicas, the organic species are also randomly distributed throughout the silica framework. However, in contrast to the OFMS's, larger organic species can be easily accommodated within the pores of these materials. More importantly, the *local* porosity around the organic groups can be precisely controlled and multiple organic groups can be spatially positioned using molecular imprinting techniques. Figure 10 schematically describes several orientations that the organic functionality can occupy in the as-synthesized material. The two-point monomer-imprint-monomer organic of Katz and Davis is shown as an example. In orientation 1, the organic moiety resides in relatively large pores, a situation that might occur if the organic monomer-imprint-monomer species were surrounded by ethanol within the pores during synthesis. In this situation, the imprint simply serves to spatially position two amino groups a specific distance apart after deprotection. The organic moiety in orientation 3 resides within a nanocavity created by the monomer-imprint-monomer functional group during synthesis. After deprotection, the two amine groups will be

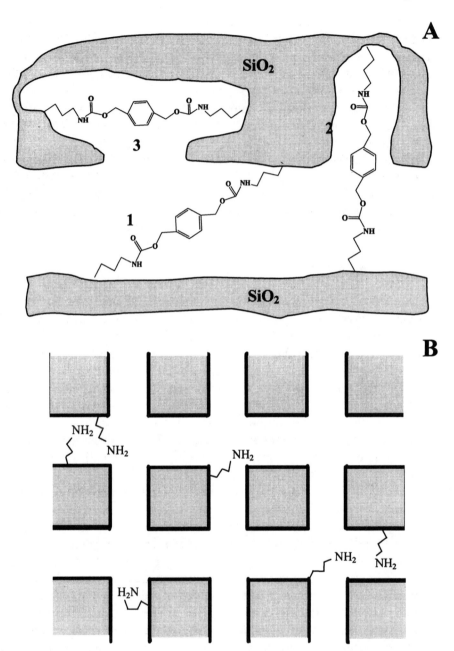

Figure 10. Schematic of organic groups and pore structure of MI silica (**A**) and OFMS (**B**). In the MI silica, the imprint creates porosity around the organic species tethered to the silica. In the OFMS, the organic groups are randomly distributed in micropores and no additional porosity is created.

precisely positioned a specific distance from each other and the space between them will be a nanocavity of the same shape and size as the imprint. Here the imprint serves both to position the organic groups and create a unique nanocavity. Orientation 2 is a hybrid of orientations 1 and 3, with the monomer-imprint monomer functional group partially creating new porosity. The degree of order of the local porosity around the organic functional groups can be maximized using procedures that favor orientations 3 and 2 over 1. Clearly, the ability to control the nanoporosity around an organic group will be greater when multipoint monomer-imprint pairs are developed. Furthermore, the results of Katz and Davis suggest that the acid catalyzed sol-gel synthesis used in the preparation of their materials resulted in silica materials with relatively small pores, favoring the formation of organic residues in orientations 2 and 3.

The properties of OFMS's and MI silicas are summarized and compared in Table 2.

Table 2. Descriptive summary of OFMS's and MI Silica.

	MI Silica	OFMS
Ordered	Organic orientation and local porosity around the organic on the Å scale	Silicon-oxygen bonds and pores: crystalline framework
Disordered	Silicon-oxygen bonds: amorphous framework	Organic group location and orientation
Summary	Potential for well-defined porosity around the organic groups and preparation of nanocavities with multiple, spatially positioned binding sites.	Ordered long range porosity with potential for incorporation of many different types of randomly located and oriented organic groups within a single framework structure.

4.3   Future Outlook for OFMS's and MI Silicas

To date, there have been a very limited number of publications on OFMS's and all of these reports have concerned the functionalization of molecular sieves with the *BEA topology.[4,8-10,30] The ability to prepare OFMS's with other crystalline frameworks has been hampered by the lack of materials that have large pore systems capable of accommodating the tethered organic groups within the micropores and that can be made porous by means other than calcination. Suitable candidates for new host structures include several of the new large pore molecular sieves synthesized recently using the fluoride route.[31] Other possibilities include functionalization of large pore materials that can be prepared in the absence of a structure-directing agent such as zeolites of the FAU topology. Because these materials do not require calcination, they would be reasonable candidates as hosts for tethered intracrystalline organic moieties. Preliminary investigations into the preparation of OFMS's using low silica, aluminosilicate zeolites such as NaY have resulted in limited success in providing OFMS materials.[32]

In contrast to the slow development of OFMS's, the number of publications concerning molecularly-imprinted materials of all types (silicas, polymers, *etc.*) continues to grow. Although additional publications using the methodologies of Katz and Davis have not yet appeared, the potential for creating complex materials with multi-point binding sites using these new synthetic techniques appears to be significant. Currently, some synthetic hurdles must be overcome before this new methodology can be applied for widespread synthesis of catalysts. For example, more effective deprotection strategies must be developed. The multi-point materials of Katz and Davis could only be partly deprotected, possibly due to incorporation of the organic imprints in completely isolated pores. However, it is expected that the work of Katz and Davis will provide a springboard for the future development of inorganic-organic hybrid catalysts with multi-point binding sites and well-controlled porosity around the active center.

ACKNOWLEDGEMENTS

Our work on OFMS's was supported by Akzo Nobel.

REFERENCES AND NOTES

1.   Moller, K.; Bein, T. *Chem. Mater.* **1998**, *10*, 2950.
2.   Venuto, P. *Micropor. Mater.* **1994**, *2*, 297.
3.   Chen, N. Y.; Degnan, T. F.; Smith, C. M. *Molecular Transport and Reaction in Zeolites*, VCH: New York, 1994.
4.   Jones, C. W.; Tsuji, K.; Davis, M. E. *Nature* **1998**, *393*, 52.
5.   Katz, A.; Davis, M. E. *Nature* **2000**, *403*, 286.
6.   Lobo, R. F.; Zones, S. I.; Davis, M. E. *J. Inclus. Phen. Mol. Recog. Chem.* **1995**, *21*, 47.
7.   Jones, C. W.; Tsuji, K.; Takewaki, T.; Beck, L. W.; Davis, M. E. *Micropor. Mesopor. Mater.* **2001**, *47*, 57.
8.   Jones, C. W.; Tsuji, K.; Davis, M. E. *Micropor. Mesopor. Mater.* **1999**, *33*, 223.
9.   Jones, C. W.; Tsapatsis, M.; Okubo, T.; Davis, M. E. *Micropor. Mesopor. Mater.* **2001**, *42*, 21.
10.   Tsuji, K.; Jones, C. W.; Davis, M. E. *Micropor. Mesopor. Mater.* **1999**, *29*, 339.
11.   Bhatia, S., *Zeolite Catalysis: Principles and Applications*, CRC Press: Boca Raton, Florida, 1990.
12.   Weisz, P. B.; Frilette, V. J.; Maatman, R. W.; Mower, E. B., *J. Catal.* **1962**, *1*, 307.

13. Brechtelsbauer, C.; Emig, G. *Appl. Catal. A.* **1997**, *161*, 79.
14. Additional control experiments also support this conclusion. In particular, nitrogen physisorption analysis indicates that nearly all the phenyl functional groups are contained within the micropores of the solid and they therefore should behave as shape-selective sites. See reference 9 for more details.
15. International Zeolite Association Website: http://www.iza-online.org/
16. Davis, M. E.; Katz, A.; Ahmad, W. R. *Chem. Mater.* **1996**, *8*, 1820.
17. Morihara and coworkers have developed an extensive body of work on imprinted aluminum sites on the surface of silicates (see ref. 16). These will not be considered here, since they do not concern organic/inorganic hybrids.
18. Heilmann, J.; Meier, W. F. *Z. Naturforsch.* **1995**, *50b*, 460.
19. Hunnius, M.; Rufinska, A.; Maier, W. F. *Micropor. Mesopor. Mater.* **1999**, *29*, 389.
20. Pinel, C.; Loisil, P.; Gallezot, P. *Adv. Mater.* **1997**, *9*, 582.
21. Ahmad, W. R.; Davis, M. E. *Catal. Lett.* **1996**, *40*, 109.
22. Dai, S.; Burleigh, M. C.; Shin, Y.; Morrow, C. C.; Barnes, C. E.; Xue, Z. *Adv. Mater.* **1999**, *38*, 1235.
23. Sasaki, D. Y.; Alam, T. M. *Chem. Mater.* **2000**, *12*, 1400.
24. Wulff, G.; Heide, B.; Helfmeier, G. *J. Am. Chem. Soc.* **1986**, *108*, 1089.
25. Wulff, G.; Heide, B.; Helfmeier, G. *React. Polym.* **1987**, *6*, 299.
26. Hwang, K.-O.; Yakura, Y; Ohuchi, F. S.; Sasaki, T. *Mater. Sci. Eng.* C3 **1995**, 137.
27. Hwang, K.-O. and Sasaki, T. *J. Mater. Chem.* **1998**, *8*, 2153.
28. Reactions were conducted at 80°C in acetonitrile with a 2:1 ratio of malononitrile to isophthaldehyde.
29. Jones, C. W.; Davis, M. E. unpublished data, 1999.
30. Shin, Y.; Zemanian, T. S.; Fryxell, G. E.; Wang, L.-Q.; Liu, J. *Micropor. Mesopor. Mater.* **2000**, *37*, 49.
31. Camblor, M.A.; Barrett, P. A.; Diaz-Cabanas, M.-J.; Villaescusa, L.A.; Puche, M.; Boix, T.; Perez, E.; Koller, H. *Micropor. Mesopor. Mater.* **2001**, *48*, 11.
32. Jones, C. W. Ph.D. Thesis, California Institute of Technology, 1999.

Chapter 13

STRATEGIES FOR THE DESIGN AND SYNTHESIS OF HYBRID MULTIFUNCTIONAL NANOPOROUS MATERIALS

Jun Liu,* Yongsoon Shin, Li-Qiong Wang, Gregory J. Exarhos, Jeong Ho Chang, Glen E. Fryxell, Zimin Nie, Thomas S. Zemanian, and William D. Samuels
Pacific Northwest National Laboratory, Battelle Boulevard, Richland, WA 99352
* Sandia National Laboratory, Biomolecular Materials and Interfaces Department, Albuquerque, NM 87185-1413

Keywords: self-assembled monolayers, supercritical fluids, functionalized silicas, zeolites, molecular imprinting, site-and-shape selective catalysis

Abstract: This chapter discusses the design and synthesis of multifunctional active sites in ordered nanoporous materials. First, the formation of homogeneous molecular monolayer structures is described. Hybrid nanoporous materials modified with functional molecules and groups are widely investigated for many applications. The molecular chain conformations depend on the surface roughness of the pore channels. A step-wise growth model has been proposed to account for the step-wise pore dimension change. This paper also discusses the use of supercritical fluids as delivery media to improve the effectiveness of surface functionalization. This technique has been used successfully to synthesize size-exclusive microporous acid catalysts. Finally, the formation of architectured monolayer molecular structures is discussed. The use of imprinting or lithograph techniques allows the synthesis of hierarchical porous materials with tunable size-and-shape selective microporosity.

1.   INTRODUCTION

Many articles, including quite a few in this book, have extensively reviewed the history and progress in the development of ordered nanoporous materials based on self-assembly principles of liquid crystalline structures.[1-6] Many different types of nanoporous materials have been reported,[7-12] including alumina, zirconia, titania, niobia, tantalum oxide, manganese oxide[13] and metals.[14] The experimental approach includes direct ionic interaction,[1] mediated ionic interaction,[15] neutral hydrogen bonding[16,17] and non-aqueous processes for semi-conducting materials.[18] Simple cationic, anionic and neutral surfactants, as well as large block copolymers have been used.[19,20] Although most materials are made in powdered forms, oriented nanoporous films on various substrates,[21,22] free-standing films,[23] spheres,[24,25] fibers,[26,27] and single crystalline nanoporous materials in which all the pre-channels are aligned,[28] been extensively fabricated. Non-powder materials are prepared either through controlled nucleation and growth[12-16,29] or by rapid evaporation, as in fiber drawing,[17,18] dip coating[30,31] and spin coating.[32]

Ordered nanoporous materials are attractive for many applications, including catalysis. They have very high surface areas (>1000 m^2/g), ordered pore structures (mostly hexagonal packed cylindrical pore channels) and extremely narrow pore-size distributions. Through careful control of the pore size and shape, and the surface chemistry, it is possible to design and synthesize nanoporous materials with novel catalytic properties, and with unique size-and-shape selectivity.

The most widely-used approach in preparing nanoporous catalytic materials is the introduction of functionality into the nanoporous channels. Some excellent review articles have become available in this area.[33,34] The purpose of this chapter is to discuss the assembly of architectured molecular structures and functional groups. We intend to illustrate how key material parameters such as the pore channel dimensions and the surface functionalities can be adjusted. Furthermore, we demonstrate how multifunctional sites, with potential for size-and-shape selectivity, can be constructed.

2.   HOMOGENEOUS MOLECULAR STRUCTURES IN NANOPOROSITY

In the literature, direct silanation has been used to introduce functional molecules and functional groups into nanoporous silica.[35-40] In this approach, various silane molecules hydrolyze and react with the surface silanol groups, becoming immobilized on the silica support. Therefore the existence of free silanol groups is critical. Direct silanation works very well for nanoporous materials prepared by a neutral surfactant route,[37,38] in which the surfactant molecules are linked to the silica substrate only through hydrogen bonding.

The surfactant can be removed easily by solvent-extraction techniques. Since these materials are not calcined at a high temperature, as required in most other preparation methods, a high concentration of surface silanols is available for the silanation reaction. Several groups have developed a co-condensation process, which combines the synthesis of the nanoporous materials and the functionalization into one step.[41]

Under optimized conditions, very high density and high quality functional monolayers can be constructed.[39,40] As illustrated in Figure 1, we first introduce one to two monolayers of water molecules onto the substrate internal surfaces. Subsequently, the organosilanes react with the water molecules during hydrolysis and finally become uniformly distributed on the substrate. As will be discussed in more detail later, the formation of these molecular monolayers in the nanoporous materials is indeed related to the preparation of self-assembled molecular monolayers (SAMs) on flat silica substrates.

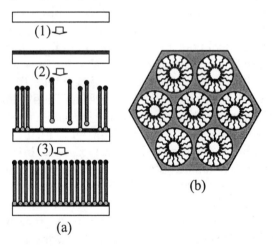

(a)

Figure 1. (a) Three step formation of high density molecular layers in nanoporous materials. (1) Application of one to two layers of water molecules. (2) Reaction of silane molecules with water. (3) Intermolecular condensation. (b) Schematic illustration of molecular layers in nanoporosity Figure 1b is reproduced with permission from *Science* **1997**, *276*, 923. Copyright 1997 Science Publishing.

SAMs are widely used to modify surfaces and interfacial properties such as wetting, adhesion and friction.[42,43] They are also used to mediate the molecular recognition processes and to direct oriented crystal growth.[44] The molecular arrangement and the chain conformation on flat substrates have been extensively studied by many state-of-the-art techniques, such as atomic force microscopy

(AFM), contact angle measurement and small angle scattering.[44] Due to their technological importance, the mechanisms of SAM formation from silane molecules on smooth surfaces, silica spheres and gels has been extensively studied. Several steps have been observed kinetically:[45] 1) Rapid reaction of silane molecules with residual or adsorbed water to form hydrolyzed monomers or oligomers, and aggregation of such monomers and oligomers; 2) Diffusion and adsorption of the silane monomers and oligomers onto the substrate; 3) Chemical adsorption to the surface, elimination of water, and the formation of Si-O-Si bonds with the substrate. Island and domain structures are widely observed.

Maoz and Sagiv[46] reported that dense molecular arrays form when the adsorption is carried out from solutions above certain critical concentrations. Less dense arrays are formed when adsorption is carried out in solutions below the critical concentration. A fractal growth model was proposed by Schwartz *et al.* to account for the self-similar domain structures formed at different reaction times.[47]

However, a careful study of monolayer structures and the kinetics of monolayer growth suggested that the molecular assembly process is not as simple as reported in the literature. Nanoporous materials prepared with surfactants contain uniform cylindrical pore channels, but both nitrogen adsorption studies and transmission electron microscopy suggest that the internal surfaces are not atomically smooth. For example, hexagonal silica with a 10 nm pore diameter was shown to have a surface roughness of more than 1 nm.[48] Based on pore size analysis and the change of pore size when the surface is gradually filled with short and long chain silanes, the step growth model in Figure 2 was proposed,[48] using aminopropyltrimethoxysilane (APS) as an example. In this model, the pore channels are regarded as modulated structures with wide and narrow pore regions. Initially, silane molecules are deposited in the wide pore region and reduce the pore radii in this region by a number corresponding to the chain length (about 7-8 Å) (**a** and **b**). No deposition occurs in the narrow pore region. After the wide pore region is filled, molecules begin to deposit in the narrow pore region, reducing the pore radii in this region by about 5 Å (**c**). Up to this stage, the pore size in the wide pore region does not decrease further. In this process, the layer thickness in the narrow pore region (5 Å) is less than the expected chain length, indicating that the packing in these regions is not efficient. This result is expected due to the positive surface curvature in these regions. When the surfaces in the narrow pore region are filled, a second layer begins to deposit in the wide pore region and reduces the pore radii in this region by another 7 Å (roughly corresponding to the chain length) (**d**). These results also suggest that deposition does not stop at the second layer. However, the third layer is not oriented, and its thickness is only about 3 Å.

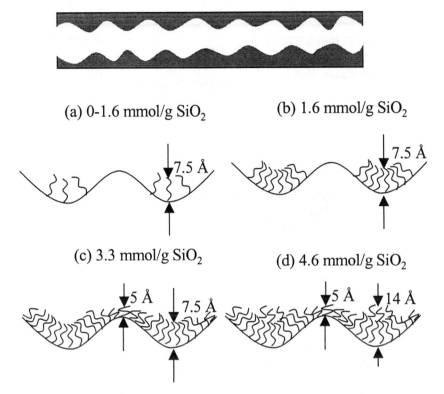

(a) 0-1.6 mmol/g SiO$_2$ (b) 1.6 mmol/g SiO$_2$

(c) 3.3 mmol/g SiO$_2$ (d) 4.6 mmol/g SiO$_2$

Figure 2. Step growth model for molecular assembly in nanoporosity. The upper figure is an idealized pore channel with modulated surface roughness. Frames (a) to (c) show molecular conformations at different surface coverages and the corresponding layer thicknesses. Adapted with permission from *J. Phys. Chem.* A **2000**, *104*, 8328. Copyright 2000 American Chemical Society.

In order for step-growth to take place, the molecules deposited on the surface must have considerable mobility. This means that chemical condensation formed through interactions with the water layers on the surface, and through intermolecular condensations. This is why we believe that the process we use is quite different from direct silanation methods, and is more similar to traditional SAM formation.

The effect of surface roughness is more evident in nanoporous materials with large irregular pore structures (Figure 3). For uniform pore structures, the molecular layer thickness is directly related to the chain length if the initial pore diameter is significantly larger than the molecular size, in agreement with previous studies.[49,50] However, for irregular pore channels, molecular layer thickness is a constant, much larger than the chain length and independent of chain length. This suggests that these molecular chains have very disordered multilayer structures.

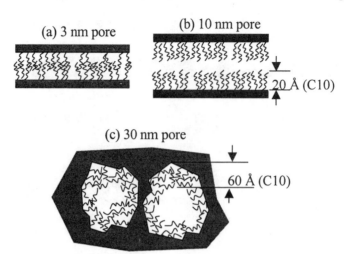

Figure 3. Molecular chain conformation of decyltrimethoxysilane (C10) in different pores. (a) Jammed porosity, with pore diameter similar to the chain length; (b) Ordered monolayer in smooth pore channels; (c) Disordered multilayers in irregular pores. Adapted with permission from *J. Phys. Chem.* A **2000**, *104*, 8328. Copyright 2000 American Chemical Society.

3. SUPERCRITICAL FLUID DEPOSITION OF MOLECULAR MONOLAYERS

Figure 3a illustrates that when the pore size becomes too small, the diffusion and delivery of the functional molecules become difficult. We investigated the use of a supercritical fluid (SCF) instead of an ordinary solvent as the delivery and reaction medium.[51] The choice of SCF is not surprising because supercritical fluid techniques have already been used in aerogel chemistry, surface treatment and in extraction of natural products.[52] SCFs have also been used to deposit metals on a substrate.[53] SCFs offer a unique environment to perform chemical reactions because of their liquid-like solvation properties and gas-like physical properties (viscosity, diffusivity).[54] When used to introduce functional molecules into nanoporosities, SCFs offer several advantages: 1) More favorable reaction conditions for molecular diffusion. The low density, low viscosity, high diffusivity and low surface tension of SCFs make them ideal media for performing silanation of the internal surfaces of porous materials; 2) CO_2 is environmentally benign, non-toxic, non-flammable and inexpensive. The mild critical conditions for CO_2 ($T_c = 31.1$ °C, $P_c = 73.8$ bar) can be attained easily, and are unlikely to cause degradation of porous substrates. By direct pressure pumping, silanes are readily delivered to the internal pore surface. Similarly, when the pressure is decreased, unreacted silanes and by-products are forced out of the inner volume of the porous substrate. In principle, the silanation process can be accomplished in a few minutes; 3) No secondary organic waste is generated because the supercritical process does not use an organic solvent; 4) The SCF process improves the quality and chemical stability of the monolayers on mesoporous silica.

We also used the supercritical process to functionalize microporous materials. Because of the small pores in microporous materials, delivery and deposition of functional molecules without blocking the pore channels by external deposition is difficult. Previously, Jones *et al.* developed an *in situ* synthesis technique to prepare acidified zeolites which showed good size exclusive selectivity.[55] We used zeolite beta (from Zeolyst)[56,57] as an example. Tris(methoxy)mercaptopropylsilane (TMMPS) was deposited into the microporous channels in a supercritical CO_2 ($SCCO_2$) environment. For comparison, TMMPS-functionalized mesoporous silica was also studied. The thiol groups of TMMPS on the zeolite and mesoporous silica were oxidized to SO_3H groups with H_2O_2. We used reactions similar to those reported by Jones *et al.*,[55] the catalytic ketalization (acetalization) of cyclic ketone (aldehyde) by ethylene glycol, to demonstrate the size selectivity of $SCCO_2$-sulfonated zeolite beta. The catalytic reaction of cyclohexanone (HEX) with glycol forms 2,2-pentamethylene-1,3-dioxolan (cyclic ketal), while the reaction of pyrenecarboxaldehyde (PYC) with glycol forms the corresponding acetal.[58] The size-selective catalytic properties of the supercritically modified zeolite compare well to similar materials prepared by an *in situ* silanation process reported by Jones *et al.*,[55] which demonstrates that the active functional groups were delivered to the internal pore surfaces of the microporous material and remain accessible to molecules that can enter the pore channels. The reaction kinetics were faster under similar conditions as a result of the high density of active groups due to enhanced diffusivity of the functional molecules in the micropore channels in SCF. The supercritical process can potentially simplify the preparation of new materials and may open up new opportunities for commercial zeolites.

Figure 4 compares the functionalized 6 nm nanoporous silica and the SCF-functionalized zeolite beta as acid catalysts. As shown in Figure 4a, the functionalized nanoporous silica has larger pore channels. Both PYC and HEX can access the acid sites, and no selectivity was observed. Similarly in Figure 4b, both large amine molecules (methyldioctylamine MDOA) and small amine (triethylamine, TEA) molecules can freely access the pores and neutralize the acid sites. Therefore when the amine molecules were introduced, the catalytic reactions stopped immediately. On the contrary, as shown in Figure 4c and 4d, SCF-functionalized zeolite has size selectivity. Faster reaction kinetics were observed for HEX, while PYC showed very limited conversion. In the same way (Figure 4d), only small amine molecules can access and poison the active sites and stop the reaction. The addition of large amines had no effect on the catalytic reactions.

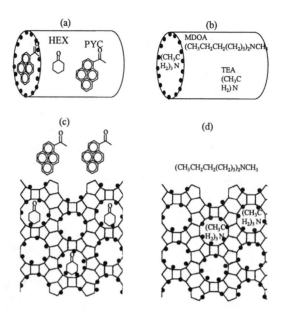

Figure 4. Comparison of functionalized nanoporous silica and SCF functionalized zeolite as acid catalysts: (a) Access of both PYC and HEX to the acid sites in 6 nm silica. Similar conversions of PYC and HEX were observed; (b) Access of large (MDOA) and small (TEA) amines to the active sites of 6 nm silica. All acid sites were neutralized by both amines and the reactions stopped immediately; (c) Access of small HEX to the active site in zeolite and exclusion of large PYC from the microporosity; (d) Access of small TEA to the acid sites in zeolite. Large MDOA has no effect on the reaction.

4.   DESIGNING MOLECULAR ARCHITECTURES IN NANOPOROSITY

Recently we developed a multistep assembly approach to control the three-dimensional architecture on the molecular level.[59] This approach is based on widely investigated molecular imprinting techniques[60,61] and involves the formation of heterogeneous, multifunctional molecular monolayers. Several key steps are involved: 1) Deposition of target molecules of desired size and shape onto internal pore surfaces; 2) Formation of an inert molecular coating in the areas not occupied by target molecules; and 3) Removal of target molecules to form desired patterns in the nanoporous materials. Except for the small length scale, this molecular imprinting approach is similar to the lithographic techniques used in the microelectronics industry. Here the patterns formed are on the order of molecular sizes, and chemical, rather than optical methods are used to form the patterns.

As an example, triangular molecules (2,4,6-tris(*p*-formylphenoxy)-1,3,5-triazine, tripods) and linear molecules (2,4-bis(*p*-formylphenoxy)-6-methoxy-1,3,5-triazine, dipods) were first bound with APS through the terminal aldehyde groups on the target molecules. The tripod chemistry on silica substrates was first reported by Tahmassebi *et al.*[62-63] These silane-bound target molecules were deposited on the internal surfaces of 10 nm nanoporous silica. The areas not occupied by the target molecules were then coated with a layer of long chain or short chain silanes. Subsequently the template molecules were selectively removed from the substrates using a solvent extraction technique to create triangular and linear patterns in the monolayer coating, Figure 5.

This approach creates a hierarchical porous material with several novel properties:

1) The material contains two level of porosities, nanoporosity formed by the surfactant and microporosity formed by the target molecules. In the last few years, there has been great interest in the synthesis of ordered nanoporous materials with zeolite-like microporosities in the walls. Sometimes binary surfactant systems have been used.[3] Unfortunately, phase separation in the surfactant system causes the formation of separate nanoporous and microporous regions. The current approach provides an unusual example of hierarchical porous ceramics with integrated nano- and microporosities.

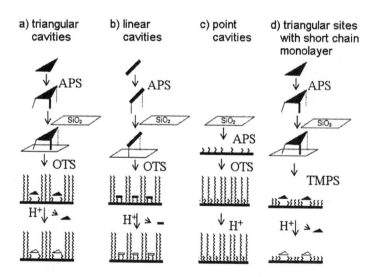

Figure 5. Molecular imprinting approach to create three dimensional patterns in nanoporosity. Reproduced from *Angew. Chem., Int. Ed.* **2000**, *39*, 2707. Copyright 2000 Wiley Publishing.

2) Very good size-and-shape selectivity is observed. We tested the different adsorption behavior of the tripod and dipod molecules. When the size and shape of the absorbate molecules match those of the microporosities, adsorption is greatly enhanced, Figure 6. The ability of the substrate to differentiate between dipod and tripod molecules is remarkable, even though the difference between them is not great. A tripod molecule has three benzaldehyde arms, while a dipod has two. The rest of the molecular structure is similar. In principle, the dipod molecules can easily fit into the triangular cavities and conform to the majority of the cavities nicely. The two aldehyde groups can also bind to the two amine groups at the corner of the cavity (by hydrogen bonding or forming a Schiff base). We would expect the dipod molecules to adsorb on triangular as well as linear cavities. The large difference in the adsorption behavior of dipod molecules on triangular and linear cavities demonstrates that the shapes of the cavities are very important.

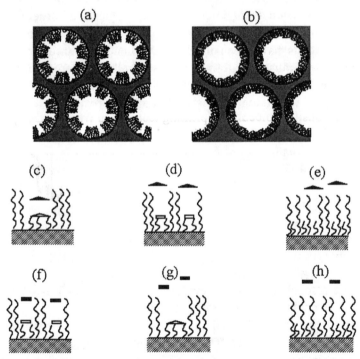

Figure 6. Tunable hierarchical materials with good size-and-shape selectivity: (**a**) Microporosities in the open position; (**b**) Microporosities in the closed position; (**c**) Enhanced adsorption of tripods on triangular microporosities; (**d**) and (**e**) Reduced adsorption of tripods on linear and point microporosities; (**f**) Enhanced adsorption of dipods on linear microporosities; (**g**) and (**h**) Reduced adsorption of dipods on triangular and point defects. Adapted with permission from *Angew. Chem., Int. Ed.* **2000**, *39*, 2707. Copyright 2000 Wiley Publishing.

3) The selectivity of the microcavities is directly related to the three-dimensional nature of the long-chain molecule coatings. Mere one-to-one correspondence of the binding sites does not give any selectivity. Materials were prepared using the same procedure without the long-chain molecule monolayer coating or with a coating layer made of small molecules (hexamethyldisilazane or trimethoxy-propylsilane (TMPS)). These materials contain exactly the same binding sites (amine groups from the APS) arranged in the same way as in the long-chain monolayers. No selectivity was observed on any of these materials without the long-chain capping. If selectivity were solely attributed to matching of the binding sites, the materials without capping or with the short-chain capping layer should have selectivities similar to those of the long-chain materials. Therefore, we can conclude that the steric effect of the cavities in the long-chain monolayers largely contributes to the selectivity of such materials.

4) Unlike in zeolites, in which the pores are formed by inorganic cages, the microporosities discussed in this chapter are formed by flexible long-chain molecules. These long-chain molecules can assume different conformations depending on the solvent conditions. In a good solvent, the chains are extended and the microporosities are open and accessible. In a poor solvent, the chains are folded and the microporosities are not accessible.

Previously, molecular imprinting techniques were used to prepare size-and-shape selective microporous silicas, which have rigid frame structures.[64] The development of tunable (or soft) microporous channels is very interesting. In nature, soft lipid membranes are known to perform exceptionally selective transport and separation.[65] These membranes contain microporous channels and pores. These biological systems self-regulate materials transport through conformational changes of membrane proteins that precisely control the ion and molecular permeability across the membrane surfaces. In a primitive way, the soft microporous materials we synthesized demonstrate similar properties of self-regulation through molecular conformational change.

We also obtained some evidence that size-selective catalysis can be performed over the soft microporosities. Preliminary experiments were conducted using Knoevenagel condensation between malononitrile and benzaldehyde or 3-pentanone.[66] The use of amino groups immobilized on silica as weak base catalysts for Knoevenagel condensations was previously reported by Angeletti *et al.*[67] and by others.[68,69] We do not have enough data pertaining to the exact reaction pathways,[64-70] but we have shown that the large cavities (triangular and linear) give a higher yield than the point cavities for the condensation between malononitrile and benzaldehyde.

ACKNOWLEDGEMENT

This work is supported by the Office of Basic Energy Sciences of the Department of Energy. Pacific Northwest National Laboratory is operated by Battelle Memorial Institute for the US Department of Energy under Contract DE-AC06-76RL01830.

REFERENCES

1. Beck, J. S.; Vartuli, J. C.; Roth, W. J.; Leonowicz, M. E.; Kresge, C. T.; Schmitt, K. D.; Chu, C. T-W.; Olson, D. H.; Sheppard, E. W.; McCullen, S. B.; Higgins, J. B.; Schlenker, J. L. *J. Am. Chem. Soc.* **1992**, *114*, 10834.
2   Kresge, C. T.; Leonowicz, M. E.; Roth, W. J.; Vartuli, J. C.; Beck, J. S. *Nature* **1992**, *359*, 710.
3. Beck, J. S.; Vartuli, J. C. *Curr. Opin. Solid State Mater. Sci.* **1996**, *1*, 76.
4. Liu, J.; Kim, A. Y.; Wang, L.-Q.; Palmer, B. J.; Chen, Y. L.; Bruinsma, P.; Bunker, B. C.; Exarhos Graff, G. L.; Rieke, P. C.; Fryxell, G. E.; Virden, J. W.; Tarasevich, B. J.; Chick, L. A. *Adv. Colloid Interface. Sci.* **1996**, *69*, 131.
5. Raman, N. K.; Anderson, M. T.; Brinker, C. J. *Chem. Mater.* **1996**, *8*, 1682.
6. Zhao, D.; Yang, P.; Hua, Q.; Chmelka, B. F.; Stucky, G. D. *Curr. Opin. Solid State Mater. Sci.* **1998**, *3*, 121.
7. Vandry, F.; Khodabandeh, S.; Davis, M. E. *Chem. Mater.* **1996**, *8*, 1451.
8. Bagshaw, B. A.; Pinnavia, T. J. *Angew. Chem., Int. Ed. Engl.* **1996**, *35*, 1102.
9. Knowles, J. A.; Hudson, M. J. *J. Chem. Soc., Chem. Comm.* **1995**, 2083.
10. Schmidt R.; Akporiaye, D.; Stocker, M.; Ellestad, O. H. *J. Chem. Soc. Chem. Comm.* **1994**, 1493.
11. Antonelli, D. M.; Ying, J. Y. *Chem. Mater.* **1996**, *8*, 874.
12. Antonelli, D. M.; Ying, J. Y. *Angew. Chem. Int. Ed. Engl.* **1995**, *34*, 2014.
13. Tian, Z.-R.; Tong, W.; Wang, J.-Y.; Duan, N.-G.; Krishnan, V. V.; Suib, S. L. *Science* **1997**, *276*, 926.
14. Attard, G. S.; Barlett, P. N.; Coleman, N. R. B.; Elliott, J. M.; Owen, J. R.; Wang, J. H. *Science* **1997**, *278*, 838.
15. Huo, Q.; Margolese, D. I.; Ciesla, U.; Feng, P.; Gier, T. E.; Sieger, P.; Leon, R.; Petroff, P. M.; Schuth, F.; Stucky, G. D. *Nature* **1994**, *368*, 317.
16. Tanev, P. T.; Pinnavaia,T. J. *Science* **1995**, *267*, 865.
17. Bagshaw, S. A.; Prouzet, E.; Pinnavaia, T. J. *Science* **1995**, *267*, 865.
18. MacLachlan, M. J.; Coombs, N.; Ozin, G.A. *Nature* **1999**, *397*, 681.
19. Zhao, D.; Feng, J.; Huo, Q.; Melosh, N.; Fredrickson, G. H.; Chmelka, B. F.; Stucky, G.D. *Science* **1998**, *279*, 548.
20. Yang, P.; Zhao, D.; Margolese, D. I.; Chmelka, B. F.; Stucky, G. D. *Nature* **1998**, *396*, 152.
21. Yang, H.; Kuperman, A.; Coombs, N.; Mamiche-Afrara, S.; Ozin, G. A. *Nature* **1996**, *379*, 703.
22. Aksay, I. A.; Trau, M.; Manne, S.; Honma, I.; Yao, N.; Zhou, L.; Fenter, P.; Eisenberger, P. M.; Gruner, S. M. *Science* **1996**, *273*, 892.
23. Yang, H.; Coombs, N.; Sokolov, I.; Ozin, G. A. *Nature* **1996**, *381*, 589.
24. Hua, Q.; Feng, J.; Scheth, F.; Stucky, G. D. *Chem. Mater.* **1997**, *9*, 14.

25. Lu, Y.; Fan, H.; Stump, A.; Ward, T. L.; Rieker, T.; Brinker, C. F. *Nature* **1999**, *398*, 223.
26. Bruinsma, P. J.; Kim, A. Y.; Liu, J.; Baskaran, S. *Chem. Mater.* **1998**, *10*, 2507.
27. Yang, P.; Zhao, D.; Chmelka, B. F.; Stucky, G. D. *Chem. Mater.* **1998**, *10*, 2033.
28. Antonelli, D. M.; Nakahira, A.; Ying, J. Y. *Inorg. Chem.* **1996**, *35*, 3126.
29. Yang, H.; Coombs, N.; Ozin, G. A. *Nature* **1997**, *386*, 692.
30. Lu, Y.; Gangull, R.; Drewien, C. A.; Anderson, M. T.; Brinker, C. F.; Gong, W.; Guo, Y.; Soyez, H.; Dunn, B.; Huang, M. H.; Zink, J. I. *Nature* **1997**, *389*, 364.
31. Zhao, D.; Yang, P.; Melosh, N.; Feng, J.; Chmelka, B. F.; Stucky, G. D. *Adv. Mater.* **1997**, *9*, 1380.
32. Baskaran, S.; Liu, J.; Domansky, K.; Kohler, N.; Li, X.; Coyle, C.; Fryxell, G. E.; Thevathasan, S.; Williford, R. E. *Adv. Mater.* **2000**, *12*, 291.
33. Moller K., Bein T. *Chem. Mater.* **1998**, *10*, 2950.
34. Stein, A; Meble, B. J.; Schroden, R. C. *Adv. Mater.* **2000**, *12*, 1403.
35. Cauvel, D.; Renard, G.; Brunel, D. *J. Org. Chem.* **1997**, *62*, 749.
36. Burkett, S. L.; Simms, S. D.; Mann, S. *J. Chem. Soc., Chem. Comm.* **1996**, 1367.
37. Mercier, L.; Pinnavaia, T. J. *Adv. Mater.* **1997**, *9*, 500.
38. Mercier, L.; Pinnavaia, T. J. *Envir. Sci. Technol.* **1998**, *32*, 2749.
39. Liu, J.; Feng, X.; Fryxell, G. E.; Wang, L.-Q.; Kim, A. Y.; Gong, M. *Adv. Mater.* **1998**, *10*, 161.
40. Feng, X.; Fryxell, G. E.; Wang, L.-Q.; Kim, A. Y.; Kemner, K.; Liu, J. *Science* **1997**, *276*, 923.
41. Lim, M. H.; Blanford, C. F.; Stein, A. *Chem. Mater.* **1998**, *10*, 467.
42. Whitesides, G. M. *Sci. Am.* **1995**, *273*, 146.
43. Ulman, A. *Chem. Rev.* **1996**, *96*, 1533.
44. Bunker, B. C.; Rieke, P. C.; Tarasevich; B. J.; Campbell. A. A.; Fryxell. G. E.; Graff. G.L.; Song. L.; Liu. J.; Virden. J. W.; McVay, G. L. *Science* **1994**, *264*, 48.
45. Bierbaum, K.; Kinzler, M.; Woll, C. H.; Grunze, M. *Langmuir* **1995**, *11*, 512.
46. Maoz, R.; Sagiv J. *J. Colloid Interface Sci.* **1984**, *100*, 465.
47. Schwartz, D. K.; Steinberg, S.; Israelachvili, J.; Zasadzinski, J. A. N. *Phys. Rev. Lett.* **1992**, *69*, 334.
48. Liu, J.; Shin, Y.; Fryxell, G. E.; Wang, L. Q.; Nie, Z; Jeong Ho Chang, J. H.; Fryxell, G. E.; Samuels. W. D.; Exarhos. G. J. *J. Phys. Chem. A* **2000**, *104*, 8328.
49. Kruk, M.; Jaroniec, M.; Sayari, A. *Langmuir* **1997**, *13*, 6267.
50. Kruk, M.; Jaroniec, M.; Sakamoto, Y.; Terasaki, O.; Ryoo, R.; Ko, C. H. *J. Phys. Chem. B* **2000**, *104*, 292.
51. Shin, Y.; Zemanian, T. S.; Fryxell, G. E.; Wang, L.; Liu, J. *Micropor. Mesopor. Mater.* **2000**, *37*,49.
52. See articles in "Supercritical Fluids", *Chem. Rev.* Noyori, R. (Ed) **1999**, *99*.
53. Watkins, J. J.; Blackburn, J. M.; McCarthy, T. J. *Chem. Mater.* **1999**, *11*, 213.
54. Watkins, J. J.; MacCarthy, T. J. *Chem. Mater.* **1995**, 7.
55. Jones, C. W.; Tsu, K.; Davis, M. *Nature* **1998**, *52*, 393, and references cited therein.
56. van der Waal, J. C.; Rigutto, M. S.; van Bekkum, H.; *J. Chem. Soc., Chem. Comm.* **1994**, 1241.
57. Camblor, M. C.; Corma, A. C.;Valencia, S. *Chem. Commun.* **1996**, 2365.
58. Solomons, T. W. G. *Organic Chemistry*, 3rd ed., Wiley: New York, 1984, pp.720-725.
59. Shin, Y.; Liu, J.; Wang, L.-Q.; Nie, Z; Samuels, W. D.; Fryxell, G. E.; Exarhos, G. J. *Angew. Chem., Int. Ed. Engl.* **2000**, *39*, 2707.
60. Wulf, G. *Angew Chem., Int. Ed.* **1995**, *34*, 1812.
61. Davis, M. E.; Katz, A.; Ahmad, W. R. *Chem. Mater.* **1996**, *8*, 1820.
62. Tahmassebi, D. C.; Sasaki, T. *J. Org. Chem.* **1994**, *59*, 679.
63. Hwang, K. O.; Yakura, Y.; Ohuchi, F. S.; Sasaki, T. *Mater. Sci. Eng. C* **1995**, *3*, 137.

64.   Katz, A.; Davis, M.E. *Nature* **2000**, *403*, 286.
65.   Gennis, R. B. *Biomembranes, Molecular Structure and Function*, Springer-Verlag: New
        York, 1989; Chapter 8, pp. 270-322.
66.   Carey F. A., Sundberg R. J. *Advanced Organic Chemistry B: Reactions and Synthesis*, 2nd
        ed. Plenum: New York, **1983**; pp. 57-59.
67.   Angeletti, E.; Capena, C.; Martinetti, G.; Venturello, P. *J. Chem. Soc., Perkin Trans. I*,
        1989, 105.
68.   Choudary, B. M.; Kantam, M. L.; Sreekanth, P.; Bandopadhyay, T.; Figueras, F.; Tuel,
        A. *J. Mol. Catal. A.* **1999**, *142*, 361.
69.   Rodriguez, I.; Iborra, S.; Corma, A.; Rey, F.; Jorda, J. L. *Chem. Commun.* **1999**, 592.
70.   House, H. O. *Modern Synthetic Reactions*, 2nd ed.; W. A. Benjamin Inc.: Menlo Park, CA,
        1972, p. 648.

Chapter 14

QUANTITATIVE RELATIONS BETWEEN LIQUID PHASE ADSORPTION AND CATALYSIS

Dirk E. De Vos,[+] Gino V. Baron,[*] Frederik van Laar,[+] Pierre A. Jacobs[+]

[+] *Centre for Surface Chemistry and Catalysis, K. U. Leuven, Kasteelpark Arenberg 23, 3001 Leuven, Belgium*
[*] *Dienst Chemische Ingenieurstechnieken, Vrije Universiteit Brussel, Pleinlann 2, 1050 Brussels, Belgium*

Keywords: physisorption, liquid chromatography, Ti zeolites, phthalocyanines

Abstract: Numerous literature examples illustrate the profound effects that changing the hydrophobic properties of a catalyst can exert on its performance. However, there is a need for quantitative data that firmly establish the relationship between adsorption and catalytic characteristics. The liquid chromatographic determination of adsorption constants K is a straightforward method for the accurate determination of intraporous concentrations of organic molecules in zeolites. These constants reflect not only the polarity of the catalyst, but also the effect of the solvent on the position of the adsorption equilibrium. Other methods to evaluate the polarity of porous catalysts such as zeolites are critically evaluated. Finally, the relationship between catalytic behavior and sorption characteristics (K values) is discussed for three cases: (1) TS-1, (2) Ti-Beta, and (3) a Y zeolite containing a macrocyclic complex.

1.   INTRODUCTION

Progress in homogeneous catalysis is usually achieved by the design of new ligands that provide a modified, optimum coordination for the active metal centre. In contrast, the activity of a heterogeneous catalyst in liquid phase reactions is determined not only by the nature of the reactive center or by its immediate coordination environment. There are other factors that may play a decisive role, for instance the dimensions of the pores giving access to the active site, or the hydrophobic or hydrophilic nature of the matrix around the catalytic site.

Catalytic scientists are increasingly using the hydrophobic/hydrophilic properties of catalyst supports as a tool to improve catalyst activity and selectivity. In qualitative terms, increased hydrophobicity favors the adsorption of apolar compounds; this may lead to increased reactant concentration at the active site, or to a modified ratio of reactants that results in improved selectivities.

Particularly with Ti catalysts for selective oxidation, this has been a popular research theme. For TS-1, there is general agreement that the high epoxidation activity is partly caused by the hydrophobic nature of the framework, which favors the adsorption of the olefin rather than of hydrogen peroxide or water.[1] With such a material, peroxide decomposition and hydrolytic epoxide degradation are negligible. While the initial Ti-Beta materials contained a substantial amount of Al, an essentially Al-free and therefore more hydrophobic Ti-Beta has been obtained via the fluoride method.[2,3] Evidently, the yield based on hydrogen peroxide and the epoxide selectivity are better with the Al-free Ti-Beta (Ti-Beta (F)) than with the Al-containing Beta catalyst (Ti-Beta (OH)).

Similar trends are observed with non-crystalline materials. A Ti-containing mesoporous material of the MCM-41 type is as such a rather poor olefin epoxidation catalyst if H_2O_2 is used as the oxygen source. Several groups have explored the effects on the catalytic properties of surface treatments with silylating agents such as *N*-trimethylsilylimidazole or (3-chloropropyl)-dimethoxymethylsilane.[4-6] There is agreement that these modifications result in better yields on a peroxide basis. Finally, attention has also been devoted to the hydrophobization of Ti-Si aerogels. This is easily achieved in the synthesis by replacing part of the tetraalkoxysilane with the corresponding alkyltrialkoxysilane. Such aerogels are among the most active Ti-based materials for epoxidation with alkylhydroperoxides (alkyl = *t*-butyl, cumyl).[7,8]

Apart from the chemistry of Ti catalysts, there are numerous other cases in which the effects of catalyst hydrophobicity or hydrophilicity on a heterogeneously-catalyzed reaction have at least qualitatively been documented:

- In the esterification of polyols with fatty acids, zeolites such as H-Beta or mesoporous molecular sieves modified with sulfonic acids have been employed. The zeolites tend to adsorb the polyol, leading to high selectivity for the monoester, with rather moderate activity. On the other hand, sulfonic acid siliceous mesoporous materials are more hydrophobic and allow much faster reaction of the polyol with the apolar fatty acid phase.[9-11]

- Composite Zr phosphonate catalysts have been used in the esterification of acetic acid with ethanol in the liquid phase. When the materials comprise an acidic function and a hydrophobic group in the same crystal, the activity per acid group is higher than with sulfonic acids.[12]
- The reverse reaction, the hydrolysis of water-insoluble esters, was studied in a liquid bi-phasic water-toluene mixture. Treatment of the external surface of H-ZSM-5 with octadecyltrichlorosilane results in a 60-fold rate increase in comparison with non-treated H-ZSM-5.[13]
- Defect-free, hydrophobic acid Beta zeolites are better catalysts for the acetalization of glucose with *n*-alkanols than more hydrophilic zeolite catalysts.[14]
- While Co-phthalocyanines are excellent catalysts for autoxidation of thiols, their activity can be further increased by immobilization in supports that have been made more hydrophobic, *e.g.*, zeolite X, exchanged on the outer surface with tetrabutylammonium ions, or a hydrotalcite-type anionic clay intercalated with dodecylsulfate anions.[14]
- Adsorption of a lipase on hydrophobic polypropylene markedly increases the rate of triolein hydrolysis in comparison with the dissolved lipase.[16]

In the above-mentioned cases, it seems reasonable to assign increased activities, selectivities, *etc.*, to a changed polarity of the catalyst surface, and resulting changes in adsorption of reactants. However, there are surprisingly few data that actually quantitatively support the relationship between adsorption of reactants and catalytic behavior in liquid phase reactions.

There is another important reason to study in detail the adsorption of reactants and products on heterogeneous catalysts. In many cases, it is hard to distinguish between adsorption and diffusion effects, particularly when microporous catalysts are involved. For instance, if a light hydrocarbon, *e.g.*, C_6, is more reactive than a heavier molecule such as C_9 in a zeolite reaction, this is commonly ascribed to the faster diffusion of the smaller reactant molecule. However, it is also possible that in the case of the reaction with C_9, the strong adsorption of the product hinders adsorption of fresh reactant. Distinction between such adsorption and diffusion phenomena is only possible if the adsorption behavior of reactants and products is known.

In this chapter, we first discuss the chromatographic approach to determination of adsorption constants. This type of measurement is then illustrated with some data for Ti catalysts and complex-containing Y zeolites. Next, the chromatographic approach is compared to alternative methods to quantify adsorption on zeolite materials. Finally, we demonstrate the value of sorption data in understanding liquid phase catalysis.

2.   CHROMATOGRAPHIC DETERMINATION OF ADSORPTION CONSTANTS

2.1   Generalities

Many methods have been developed to determine pure and mixture gas phase adsorption equilibria and a variety of commercial instruments are available.[17,18]   Thermodynamic theory describing the phenomena is well developed and a range of isotherm equations are available for data treatment and prediction for multicomponent systems. Data can be obtained in catalytically relevant conditions at high temperature and pressure and be used to model multicomponent reactions such as hydroconversion.[19,20]

Far less data are available on adsorption of polar and apolar molecules and their mixtures in the liquid phase on microporous catalytic materials such as zeolites. In the gas phase, especially at low loading, most molecules essentially interact directly with the walls of the pore system. In the liquid phase, the pores of the adsorbent are filled with a mixture of molecules. In these conditions of high loading, non-idealities like adsorbate-adsorbate interactions and surface heterogeneity become more important.[21-24] For the modelling of multicomponent adsorption of liquid mixtures, it is often necessary to introduce cross correlation coefficients between components and complex expressions for the activity coefficients to obtain a good fit between model and experimental adsorption data.[22,25]   It is thus clear that adsorption in the liquid phase is a complex subject and that liquid phase adsorption properties cannot be extrapolated easily from gas phase adsorption isotherms. For materials such as ion-exchange resins used as acid catalysts, certain components can induce strong swelling of the material resulting in highly selective sorption.[26] The inherent complexity and the experimental difficulties explain why data on competitive adsorption between organic molecules with different polarity and carbon number at temperatures and pressures relevant to catalysis are lacking in literature.

Moreover, most of the measurements reported in the literature have been obtained for dilute solutions, such as activated carbon/water systems for wastewater or drinking water treatment. Isotherm equations similar to those in gas phase systems (linear, Langmuir, Freundlich, *etc.*) can be used and a vast amount of data have been reported. Given the strong adsorption of components of interest and the very weak interactions of water with these carbonaceous materials, this is a reasonable approach, resembling the situation in gas phase at low coverage. The fitted equations can then be used to design separation equipment.

Liquid phase adsorption equilibria are most often determined by a simple batch test. A known quantity of solid is contacted with a mixture of known composition under controlled conditions of temperature and pressure. After equilibration (minutes to days), a sample of the liquid is analyzed, and from a material balance the concentrations in the adsorbed phase are calculated. When intraparticle or intracrystal transport is sufficiently slow, analysis of samples taken after different equilibration times in a well-mixed system allows for extraction of diffusion or transport parameters.[27] If selective adsorption occurs, then the external bulk phase concentration will change significantly after equilibration and can then be measured. When the adsorption is less selective, then the change is small and to obtain sufficient accuracy, the free liquid phase volume has to be minimized. Mixing of the suspension may then be difficult, and taking a sample may give problems if the solid is in the form of fine crystals, which is often the case for zeolites. Centrifuging the sample or filtration is then necessary before injection into analytical equipment. One possible solution is to use headspace chromatography when the vapor-liquid equilibrium and the vapor pressure allow such measurement. Evaporation of the solvent is more often the real problem, and in some cases data are needed at temperatures above the boiling point at ambient pressure. This requires operation under pressure, further complicating the measurement.

An alternative approach, which avoids most of the aforementioned problems, is to use a chromatographic method in analogy to gas phase measurements. Due to its simplicity and accuracy, the liquid phase chromatographic method has often been applied to study adsorption and diffusion properties of a range of components on several adsorbents.[28-33] A conventional HPLC setup can be adapted to perform these measurements, and enables working temperatures exceeding the boiling temperatures of the fluids used as mobile phases. A liquid phase mixture, similar to the reaction fluid, is passed over the column packed with the catalyst. This can be just the solvent or one of the components of the mixture, or a more complex mixture. A tracer, either one of the components of the reaction mixture (solvent, product or reactant), or any other probe molecule, is then injected and the retention time of the peak determined, as recorded by a suitable detector at the outlet of the column. The retention time gives an indication of the strength of adsorption and can, with the help of a model, yield the parameters in the isotherm equation. A small quantity of the catalyst is sufficient $(0.5 - 1.0$ g$)$ and only limited amounts of the different mixture components and solvents are needed.

2.2   Experimental Setup

Basically, any standard HPLC setup can be used (*e.g.*, Hewlett Packard HP1100 used in our work) with a refractometer detector, UV-VIS detector or Mass Spectrometer (such as those used in GC/MS or LC/MS systems). The refractive index detector is certainly the most universal and generally shows high sensitivity, but it requires excellent temperature control (*e.g.*, HP1037A) and should have minimal cell volume. Catalyst crystals (down to 0.1 µm) or particles (up to 100 µm) are carefully packed into standard stainless steel 1/4" diameter columns with a length of 10 to 100 mm, fitted with stainless steel filter frits on both ends. The mobile phase (solvent or mixture) is pumped through the column filled with adsorbent crystals by an isocratic pump (see Figure 1).

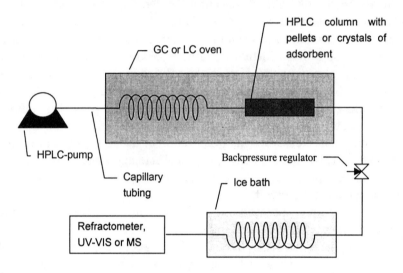

Figure 1. Experimental setup.

Flow rates ranging from 0.1 to 4 mL/min are applied. Flow rate and column length are adapted to obtain adequate retention times (several minutes). Since high pressures can easily be generated with the HPLC pumps, pressure drop is not an issue and working directly with the crystals is possible. This avoids macropore diffusion limitations and the resulting peak broadening. In some cases, intracrystalline diffusion coefficients can be extracted from the response curves,[18] but in many instances, axial dispersion makes this measurement problematic. Particles can be used, preferentially as small as possible, and flow rates have to be adjusted to allow macropore diffusion to occur. Fairly symmetric ("Gaussian") peaks are a good indication of proper operation.

The column is connected to the injector and detector with 1/16" capillary tubing (with an internal diameter of 0.17 mm) to avoid dead volumes which require correction of the retention time and lead to peak broadening as well. The inlet pressure of the system needs to be constantly monitored. In order to ensure liquid phase operation throughout the column when performing measurements above the boiling temperature of the fluid mixture, a backpressure regulator is placed at the end of the column and set to a sufficiently high pressure. The column and a length of tubing coming from the injector are placed in a suitable oven (such as the column oven of an HPLC- or GC-type hot air oven) to ensure precise temperature control. About 1 m of the tubing at the outlet of the HPLC column is passed through an ice bath, in order to avoid vapor bubbles entering the refractometer.

A small pulse (typically 0.1 to 10 μL) of the tracer component is injected into the mobile phase using a manual injector or preferably an autoinjector, allowing automated sequential measurements. The response to this perturbation is recorded at the outlet of the column with the detector. After subtraction of the baseline, the first moment of the response curve is determined by integration. The dead time of the system is determined (without column) at different flow rates and temperatures and is subtracted from the retention time.

2.3   Calculation of the Partition Coefficients

A liquid is passed over the column filled with adsorbent crystals. The inlet volumetric flow rate is held constant. The porosity of the column is given by ε. The column mass balance is then given by:

$$ - \varepsilon D_{ax} \frac{\partial^2 c}{\partial z^2} + \varepsilon \frac{\partial(uc)}{\partial z} + \frac{\partial}{\partial t}\left(\left(\varepsilon + (1-\varepsilon)\frac{\partial q}{\partial c}\right)c\right) = 0 \qquad (1)$$

in which D_{ax} represents the axial dispersion coefficient, u the interstitial column velocity, c the bulk phase concentration of a given component in the solvent, q the adsorbed phase concentration, and z the axial co-ordinate.

In the case of linear equilibrium, valid at low concentrations of the tracer compound, the relation between the concentration c in the mobile phase and q in the adsorbed phase (defined as the total volume within the crystal boundary, including both solid and pore space) can be written as:

$$ q=Kc \qquad (2)$$

where K is the "partition" coefficient. High values of K (dimensionless) signify that the component is enriched inside the pores of the material. This results in:

$$ - \varepsilon D_{ax} \frac{\partial^2 c}{\partial z^2} + \varepsilon \frac{\partial(uc)}{\partial z} + \frac{\partial}{\partial t}\left((\varepsilon + (1-\varepsilon)K)c\right) = 0 \qquad (3)$$

Other equilibrium equations can be used, and even mixture equilibria,[32] but we will restrict ourselves here to linear equilibrium. Under conditions of constant temperature and pressure, the interstitial velocity is constant in time. However, the velocity is still a function of the fluid density. For a mass flow rate $\dot{m}$ at the inlet of the column, the interstitial column velocity u is then given by:

$$u = \frac{\dot{m}}{\rho \varepsilon \Omega} \tag{4}$$

where Ω is the cross section of the column and ρ is the density of the mobile phase. The density is related to both the pressure and the temperature of the mobile phase. Several empirical and semi-empirical estimation methods for the liquid density as a function of temperature and pressure are available.[34-37] However, most of these estimation methods fail at near-critical temperatures. Above critical temperatures, estimation methods are lacking.

As a result of the pressure gradient, the boiling point of the mobile phase decreases along the length of the column. A transition from liquid to vapor phase may occur at some point, and hence a density calculation or, even better, a phase equilibrium calculation must be performed to ensure liquid phase operation. The pressure difference over the column is, however, normally small enough to give constant liquid density.

The mass balance over the column (eq. 3) is solved by transformation to the Laplace domain and application of the Ver der Laan theorem. The first moment of the response curve μ or retention time is then given by:[17]

$$\mu = \frac{L}{u_m}\left(\varepsilon + (1 - \varepsilon)K\right) \tag{5}$$

The partition coefficient K can easily be calculated from the column retention time μ and values of column length L, average superficial (empty tube) velocity $u_m = \dot{m}/\rho\Omega$ and external or interparticle porosity ε. Instead of calculating the mean velocity u_m, it can sometimes also be determined in an experimental way: a non-selectively-adsorbing component is injected into the system. For such a component, it is known that the partition coefficient K is equal to the microporosity, since the pores are filled with the same fluid as in the bulk external phase. The mean velocity u_m can then be calculated, eq. 5. The external porosity ε can also be determined from the retention time with a non-adsorbing component which does not enter the pore system ($K = 0$), or calculated from pycnometric measurements.

3. ILLUSTRATIVE LIQUID PHASE ASDSORPTION DATA

3.1 TS-1

Figures 2 and 3 show data for adsorption of several 1-alkenes and 2-alkanols on TS-1. Similar data have been obtained for adsorption of alkanes and epoxides.[38] It is clear from the graphs that the partition coefficients within a homologous series increase sharply as the alkyl chains get longer. Moreover, the solvent has a major effect on the K value. When methanol is the carrier, intraporous concentrations of 2-alkanols or 1-alkenes are clearly much higher than when a less polar solvent is used.

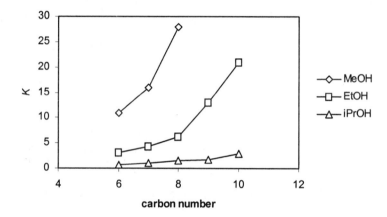

Figure 2. K values for adsoption of 1-alkenes on TS-1 in different solvents.

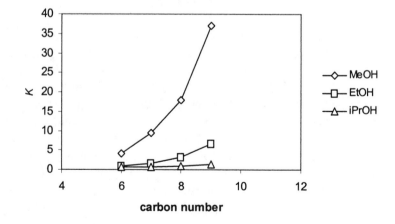

Figure 3. K values for adsorption of 2-alkanols on TS-1 in different solvents.

3.2   Ti-Beta Zeolites

For Ti-Beta samples, similar adsorption trends are found as for TS-1: (1) the adsorption constants increase with increasing carbon number; and (2) the constants are strongly solvent dependent. In Figure 4, the effect of the Ti content on the polarity of the molecular sieve is shown. The K-values drop with increasing amount of titanium. For a similar sample with Al and Ti (Ti/Si = 0.04), a K value of 7 is obtained, in comparison with 22 for the Al-free sample.

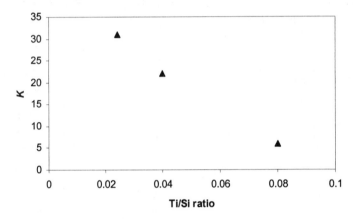

Figure 4. K values for 1-methyl-1-cyclohexene in methanol over different Ti-Beta samples.

3.3   Fe-Phthalocyanines in Zeolite Y

A third example concerns a faujasite NaY zeolite which contains catalytically active phthalocyanine complexes.[39-41] In particular, encapsulated Fe-phthalocyanines (FePc) are highly active for the oxyfunctionalization of alkanes such as cyclohexane with *tert*-butylhydroperoxide as the oxidant. While several methods to incorporate porphyrin or phthalocyanine complexes into zeolites have been reported, we prefer the *in situ* synthesis of phthalocyanines starting from four molecules of dicyanobenzene. This method allows the even distribution of complexes throughout the lattice, and the gradual variation of the phthalocyanine content even in zeolites with widely varying Si/Al ratios.

Table 1 contains data for adsorption on FePc-US-Y, and on the parent US-Y zeolite. The latter has a Si/Al ratio of 130, and thus would be expected to be a relatively apolar material. The FePc-Y material contained 10^{-4} mole complexes per gram, or one complex per six supercages.

Table 1. Adsorption constants K for US-Y zeolite and FePc-containing US-Y zeolite.

carrier solvent	US-Y cyclohexane	US-Y acetone	FePc-US-Y cyclohexane
cyclohexane	--	0.7	--
cyclohexanol	3.6	0.8	13
cyclohexanone	2.2	0.8	11
'BuOOH	7.5	1.2	28

Since cyclohexane is the least polar compound in the reaction mixture, it was chosen as the carrier solvent, as well as acetone, which is often used as a reaction solvent with these catalysts. A striking observation is the effect of the carrier solvent: while in cyclohexane all compounds are clearly enriched within the porous structure, the situation is almost reversed in acetone, and only 'BuOOH is to some extent accumulated in the pores. Second, decoration of the US-Y material with phthalocyanines increases its affinity for oxyfunctionalized organic compounds. Whether this is due to an increased overall polarity of the material, or to specific interactions between the aromatic Pc molecule and the adsorbates, cannot be decided based on *K* values alone.

In conclusion, *K* values from chromatographic measurements not only reflect the polarity of the adsorbent; they also enable one to look at the effect of a solvent on the intraporous concentrations of various adsorbates.

4. OTHER APPROACHES TO QUANTIFY ADSORPTION PROPERTIES OF ZEOLITES

Anderson and Klinowski define the **hydrophobicity *h*** of a zeolite as the ratio of its weight losses at 150°C and 400°C in a thermogravimetric experiment.[42] The larger the fraction of physisorbed water that is easily desorbed, *i.e.*, at temperatures below 150°C, the higher *h*. However, since the amount of physisorbed water may vary widely, *e.g.*, between 2 and 25 weight percent, the measurement is rather inaccurate, particularly for strongly dealuminated zeolites.

An alternative parameter, the **hydrophobicity index (*HI*)**, was defined as follows:[43]

$$HI = X_{toluene} / X_{H_2O} \qquad (6)$$

$X_{toluene}$ and X_{H_2O} represent the mass of toluene or water adsorbed per mass of dry catalyst in the competitive adsorption of gaseous toluene and water. Typically, breakthrough curves are measured in a gas phase reactor setup. As an alternative to toluene, *n*-octane or other alkanes may be used. This method can adequately describe the effect of temperature on the equilibrium concentrations of water and hydrocarbon in the zeolite.

Combined **adsorption-microcalorimetry** experiments were performed to characterize various Ti-Beta samples.[3] A classical volumetric apparatus was used to measure the amounts of water, toluene or *n*-hexane adsorbed on an outgassed sample. When the apparatus is well-calibrated, quantitative heats of adsorption can simultaneously be obtained. Several types of information can be retrieved from such measurements. For instance, the adsorption heat for toluene at low coverage increases from 60 kJ mol^{-1} for Ti-Beta (F) to 68 kJ mol^{-1} for Ti-Beta (OH). Differences are clearer for the adsorption of water. On Beta (F), water displays a type III isotherm, which typifies a stronger affinity of the adsorbate molecules for each other than for the adsorbent. For Ti-Beta (F), the heat of adsorption is initially high, but decreases sharply after adsorption of approximately one molecule of water per Ti. This indicates that each Ti site coordinatively binds one water ligand. Finally, a clear type II isotherm is observed for water on Ti-Beta (OH). Moreover, complex experiments can be performed, such as pre-exposure of the zeolite to toluene, followed by a short outgassing and exposure to water. These data were related to the properties of these zeolites in epoxidation of highly hydrophobic molecules such as oleic acid with H_2O_2.

Finally, **competitive adsorption** on Ti-Beta catalysts was studied with mixtures of alcohols and the carbonyl compound 4-methylcyclohexanone.[44] These experiments are relevant for the selective Meerwein-type reduction of 4-methylcyclohexanone, in which an alcohol is used as the hydrogen donor. In the adsorption experiments, 1,3,5-triisopropylbenzene was used as a solvent, since it is size-excluded from the 12-membered ring pores of the zeolite. Binary adsorbate mixtures were then added. The experiments showed that longer chain alcohols, *e.g.*, 2-heptanol *vs.* 2-butanol, had a greater affinity for the Ti-Beta zeolite. The general advantages and disadvantages of this batch adsorption approach were discussed in section 2.1.

5.   APPLICATION OF SORPTION DATA TO CATALYSIS

5.1   Olefin Epoxidation with TS-1

It is generally known that the nature of the solvent strongly affects the rates of epoxidation or other oxyfunctionalization with TS-1 and H_2O_2. In the case of methanol as the solvent, a widespread interpretation is that methanol coordinates on the Ti sites. For instance, Bellussi *et al.* proposed the following Ti-hydroperoxo-MeOH adduct (**a**):[45]

The electrophilicity of one of the oxygen atoms in the peroxide group is enhanced by formation of a hydrogen bond with a coordinated methanol ligand. This scheme possibly accounts for the high epoxidation rates with, for example, 1-hexene in methanol. In analogy to the methanol-coordinated Ti species, coordination of water on the Ti sites has been proposed as well (species **b**).[2] However, it is difficult to directly prove the coordination of methanol or water on Ti by spectroscopic means.

Another complex issue with TS-1 catalyzed oxidations is the effect of increasing alkyl chain length on the oxidation rate. Early observations indicated that in methanol, the reactivity of the olefins decreases with increasing chain length:[46]

1-butene > 1-pentene > 1-hexene > 1-octene

Similar results were obtained for alkane hydroxylation in methanol:[47]

n-hexane > *n*-heptane > *n*-nonane

Note that in these reactions, rather small catalyst quantities were used, and that reactions were analyzed when a major part of the oxidant was consumed. Thus, in some cases, the catalyst may already have been subject to deactivation. These data were interpreted in terms of diffusion: a molecule with a longer chain should diffuse more slowly into the zeolite pores, and hence the oxidation rate should be lower. However, we obtained several sets of data under truly initial conditions;[38] these data are not in agreement with the previously reported data at higher olefin conversion. First, when acetone is used as a solvent, 1-octene reacts much faster than 1-hexene, and second, in a competitive experiment with equimolar amounts of olefin, 1-nonene is clearly epoxidized faster than 1-hexene.

Although diffusion undoubtedly plays a role in some reactions in TS-1 when crystals are large, or for larger, sterically hindered substrate and product molecules, it is not the only effect that governs the relative reactivities of olefins. We should not forget that slowly diffusing species have dimensions close to the channel dimensions and hence will also experience strong adsorption effects.

Adsorption and diffusion effects are often linked in zeolites. Based on extensive sets of sorption data, and on a large number of kinetic experiments, we rationalized the behavior of TS-1 catalyzed epoxidations as follows:[38]

1. *The superior rates for olefin epoxidation in methanol are primarily due to high intraporous olefin concentration.* Methanol forms a highly polar, hydrogen-bonded solvent network around the zeolite pores. Apolar molecules, which are not easily accommodated in such a network, are driven into the pores of the catalyst. Less polar solvents such as ethanol or 1-propanol lead to less olefin accumulation in the pores, and hence, lower reaction rates. The rates in the initial phase of the reaction (r_{IN}), for instance for 1-hexene in various solvents, parallel the adsorption constants K, as illustrated in Table 2.

Table 2. Rates for 1-hexene epoxidation with TS-1 at 308 K (expressed as mmol converted per g catalysts per hr) during the initial phase of the reaction, and adsorption constants, K.

	r_{IN} (mmol g^{-1} hr^{-1})	K
methanol	42.8	11
ethanol	14.4	3.1
1-propanol	7.2	0.7

2. *Catalyst deactivation can be caused by strong adsorption of epoxide products on the TS-1 catalyst.* The K values for epoxides are comparable to or even larger than those of the corresponding olefins. This somewhat conflicts with the intuitive idea that oxyfunctionalization of a product increases its polarity. In the specific example of 1-decene, this means that the 1,2-epoxydecane product is even more strongly adsorbed on TS-1 than the reagent olefin, particularly in methanol. Therefore, long α-olefins are epoxidized at a high initial rate, but deactivation may be fast, eventually leading to higher epoxide yields with 1-hexene than with 1-nonene.

3. *The preference for olefins in competitive oxidations follows the order predicted by the adsorption constants.* Particularly at low reaction temperatures, differences between the adsorption constants of, for example, 1-hexene and 1-nonene, are large. As a consequence, 1-nonene competitively inhibits adsorption of 1-hexene, and hence suppresses its epoxidation. The ratio epoxynonane/epoxyhexane decreases with increasing temperature, in accordance with a weaker adsorption at higher temperature.

Table 3. Epoxidation of 1-hexene by TS-1 and Ti-Beta

Entry	Catalyst	Yield (%)
1	TS-1	7
2	Ti-Beta (TiO$_2$/SiO$_2$: 0.024)	4.6
3	Ti-Beta (TiO$_2$/SiO$_2$: 0.04)	3.7
4	Ti-Beta (TiO$_2$/SiO$_2$: 0.08)	2.3
5	Ti, Al-Beta (TiO$_2$/SiO$_2$: 0.04)	0.8

Reaction conditions: 12.5 mg catalyst, 2.1 mmol 1-hexene, 0.6 mmol H$_2$O$_2$ and 1.16 mL CH$_3$CN. Reaction temperature 323 K; samples were taken after 2 hr; yields on substrate basis.

5.2 Epoxidation with Ti-Beta

From a catalytic point of view, it may seem desirable to have as many active sites as possible in order to increase the productivity of the catalyst. However, the opposite trend emerges from the data on the epoxidation of 1-hexene with various Ti-Beta zeolites, Table 3. Within the series of Ti-Beta zeolites synthesized by the fluoride method (entries 3-5), a lower Ti content leads to higher hydrophobicity, hence higher intraporous olefin concentration and a higher epoxide yield.

Entries 3 and 5 allow a comparison between two Ti-Beta zeolites with equal Ti contents, with or without Al in the framework. Clearly, the more hydrophilic Al-containing structure is a less active material, which is in line with a lower olefin enrichment in the pores.

Note that the Ti-Beta catalyst can also be used to oxidize more bulky olefins, such as cyclohexene. In the epoxidation of cyclohexene over various Ti-Beta zeolites, similar trends were observed as for 1-hexene epoxidation. Of course, cyclohexene epoxide yields are negligible with TS-1.

5.3 Alkane Hydroxylations with FePc in US-Y

In the hydroxylation of cyclohexane with 'BuOOH as the oxidant, the polarity of reagents and products varies widely. The adsorption data of Table 1 show that the intraporous concentration of cyclohexane must be relatively low in comparison with that of the 'BuOOH oxidant. Moreover, since cyclohexanol and cyclohexanone have higher K values than cyclohexane, product desorption is expected to be difficult. Major consequences for the overall catalytic reaction are:

1. *A relatively large fraction of the peroxide is lost by decomposition.* Since the intraporous concentration of ^{t}BuOOH is high, a peroxide-activated FePc has a high probability to react with another molecule of ^{t}BuOOH, rather than with the cyclohexane substrate. Consequently, typical yields on a peroxide basis may be as low as 11%.

2. *Adsorption of the reaction products leads to fast deactivation.* Since the products tend to be strongly adsorbed, the initial activity decreases rapidly, due to low intraporous cyclohexane concentrations. In order to improve the oxidant efficiency, and to avoid deactivation, several approaches have been successfully followed, such as the use of carbon black as an adsorbent for the phthalocyanines instead of zeolite Y,[48] and the incorporation of the zeolite in a hydrophobic polymeric membrane.[39]

6.   CONCLUDING REMARKS

The three examples described here demonstrate that chromatographic determination of partition coefficients is an easily accessible method to study quantitatively the complex phenomena of adsorption in the liquid phase. The chromatographic method is not only helpful in quantifying the polarity of zeolites or other porous supports; it also takes into account the competitive adsorption of a more or less polar solvent. The latter point is a main advantage in comparison with indices such as the hydrophobicity *h* or Weitkamp's hydrophobicity index *HI*.[42,43]

Nevertheless, one must recognize some limitations of the method, and some possibilities for further improvement. A serious disadvantage is that the chromatographic approach is essentially a perturbation method; this means that only a small amount of adsorbate interacts with a large mass of catalyst. Considering that catalytic reactions are often performed with 5-20 g of substrate per g of catalyst, it becomes clear that the use of *K* values in interpretation of catalytic data is essentially an extrapolation from simple to more complex multicomponent adsorption. A second point is that *K* values reflect the average behavior of all sites in an adsorbent. For instance, a Ti-Beta (F) zeolite contains essentially hydrophobic channels, decorated with hydrophilic Ti centers. While it would be interesting to study separately the properties of these adsorption sites, the chromatographic method does not allow one to distinguish between various sites, and a lumped K value is obtained. Site differentiation may be more adequately studied by adsorption microcalorimetry at different adsorbate loadings, as was shown by Corma and co-workers.[3]

Although measuring different binary systems of components with the method described here may yield most of the qualitative and some quantitative information to analyze a given system, true quantitative multicomponent measurements and descriptions are to be preferred. A viable experimental approach is the use of mass-selective detectors and/or isotopes in chromatographic adsorption experiments. The multicomponent analysis can now only be performed routinely in gas phase systems, and remains an extremely challenging goal in liquid phase, requiring further developments in experimental techniques as well as well-founded thermodynamic models.

ACKNOWLEDGEMENTS

This research received support from the National Science Policy Office (IUAP-P4/11 Programme on Supramolecular Chemistry and Catalysis) and the Flemish F.W.O. (Research Grant to G.B.).

REFERENCES

1. Khouw, C. B.; Dartt, C. B.; Labinger, J. A.; Davis M. E. *J. Catal.* **1994**, *149*, 195.
2. Corma, A.; Esteve, P.; Martinez, A. *J. Catal.* **1996**, *161*, 11.
3. Blasco, T.; Camblor, M. A., Corma, A.; Esteve, P.; Guil, J. M.; Martinez, A.; Perdigon-Melon, J. A.; Valencia, S. *J. Phys. Chem. B* **1998**, *102*, 75.
4. Corma, A.; Domine, M.; Gaona, J. A.; Navarro, M. T.; Rey, F.; Perez-Pariente, J.; Tsuji, J.; McCulloch, B.; Nemeth, L. T. *Chem. Commun.* **1998**, 2211.
5. Bhaumik, A.; Tatsumi, T. *Catal. Lett.* **2000**, *66*, 181.
6. Bu, J.; Rhee, H. K. *Catal. Lett.* **2000**, *66*, 245.
7. Muller, C. A.; Deck, R.; Mallat, T.; Baiker, A. *Top. Catal.* **2000**, *11*, 369.
8. Kochkar, H.; Figueras, F. *J. Catal.* **1997**, *171*, 420.
9. Heykants, E.; Verrelst, W. H.; Parton, R. F.; Jacobs, P. A. *Stud. Surf. Sci. Catal.* **1997**, *105*, 1277.
10. Bossaert, W. D.; De Vos, D. E.; Van Rhijn, W. M.; Bullen, J.; Grobet, P. J.; Jacobs, P. A. *J. Catal.* **1999**, *182*, 156.
11. Diaz, I.; Marquez-Alvarez, C.; Mohino, F.; Perez-Pariente, J.; Sastre, E. *J. Catal.* **2000**, *193*, 295.
12. Segawa, K.; Ozawa, T. *J. Mol. Catal. A* **1999**, *141*, 249.
13. Ogawa, H.; Koh, T.; Taya, K.; Chihara, T. *J. Catal.* **1994**, *148*, 493.

14.   Camblor, M. A.; Corma, A.; Iborra, S.; Miquel, S.; Primo, J.; Valencia, S. *J. Catal.* **1997**, *172*, 76.
15.   Iliev, V.; Ileva, A.; Bilyarska, L. *J. Mol. Catal. A* **1997**, *126*, 99.
16.   Huang, F. C.; Ju, Y.H.; *Biotechn. Techn.* **1994**, *8*, 827.
17.   Ruthven, D.M. *Principles of Adsorption and Adsorption Processes*, Wiley: Toronto, 1984.
18.   Kärger, J.; Ruthven, D.M. *Diffusion in Zeolites and Other Microporous Solids*, Wiley: New York, 1992.
19.   Denayer, J. F. M.; Baron, G. V.; Jacobs, P. A.; Martens, J. *Phys. Chem. Chem. Phys.* **2000**, *2*, 1007.
20.   Denayer, J. F. M.; Baron, G. V.; Vanbutsele, G.; Jacobs, P. A., Martens, J. *J. Catal.* **2000**, *190*, 469.
21.   Rouquerol, F.; Rouquerol, J.; Sing, K. *Adsorption by Powders and Porous Solids*, Academic: London, 1999.
22.   Hulme, R.; Rosenweig, R. E.; Ruthven, D. M. *Ind. Eng. Chem. Res.* **1991**, *30*, 752.
23.   Larionov, O. G.; Myers, A. L. *Chem. Eng. Sci.* **1971**, *26*, 1025.
24.   Myers, A. *Proc. 3rd Int. Conf. on Fundamentals of Adsorption*, Sonhofen, Germany, May 5-9, Eds. Mersmann, A., Scholl, S.E. **1989**, 609.
25.   Chiang, A. S. T.; Lin, K. S., Fun, L.Y. *Proc. 4rd Int. Con. Fundamentals of Adsorption*, Kyoto, May 17-22, Ed. Suzuki, M. **1992**, 81.
26.   Melis, S., Markos, J., Cao, G.,. Morbidelli, M. *Fluid Phase Equilibria* **1996**, *117*, 281.
27.   Do, D. D. *Adsorption Analysis: Equilibria and Kinetics*, Imperial College: London, 1998.
28.   Brandani, S.; Ruthven, D. M. *Chem. Eng. Sci.* **1995**, *13*, 2055.
29.   Muralidharan, P. K.; Ching, C. B. *Ind. Eng. Chem. Res.* **1997**, *36*, 407.
30.   Ching, C. B.; Ruthven, D. M. *Zeolites* **1998**, *8*, 68.
31.   Lin, Y. S.; Ma, Y. H. *Ind. Eng. Chem. Res.* **1989**, *28*, 622.
32.   Claessens, R.; Baron, G. V. *Chem. Eng. Sci.* **1996**, *51*, 1869.
33.   Denayer, J. F.; Bouyermaouen, A.; Baron, G. V. *Ind. Eng. Chem. Res.* **1998**, *37*, 3691.
34.   Chueh, P. L.; Prausnitz, J. M. *AIChE J.* **1969**, *15*, 471.
35.   Chueh, P. L.; Prausnitz, J. M. *AIChE J.* **1967**, *13*, 1099.
36.   Gunn, R. D.; Yamada, T. *AIChE J.* **1971**, *13*, 351.
37.   Yen, L. C.; Woods, S. S. *AIChE J.* **1966**, *12*, 95.
38.   Langhendries, G.; De Vos, D.E.; Baron, G.V.; Jacobs, P.A. *J. Catal.* **1999**, *187*, 453.
39.   Parton, R. F.; Bezoukhanova, C. P.; Casselman, M.; Uytterhoeven, J.; Jacobs, P. A. *Nature* **1994**, *370*, 541.
40.   Parton, R. F., Bezoukhanova, C. P., Thibault-Starzyk, F., Reynders, R. A., Grobet, P. J., Jacobs, P. A. *Stud. Surf. Sci. Catal.* **1994**, *84*, 813.
41.   Parton, R. F.; Bezoukhanova, C. P.; Grobet, J.; Grobet, P. J.; Jacobs, P. A. *Stud. Surf. Sci. Catal.* **1994**, *83*, 371.
42.   Anderson, M. W.; Klinowski, J. *J. Chem. Soc., Faraday Trans* **1986**, *82*, 1449.
43.   Weitkamp, J.; Ernst, S.; Roland, E.; Thiele, G. F. *Stud. Surf. Sci. Catal.* **1997**, *105*, 763.
44.   van der Waal, J. C.; Kunkeler, P. J.; Tan, K.; van Bekkum, H. *J. Catal.* **1998**, *173*, 74.
45.   Bellussi, G.; Carati, A.; Clerici, M. G.; Maddinelli, G.; Millini, R. *J. Catal.* **1992**, *133*, 220.
46.   Clerici, M. G.; Ingallina, P. *J. Catal.* **1993**, *40*, 71.
47.   Clerici, M. G. *Appl. Catal.* **1991**, *68*, 249.
48.   Parton, R. F.; Neys, P. E.; Jacobs, P. A.; Sosa, R. C.; Rouxhet, P. G. *J. Catal.* **1996**, *164*, 1.

INDEX